Current Topics in Developmental Biology
Volume 45

Series Editors

Roger A. Pedersen and **Gerald P. Schatten**
Reproductive Genetics Division   Departments of Obstetrics–Gynecology
Department of Obstetrics, Gynecology,   and Cell and Developmental Biology
and Reproductive Sciences   Oregon Regional Primate Research Center
University of California   Oregon Health Sciences University
San Francisco, California 94143   Beaverton, Oregon 97006-3499

Editorial Board

Peter Grüss
Max-Planck-Institute of Biophysical Chemistry
Göttingen, Germany

Philip Ingham
University of Sheffield, United Kingdom

Mary Lou King
University of Miami, Florida

Story C. Landis
National Institutes of Health/
National Institute of Neurological Disorders and Stroke
Bethesda, Maryland

David R. McClay
Duke University, Durham, North Carolina

Yoshitaka Nagahama
National Institute for Basic Biology, Okazaki, Japan

Susan Strome
Indiana University, Bloomington, Indiana

Virginia Walbot
Stanford University, Palo Alto, California

Founding Editors

A. A. Moscona
Alberto Monroy

Current Topics in Developmental Biology

Volume 45

Edited by

Roger A. Pedersen
*Reproductive Genetics Division
Department of Obstetrics, Gynecology,
and Reproductive Sciences
University of California
San Francisco, California*

Gerald P. Schatten
*Departments of Obstetrics–Gynecology
and Cell and Developmental Biology
Oregon Regional Primate Research Center
Oregon Health Sciences University
Beaverton, Oregon*

Academic Press
San Diego   London   Boston   New York   Sydney   Tokyo   Toronto

Front cover photograph: Longitudinal sections through mature (stage 12) wild-type carpels. (For more details see Chapter 4, Figure 2.)

This book is printed on acid-free paper. ∞

Copyright © 1999 by ACADEMIC PRESS

All Rights Reserved.
No part of this publication may be reproduced or transmitted in any form or by any means, electronic or mechanical, including photocopy, recording, or any information storage and retrieval system, without permission in writing from the Publisher.
The appearance of the code at the bottom of the first page of a chapter in this book indicates the Publisher's consent that copies of the chapter may be made for personal or internal use of specific clients. This consent is given on the condition, however, that the copier pay the stated per copy fee through the Copyright Clearance Center, Inc. (222 Rosewood Drive, Danvers, Massachusetts 01923), for copying beyond that permitted by Sections 107 or 108 of the U.S. Copyright Law. This consent does not extend to other kinds of copying, such as copying for general distribution, for advertising or promotional purposes, for creating new collective works, or for resale. Copy fees for pre-1999 chapters are as shown on the title pages. If no fee code appears on the title page, the copy fee is the same as for current chapters.
0070-2153/99 $30.00

Academic Press
a division of Harcourt Brace & Company
525 B Street, Suite 1900, San Diego, California 92101-4495, USA
http://www.apnet.com

Academic Press
24-28 Oval Road, London NW1 7DX, UK
http://www.hbuk.co.uk/ap/

International Standard Book Number: 0-12-153145-7

PRINTED IN THE UNITED STATES OF AMERICA
99 00 01 02 03 04 EB 9 8 7 6 5 4 3 2 1

Contents

Contributors   ix
Preface   xi

1

Development of the Leaf Epidermis
Philip W. Becraft

    I. Introduction   1
   II. Morphology   2
  III. Ontogeny   5
  IV. Epidermal Identity   9
   V. Epidermal Growth   11
  VI. Interactions with Internal Tissues   16
 VII. Pattern Formation and Cellular Differentiation   18
VIII. Determined State of the Epidermis   24
  IX. The Cuticle   29
   X. Summary   32
      References   32

2

Genes and Their Products in Sea Urchin Development
Giovanni Giudice

   I. Introduction   42
  II. Histones   43
 III. Cytoskeletal Actin Genes   47
 IV. Endo16 and Other Endoderm-Specific Genes   55
  V. Ectoderm-Specific Genes   56
 VI. Genes of the Primary Mesenchyme or However Relevant to the Extracellular Matrix   59

VII. Genes for Cortical Granule Content and Genes Related to the Wnt Pathway and Gastrulation   65
VIII. Homeobox-Containing Genes and Genes for Transcription Factors That Do Not Contain Homeobox   70
IX. Tubulin Genes and Other Genes Related to Ciliogenesis and to the Mitotic Apparatus   73
X. Genes for Cyclin and Other Cell Division Related Proteins   75
XI. Genes for Kinases, a Fibroblast Growth Factor Receptor, and Metallothionein   76
XII. Genes for Heat Shock Proteins (hsps) and Ubiquitin Genes   77
XIII. Genes That Are Differentially Expressed along the Animal-Vegetal Axis   79
XIV. Genes for Arylsulfatase, Myogenic Factors, and Ribosomal Proteins, Genes with an EGF Domain, and Genes for Poly(A)-Binding Proteins   81
XV. Genes for Jelly Proteins and the Egg Receptor for Sperm   83
XVI. Genes Expressed by Coelomocytes   84
XVII. Miscellaneous Genes   84
XVIII. Proto-oncogenes and Retroposons   87
References   88

3

The Organizer of the Gastrulating Mouse Embryo
Anne Camus and Patrick P. L. Tam

I. Introduction   117
II. The Mouse Gastrula Organizers   121
III. Separate Head and Trunk Organizing Activity   131
IV. Tissue Patterning and the Specification of the Gastrula Organizer   139
V. An Organizer for All Seasons?   144
References   145

4

Molecular Genetics of Gynoecium Development in *Arabidopsis*
John L. Bowman, Stuart F. Baum, Yuval Eshed, Joanna Putterill, and John Alvarez

I. Introduction   156
II. Gynoecium Structure and Development in *Arabidopsis*   159
III. Molecular Genetics of Carpel Development   171
IV. Lessons from Analyses of Gene Expression Patterns   189
V. Conclusions   192
References   199

5

Digging out Roots: Pattern Formation, Cell Division, and Morphogenesis in Plants
Ben Scheres and Renze Heidstra

    I. Introduction   208
    II. Embryonic Pattern Formation   209
    III. Postembryonic Perpetuation of Cellular Pattern   220
    IV. Control of Cell Division during Development   225
    V. Growth and Organ Morphogenesis   234
    VI. Concluding Remarks   239
        References   240

Index   249
Contents of Previous Volumes   259

Contributors

Numbers in parentheses indicate the pages on which the authors' contributions begin.

John Alvarez (155), Department of Biological Sciences, Monash University, Clayton, Melbourne, Victoria 3168, Australia

Stuart F. Baum (155), Section of Plant Biology, University of California, Davis, Davis, California 95616

Philip W. Becraft (1), Department of Zoology and Genetics and Department of Agronomy, Iowa State University, Ames, Iowa 50011

John L. Bowman (155), Section of Plant Biology, University of California, Davis, Davis, California 95616

Anne Camus (117), Embryology Unit, Children's Medical Research Institute, Wentworthville, New South Wales 2145, Australia

Yuval Eshed (155), Section of Plant Biology, University of California, Davis, Davis, California 95616

Giovanni Giudice (41), Department of Cellular and Developmental Biology, University of Palermo, 90128 Palermo, Italy

Renze Heidstra (207), Department of Molecular Cell Biology, Utrecht University, 3584 CH Utrecht, The Netherlands

Joanna Putterill (155), School of Biological Sciences, University of Auckland, Auckland, New Zealand

Ben Scheres (207), Department of Molecular Cell Biology, Utrecht University, 3584 CH Utrecht, The Netherlands

Patrick P. L. Tam (117), Embryology Unit, Children's Medical Research Institute, Wentworthville, New South Wales 2145, Australia

Preface

Plant developmental biologists will be particularly interested in this volume of *Current Topics in Developmental Biology*. Three chapters are devoted to exciting breakthroughs in plant development. Chapter 1, by Philip W. Becraft, discusses the development of the leaf epidermis. John L. Bowman, Stuart F. Baum, Yuval Eshed, Joanna Putterill, and John Alvarez contributed Chapter 4, which considers the molecular genetics of gynoecium development in *Arabidopsis*. Ben Scheres and Renze Heidstra discuss pattern formation, cell division, and morphogenesis in plants in their cleverly titled chapter "Digging out Roots."

This volume also continues the custom of this series in addressing developmental mechanisms in a variety of experimental systems. In Chapter 2, Giovanni Giudice provides a comprehensive treatment of one of the powerful experimental systems of development—the sea urchin—which had been the subject of some concerns regarding genetic investigations. Anne Camus and Patrick P. L. Tam discuss the organizer of the gastrulating mouse embryo, an essential aspect of mammalian development.

Together with the other volumes in this series, this volume provides a comprehensive survey of major issues in the forefront of modern developmental biology. These chapters should be valuable to researchers in the fields of plant and animal development, as well as to students and other professionals who want an introduction to current topics in cellular, molecular, and genetic approaches to both developmental and plant biology. This volume in particular will be essential reading for anyone interested in plant development and plant biology, morphogenesis and embryo formation, gene regulation of development, development in invertebrates, and molecular basis of mammalian embryogenesis.

This volume has benefited from the ongoing cooperation of a team of participants who are jointly responsible for the content and quality of its material. The authors deserve full credit for their success in covering their subjects in depth, yet with clarity, and for challenging the reader to think about these topics in new ways. We thank the members of the Editorial Board for their suggestions of topics and authors and Liana Hartanto and Michelle Emme for their exemplary administrative and editorial support. We are grateful for the unwavering support of Craig Panner and Hilary Rowe in the editorial office at Academic Press in San Diego. We are also grateful to the scientists who prepared chapters for this volume and to their funding agencies for supporting their research.

Gerald P. Schatten
Roger A. Pedersen

1
Development of the Leaf Epidermis

Philip W. Becraft
Department of Zoology and Genetics and Department of Agronomy
Iowa State University
Ames, Iowa 50011

I. Introduction
II. Morphology
   A. Dicot Leaf Epidermis
   B. Monocot Leaf Epidermis
III. Ontogeny
   A. Epidermal Formation in Embryogenesis
   B. Epidermal Formation during Leaf Development
   C. Cell Lineage of the Leaf Epidermis
IV. Epidermal Identity
   A. Heteroblasty: Variation among Leaves
   B. Regional Identity within Leaves
V. Epidermal Growth
   A. Cell Division
   B. Cell Expansion
   C. Hormone Targets
   D. Relation to Plant Morphogenesis
VI. Interactions with Internal Tissues
   A. Correlations among Structures
   B. Genetic Mosaics
   C. Cytoplasmic Connections
VII. Pattern Formation and Cellular Differentiation
   A. Stomates
   B. Trichomes
VIII. Determined State of the Epidermis
   A. Determination of Epidermal Fate
   B. Epidermal Potency
IX. The Cuticle
X. Summary
   References

I. Introduction

The epidermis is the cell layer which forms the interface between a plant and its environment. As such, the epidermis is crucial for protecting the plant against environmental insults and for regulating the exchange of materials between a plant and its environment. Leaves constitute the majority of the surface area of aerial portions in most plants. Leaves are the major site of gas exchange, transpiration, and attack by insects and pathogens. Thus the leaf epidermis is of particular importance for plant protection. Plant epidermises have evolved various features to allow the performance of this protective role. The epidermis consists of epidermal ground cells interspersed with specialized structures such as stomates, trichomes, or fibers which enable the epidermis to perform its various functions. The waxy cuticle forms a barrier to moisture loss and has been implicated in resistance to insects, pathogens, frost, and UV radiation. The epidermis is also a physically tough tissue, providing mechanical support and protection. In addition, the epidermis plays key roles in development. Evidence suggests that patterns of surface reinforcement are a critical element in organogenesis and shape determination. Likewise epidermal properties are important in controlling whether leaves or floral organs remain as individual organs or fuse into compound structures.

Because of the functional significance of various epidermal components, there is potential practical application in understanding their development. In addition, the epidermis has emerged as a powerful system for studying fundamental processes of development. Questions of cell fate specification, cellular pattern formation, and cellular morphogenesis are readily addressable. Epidermal markers have allowed excellent fate-mapping experiments, interpretation of homeotic transformations, and analysis of cell interactions. The purpose of this review is to provide a broad framework in which to consider these various aspects of epidermal development. The focus will be on angiosperms and particular attention will be paid to model systems where the most information is available.

II. Morphology

General descriptions of epidermal morphology can be found in most plant anatomy texts (Fahn, 1990). Epidermal morphology is extremely variable, reflecting the vast array of environments in which plants grow. The epidermis is most commonly a single-cell layer (uniseriate) although some plants such as *Pepperomia* and *Ficus* contain a multiple, or multiseriate, epidermis (more than one cell thick). The internal cells of a multiple epidermis are commonly large unspecialized cells with no or few chloroplasts and often function in water storage. Some plants form an anatomically similar tissue called the hypodermis or subepidermis, the difference being one of ontogeny; a multiple epidermis is derived by periclinal divisions (new cell wall parallel to the organ surface) of the developing epidermis whereas the hypodermis descends from internal tissues.

The epidermis is covered by a waterproof layer of complex lipid polymers called the cuticle, which will be discussed in detail later. The cuticle has a layered structure and the surface can be variously textured. The cuticle is bound to the cellulosic cell wall by a layer of pectin. Cell wall thickness varies among plants and among cell types within an epidermis although the outer cell wall is frequently thicker than the others, the anticlinal walls being thinnest (anticlinal walls are oriented perpendicular to the leaf surface). This increases the tensile strength of the epidermis, which is important for its mechanical support role.

The epidermis consists of relatively unspecialized ground cells interspersed with specialized cell types such as stomates or trichomes. Stomates function in gas exchange, allowing uptake of CO_2 and release of O_2 for photosynthesis while restricting transpirational water loss. They consist of a pore and a pair of guard cells. In some plants additional cells called accessory or subsidiary cells are associated with the stomatal complex. Guard cells change shape in response to turgor, resulting in regulated pore opening and closure that is coupled to the photosynthetic and moisture status of the plant. Guard cells are typically the only epidermal cells to contain chloroplasts; exceptions can be found in shade plants and aquatics.

Trichomes are epidermal appendages and display tremendous variation. They can be glandular or nonglandular, single-celled or multicellular, or simple or ornately branched. They range from single-celled hairs to the multicellular glandular trichomes of carnivorous sundews (*Drosera*) that bend in response to an action potential triggered by an insect (Williams and Pickard, 1972). The trichomes of leaves generally perform protective functions, providing a mechanical barrier against insects, dead air space to slow moisture loss, or reflectivity to excessive light (reviewed in Johnson, 1975). Trichomes can also accumulate chemical protectants such as the skin irritants in stinging nettles or UV-absorbing compounds (Karabourniotis *et al.*, 1995). In soybean, trichome density was positively correlated with yield (Zhang *et al.*, 1992).

The orientation and distribution of various cells can appear random or show highly ordered patterns. In many cases, the patterns of cell types observed in the epidermis correlate with internal features such as vasculature or air spaces. In other cases, various epidermal cell types are associated in complex structures. For example, stomates are often associated with structures that aid in retarding transpirational water loss. They can be surrounded by trichomes or located in depressions called crypts that protect the stomatal pore from air flow.

Leaves may be unifacial, with morphologically identical epidermis on both sides, but are commonly bifacial. For example, many xerophytic plants contain a thicker cuticle and abundant hairs on the adaxial (upper) surface where the greatest protection from sun in required, whereas the majority of stomata are located on the abaxial (lower) surface where they will be less prone to excessive moisture loss. Conversely, floating aquatic plants contain stomata only on the upper surface.

A. Dicot Leaf Epidermis

In dicot leaves, the ground cells (often referred to as pavement cells) of the laminar epidermis usually have highly undulated, interlocking lateral walls, giving the cells a jigsaw puzzle-like appearance (Fig. 1). In some leaves such as *Arabidopsis*, the epidermis is uniform with respect to the underlying anatomy. In other plants

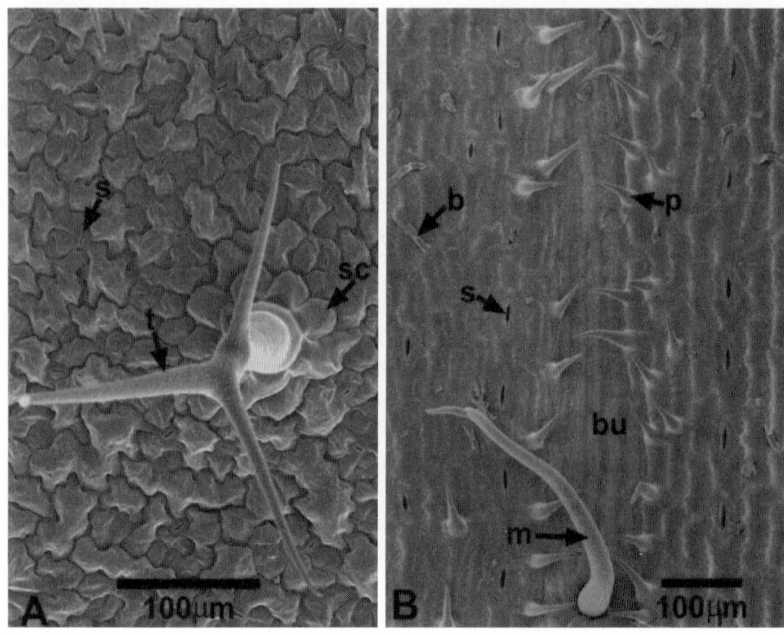

Fig. 1 SEM of leaf surfaces of (A) *Arabidopsis* and (B) maize. Abbreviations: b, bicellular hair; bu, bulliform cell row; m, macrohair; p, prickle hair; s, stomate; sc, support cell; t, trichome.

such as oak (*Quercus*), vascular bundle sheaths extend into the epidermis and are distinct, typically being elongated parallel to the vasculature.

Stomata consist of a pair of kidney-shaped guard cells. The stomata may be aligned parallel to venation or oriented randomly. They may or may not be associated with subsidiary cells (which are thought to function jointly with guard cells in the physiological control of stomate opening) and stomata are classified according to the various configurations of subsidiary cells. Stomata without anatomically distinct subsidiary cells are called anomocytic, ones with subsidiary cells elongated parallel to the guard cell are paracytic, and so on (Esau, 1977). *Arabidopsis* and other crucifers are anisocytic with the stomata surrounded by three unequally sized epidermal cells.

Dicots contain the full range of trichome variation discussed above. The leaf trichomes of *Arabidopsis* are unicellular and branched. Fibers and a few other specialized cell types occur in the leaf epidermis of some species.

B. Monocot Leaf Epidermis

In contrast to dicots, epidermal cells of monocots are typically rectangular and occur in files parallel to the leaf venation. The epidermis of some monocots—for example, the spider plant *Chlorophytum comosom*—is extremely simple, containing only stomates and long cells (the ground cells of a monocot epidermis). On the other hand, the epidermis of maize is considerably more complex than that of most dicots. The adaxial surface of an adult maize leaf contains at least 10 different cell types (Freeling and Lane, 1994). The cells occur in files, with each file composed of a particular set of cell types (Fig. 1B). Over the major (lateral) veins are cells called costal long cells. Unspecialized epidermal cells between the major veins are called intercostal long cells and are distinct from costal long cells. Intercostal long cells alternate with silica cells and cork cells (collectively referred to as short cells) in the cell files. Stomatal files consist of stomata alternating with interstomatal cells, which are unspecialized cells that resemble intercostal long cells but are shorter. Stomatal complexes consist of a pair of dumbbell-shaped guard cells associated with triangular subsidiary cells. Files of bulliform cells occur periodically. These are thin-walled cells enlarged in depth relative to the rest of the epidermis. They are thought to function in leaf unrolling during leaf growth and/or leaf rolling in response to drought-associated turgor loss. Within the bulliform rows are periodic macrohairs consisting of a large, unicellular hair surrounded by a raised multicellular base. Prickle hairs are short curved hairs that often occur alongside the bulliform files. Bicellular hairs are interspersed among the other cell types. The complexity evident in the variety of cell types and their highly ordered pattern within the grass epidermis contrast markedly with the relatively simple epidermis of *Arabidopsis*.

III. Ontogeny

A. Epidermal Formation in Embryogenesis

The epidermis is derived from the embryonic protoderm. Protoderm formation represents the earliest recognizable stage of histogenesis in plant embryo development. In *Arabidopsis*, the protoderm is formed by the fifth zygotic cell division, the fourth round of cell division of the embryo proper (Mansfield and Briarty, 1991) (Fig. 2A). All eight cells of the octet-stage embryo divide periclinally (new cell wall parallel to the embryo surface), giving rise to a surface cell and an internal cell. This is the earliest point in embryogenesis where it is geometrically possible to set apart surface and internal cells. Protoderm is an element of radial pattern in the establishment of the embryo body and mutants are known to disrupt this pattern (Mayer *et al.*, 1991). In maize the pattern of cell divisions during early embryogenesis is less regimented and protoderm formation occurs somewhat later (Randolph, 1936; Kiesselbach, 1949). A variable pattern of cell division leads to a cone-shaped embryo in which the suspensor and embryo proper are not clearly demarcated (Fig. 2B). In the rounded top of the cone comprising the embryo proper, the protoderm becomes distinct as a layer of regularly shaped surface cells that divide primarily in the anticlinal plane. Following protoderm formation in all

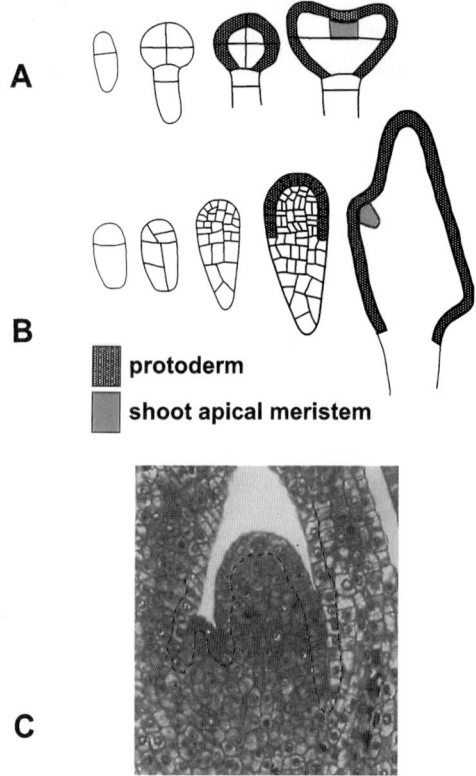

Fig. 2 Protoderm formation in *Arabidopsis* and maize. (A) In *Arabidopsis* embryogenesis, the zygote divides asymmetrically, forming a smaller apical cell and a larger basal cell. The apical cell undergoes three rounds of division to form the eight-celled embryo proper. All eight cells then divide periclinally, giving rise to a surface layer, the protoderm. The protoderm divides anticlinally, forming a stable cell lineage that includes the outer cell layer of the shoot apical meristem and persists throughout the life of the plant. After Laux and Jürgens (1997). (B) Maize embryogenesis involves a variable pattern of cell division. At the transition stage, cell division in the outer cells of the apical end of the embryo becomes restricted to the anticlinal plane, forming the protoderm. The anticlinal division pattern persists after this layer is incorporated into the shoot apical meristem (Randolph, 1936). (C) The L1 layer (demarcated by the line) of the dome-shaped maize SAM divides anticlinally and contributes the epidermis of flanking leaf primordia.

species, the division plane of surface cells is predominantly anticlinal (new cell walls perpendicular to the surface). As a consequence, daughter cells of most surface cell divisions remain within the surface layer, effectively setting the protoderm aside as an independent cell lineage (Steeves and Sussex, 1989). It should be emphasized that this lineage independence is not absolute (Stewart *et al.*, 1974; Stewart and Dermen, 1975; McDaniel and Poethig, 1988).

It appears from the pattern of embryonic cell divisions that one of the primary goals in plant embryogenesis is to establish a surface layer. In fact, it is possible that the embryo's surface is defined even prior to the onset of cell division. Citrus has been found to contain a cuticle over the zygote prior to the onset of cell division, which was interpreted to mean that the zygotic cell has epidermal identity (Bruck and Walker, 1985a). It is likely that even during embryogenesis the epidermis may function in regulating interactions between the young plant and the environment of the surrounding maternal tissues. The surface is also implicated in critical morphogenetic roles (see later), which could be another reason for quickly differentiating a surface layer of cells.

Sometime during *Arabidopsis* embryogenesis the shoot epidermis becomes symplastically isolated as revealed by dye-coupling experiments (Duckett *et al.*, 1994). Dye loaded onto cut cotyledons moved through the epidermis into the hypocotyl epidermis but there was an abrupt boundary between the hypocotyl and the root and the dye never entered internal hypocotyl cells. Thus the epidermis of the seedling shoot forms a symplastic continuum that is isolated from other tissues. In mature leaves of several other species the epidermis was found to contain symplastic connections with mesophyll cells (Erwee *et al.*, 1985), suggesting that the symplastic isolation of the shoot epidermis is transitory or is not a universal feature of plant development.

As the developing embryo forms a shoot apical meristem (SAM), the protoderm layer becomes the L1 layer, or tunica, of the meristem (or the outer tunica layer in species such as *Arabidopsis* with a two-layered tunica). The SAM is the site of all primary postembryonic growth and is responsible for establishing the shoot system of the plant, including leaves. The anticlinal pattern of cell division is retained in the tunica; thus the independent cell lineage established in early embryogenesis is maintained for the life of the plant (McDaniel and Poethig, 1988). Again it must be emphasized that this is not an absolutely separate lineage and the tunica occasionally makes contributions to the internal tissues (Stewart *et al.*, 1974; Stewart and Dermen, 1975). In these cases, the L1-derived cells are assimilated into the internal tissues, indicating that there is no fixed cell fate associated with the different lineages.

B. Epidermal Formation during Leaf Development

Leaves initiate as bulges on the flanks of the SAM. The L1 layer of the SAM becomes the surface cell layer of the leaf primordium, destined to become the leaf epidermis. The L1 layer of the SAM is generally considered undifferentiated although alternative views suggest it is already predisposed to form epidermis and therefore should be considered protoderm (Bruck and Walker, 1985a). As the leaf grows and develops, the predominance of anticlinal divisions in the surface continues although it is less stringent than in the meristem. In many plants, a gradient in developmental maturity forms in leaves, with the tip maturing before the base

(Sylvester et al., 1990; Croxdale et al., 1992; Lloyd et al., 1994). This feature allows different stages of epidermal development to be studied in a single leaf. In plants such as maize it is possible to recognize cell lineages because of the longitudinal cell files. The general pattern observed is typical of plant development in that early stages are marked by a high frequency of cell division whereas later stages show rapid cell elongation. The process of cell differentiation overlays this pattern. At the basalmost part of the growing leaf is a region called the basal meristem, characterized exclusively by cell division with no apparent differentiation. Slightly distal to this region is an area still displaying a high rate of cell division but in which some evidence of cell differentiation, such as the characteristic pattern of cell divisions that produce a stomate, can be observed. Beyond this area the rate of cell division declines and cells expand, differentiate, and mature.

C. Cell Lineage of the Leaf Epidermis

Fate-mapping studies are commonly performed by genetically marking cell lineages with a trait such as albinism. Such studies indicate that derivatives of cells in particular positions in the SAM will occupy predictable but variable portions of the mature leaf (Satina et al., 1940; Stewart, 1978; Stewart and Dermen, 1979; Poethig, 1984; Poethig and Sussex, 1985). Periclinal chimeras are plants in which the layers of the SAM are differentially marked. These show that L1 cells are destined to form the epidermis, some of the internal tissues near the leaf margins, and occasional areas of subepidermal cells scattered throughout the leaf. The marginal region of L1-derived internal cells is generally minor in dicots whereas it tends to occupy a somewhat larger area of monocot leaves (Stewart, 1978; Stewart and Dermen, 1979).

Cell lineage in plants is generally regarded as stochastic and of little consequence in the specification of cell fate (Stewart, 1978; Poethig, 1987; Bossinger et al., 1992; Schnittger et al., 1996). In periclinal chimeras one can detect groups as small as two mesophyll cells that were derived from the L1 (Stewart et al., 1974; Stewart and Dermen, 1975). Thus only one cell division occurred after the cell was contributed from the epidermis, indicating that even very late in development cell lineage does not control the developmental fate of cells.

Although lineages are variable, certain patterns emerge with relatively high probability. The width between lateral veins of maize leaves has been designated as a developmental module because of the propensity of epidermal sectors to begin and end at lateral veins (Cerioli et al., 1994). By studying reversions of an epicuticular wax mutation, it was found that over 80% of clonal sectors of one module width or greater started at a lateral vein (i.e., the inner boarder was located over the vein). Sectors of half-module widths or smaller most often started at a lateral vein or an intermediate vein. The marginal boundaries also tended to end at lateral or intermediate veins for sectors of half-module width or greater. The pattern of sector placement was consistent with a model in which a module progenitor divides longitudinally to produce a derivative to the marginal side. The two daughter cells then become progenitors for the two half-modules (demarcated by the intermediate vein). The results suggest that a consistent pattern of cell divisions occurs in founder cells to produce the developmental modules. Similarly, epidermal sectors in *Arabidopsis* leaves border the midvein with high frequency (Schnittger *et al.*, 1996).

Such regularity in sector placement suggests a typical pattern of cell division in the development of the maize leaf epidermis. However, lineage may be preserved in the organization of a tissue simply because the primordial cell pattern serves as a structural template for subsequent development (Dawe and Freeling, 1991). Thus it is also possible that the regularity in clonal relationships of the mature leaf epidermis reflects more the regular arrangement of cells in the leaf primordium than a set pattern of cell division during leaf development.

IV. Epidermal Identity

Epidermal characteristics vary from organ to organ or according to position within an organ. As such, the specification of epidermal identity is closely related to specification of organ identity and positional cues within the organ.

A. Heteroblasty: Variation among Leaves

Many plants produce leaves in a heteroblastic series where the morphology of successive leaves changes. Epidermal characteristics often show some of the most obvious differences. In maize the epidermis of juvenile leaves is covered by a dull epicuticular wax bloom (Fig. 5), the cuticle is thinner, and hairs and bulliform cells are absent. Epidermal cells are circular in transverse section and, from a surface view, are shorter and wider than adult and have weakly crenulated lateral walls. Adult leaves lack epicuticular wax, have a thick cuticle, and contain hairs and bulliform cells. Epidermal cells have rectangular cross sections, are long and narrow, and have highly crenulated side walls (Poethig, 1990; Evans *et al.*, 1994; Lawson and Poethig, 1995). The *glossy15* (*gl15*) gene specifically regulates juvenile characteristics in the epidermis (Evans *et al.*, 1994; Moose and Sisco, 1994). Loss-of-function mutations in *gl15* cause a premature acquisition of adult epidermal morphology in early leaves but do not affect other juvenile characteristics such as leaf shape. The *gl15* gene encodes a putative transcription factor related to the *APETALA2* floral gene of *Arabidopsis* (Moose and Sisco, 1996) and appears to act autonomously in epidermal cells (Moose and Sisco, 1994). Dominant gain-of-function mutations in the *teopod* and *corngrass* genes prolong the juvenile phase. The *Teopod1* and *Teopod2* mutants act non-cell-autonomously to regulate juvenile traits throughout the leaf (Dudley and Poethig, 1993) and require a functional *gl15* gene for their effect on epidermal traits (Evans *et al.,* 1994). Thus *gl15* acts downstream of the *teopod* genes to control epidermal cell identity.

Arabidopsis similarly shows heteroblastic leaf development. This is most obvious in the difference between rosette leaves and cauline leaves or bracts of the inflorescence; however, differences also occur among rosette leaves and cauline leaves (Telfer and Poethig, 1994). Leaf shape and trichome distributions change as plant development progresses. The first few rosette leaves have trichomes on the adaxial surface but not the abaxial (Telfer *et al.*, 1997), whereas in the cauline leaves the number of adaxial trichomes successively declines, becoming restricted to the leaf tip (Chien and Sussex, 1996). At least some of the differential distribution of trichomes on the adaxial versus abaxial surface can be explained by differential sensitivity to gibberellic acid (Chien and Sussex, 1996), which is involved in phase change (Evans and Poethig, 1995).

B. Regional Identity within Leaves

Epidermal identity in bifacial leaves is determined by factors that control dorsiventrality. Dorsiventrality is specified early in primordial leaf development and involves signals emanating from the shoot apical meristem (Sussex, 1955). The recently identified *phantastica* (*phan*) gene of *Antirrhinum* is involved in specifying dorsiventrality (Waites and Hudson, 1995). Mutant leaves are radially symmetrical and uniformly covered with epidermis of abaxial identity (i.e., unifacial). Some leaves are mosaics; patches of abaxial epidermis occur on the adaxial surface, or parts of the leaf are radially symmetrical while other parts show dorsiventrality. In either case, where adaxial and abaxial epidermises juxtapose, an outgrowth reminiscent of a lateral lamina occurs. The authors propose that *phan* provides a dorsalizing function that specifies adaxial (dorsal) identity. PHAN encodes a MYB-related protein that is expressed in lateral organ primordia in a pattern complementary to the *Kn1*-related *AmSTM* gene, suggesting reciprocal functions (Waites *et al.*, 1998). In maize, a dominant mutant called *Rolled1* (*Rld1*) has the effect of reversing adaxial and abaxial identity of the epidermis (Freeling, 1992) while apparently not affecting internal anatomy (unpublished observation). Thus the *Rld1* mutant uncouples dorsiventral axis specification or recognition in the epidermis from that of the rest of the leaf, implying that epidermal identity specification occurs downstream of axis specification. By analogy to heteroblasty, one might expect a factor that functions similarly to *gl15* in specifying dorsiventral epidermal identity.

Epidermal identity shows regional variation within a leaf. In dicots, the lamina, midrib, and petiole all have different epidermal characteristics, whereas in monocots, the lamina, sheath, and collar regions each display different epidermal traits. In maize, a group of mutants alter regional identities within the leaf (Freeling,

1992; Sylvester *et al.*, 1996). Most notable are dominant mutants caused by ectopic expression of homeobox genes related to *knotted1* (*kn1*). These mutants transform the identity of certain regions of blade into sheath and these transformations are reflected in altered epidermal identity (Becraft and Freeling, 1989; Becraft and Freeling, 1994; Sinha and Hake, 1994; Fowler and Freeling, 1996). In the case of *Kn1*, expression studies and genetic mosaics both show that the altered epidermal identity is genetically controlled by the internal cells (Hake and Freeling, 1986; Sinha and Hake, 1990; Smith *et al.*, 1992). Normal *kn1* transcript is restricted to the internal cells of the shoot apical meristem (Jackson *et al.*, 1994) and the KN1 protein itself is trafficked to the surface cells (Lucas *et al.*, 1995), suggesting that transport of the KN1 protein from internal to surface cells might be responsible for the control of epidermal identity by internal cells. A related mutant gene, *Lg3*, also acts internally to specify epidermal fate (Fowler *et al.*, 1996).

V. Epidermal Growth

A. Cell Division

Control of cell division plane is a fundamental and controversial problem in plant biology. Observations of plant cell division have led to a set of "rules" that govern the plane of cell division (reviewed in Lloyd, 1991). Most plant cells divide perpendicular to the long axis of the cell, in the plane of least area that will divide the cell equally in two. These observations have led to theories of minimal free energy configurations for cell plate formation and cell packing. Aspects of these theories are conveniently testable in epidermal cells.

Cell division is associated with concerted changes in cytoskeletal arrays which are thought to determine the plane of division. Thus any model for controlling division plane must account for cytoskeletal components. In studying geometric factors that could influence the plane of cell division, it was found that the arrangement of radial microtubular arrays in premitotic epidermal cells of *Datura stramonium* L. resembled the pattern of soap bubbles or sliding springs suspended within frames of similar shape to the cell (Flanders *et al.*, 1990). Radial elements assumed the shortest distance to the surrounding wall and consequently did not intercept wall junctions. This is consistent with the observation that new epidermal cell walls avoid existing three-way junctions, creating an alternating brickwork-like cell pattern (Sinnott and Bloch, 1941). The radial patterns changed as cell shape was varied from hexagonal to elongated and the correlation between cells and the mechanical models was maintained. The microtubule strands appeared to obey the same laws of geometry as the purely mechanical models. A prediction of this model is that the radial cytoskeletal strands should be under tension within the premitotic cell. Based on the rapid elastic-like retraction of radial strands following severance with a microlaser, this appears to be the case (Goodbody *et al.*,

1991). Thus it seems reasonable that geometric factors can influence cell division planes through their effects on cytoskeletal arrays.

Plant cells can obviously overcome these geometric constraints in specialized cell divisions. Asymmetric divisions precede the formation of stomata and often trichomes (reviewed in Larkin *et al.*, 1997). A graphic example is the highly asymmetric division associated with subsidiary cell formation in the stomatal complex of monocots (Fig. 3). Following karyokinesis, a crescent-shaped cell wall forms with both new ends adjoining the same longitudinal wall of the mother cell (Stebbins and Shah, 1960). Hence the mother cell is not bisected at all but a daughter cell is cleaved out from the side.

Cell divisions that would not obviously be considered specialized nonetheless might be controlled by specific programs if they do not follow the geometric principles outlined above. In grasses, the majority of epidermal cell divisions are transverse such that the daughters contribute to leaf length and the relatively infrequent

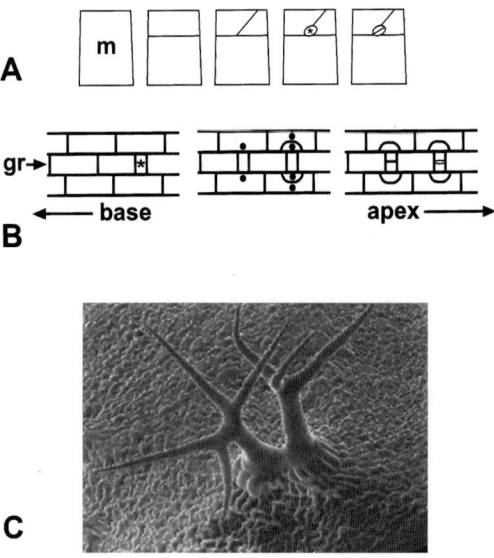

Fig. 3 (A) Cell division pattern giving rise to stomates in *Arabidopsis*. A meristemoid (m) undergoes an asymmetric division. Two additional asymmetric divisions occur in the smaller daughter of the previous division. This produces a GMC (asterisk) surrounded by three cells of the same lineage. The GMC then divides symmetrically to form a guard cell pair. After Larkin *et al.* (1997). (B) Development of monocot stomata progresses in linear files called guard cell rows (gr). Meristemoids divide asymmetrically to form a GMC (asterisk) in the smaller distal daughter. Nuclei in subsidiary mother cells in neighboring cell files migrate adjacent to the GMC and a highly asymmetric division produces a lens-shaped subsidiary cell. The GMC then divides symmetrically to form a guard cell pair. (C) Trichome cluster in *try;35S::GL1* plants indicating that *try* functions to suppress trichome formation in neighboring cells. Figure 3C courtesy of David M. Marks.

longitudinal divisions contribute width. The latter do not obey the "rules" of division because the longitudinal area is not the minimum area which can bisect a wall. The *tangled1* mutant of maize specifically affects longitudinal divisions, making them seemingly random, whereas transverse divisions appear normal (Smith *et al.*, 1996). Thus the *tangled1* gene product is required specifically for longitudinal divisions, indicating this type of division is controlled by a specific program and not simply geometric principles.

Development of the ligule on grass leaves reveals additional controls of cell division. The ligule is a fringe of epidermally derived tissue that forms at the boundary between the blade and sheath. A series of highly controlled cell divisions precede ligule formation. Initially, a relatively uniform field of cells occupies the preligular region (Sylvester *et al.*, 1990). Within that field a series of anticlinal cell divisions form a transverse band of small cells. This process involves a localized increase in the rate of cell division and the proportion of longitudinal divisions, accompanied by a decreased rate of cell elongation (Sylvester *et al.*, 1990). Within the band of small cells, a narrower band of cells undergo periclinal divisions to create a ridge that will continue to grow out from the leaf surface and form the mature ligule (Sharman, 1942; Becraft *et al.*, 1990). Thus ligule formation involves dramatic controls on the rate and orientation of cell division as well as the rate and polarity of cell expansion. Two recessive mutations called *liguleless1* and *lgl2* block initiation of the ligule (and an associated structure called the auricle) at the earliest recognizable step but have no other detectable effects on the plant (Becraft *et al.*, 1990; Sylvester *et al.*, 1990; Harper and Freeling, 1996), again highlighting the specific developmental regulation of these specialized cell divisions.

B. Cell Expansion

Discussions of plant development often focus on cell division but in fact cell expansion is the major contributor to plant growth. Epidermal cells in the basal region of a growing maize leaf are 15–20 μM in length whereas mature cells are 100–300 μM (Giles and Shehata, 1984; Palmer and Davies, 1996). Epidermal cell length in *Lolium perenne* increase from 12 to 550 μM (Schnyder *et al.*, 1990). Thus cell expansion can produce 50-fold increases in cell size. On the other hand, cytokinesis divides an existing cell volume in half and in the absence of cell expansion contributes nothing to growth. The extent and direction of cell expansion are therefore central in determining final form. It is well accepted that turgor pressure provides the driving force for cell expansion and that the interplay between turgor and cell wall extensibility determines the direction and extent of cell expansion (reviewed in Cosgrove, 1997). The major structural element of the cell wall is the cellulose microfibril. The orientation and cross-linking of microfibrils as well as cell wall thickness are hypothesized to contribute to cell wall extensibility. Because of its surface location, properties of the outer epidermal cell wall could be especially important for determining the overall pattern of growth and form of a plant or organ, a view which is the subject of controversy (see later).

Because of the polar developmental gradient in many leaves, particularly monocots, there is a restricted zone of cell elongation at the base of a growing leaf. This allows comparison of cellular properties at various stages of elongation within a single leaf. In wheat leaves there is a correlation between the orientation of epidermal cellulose microfibrils and elongation; within the zone of elongation the microfibrils are oriented transversely and near the distal end of the elongation zone they become longitudinal (Paolillo, 1995). This correlation held in several dwarf mutants where it was determined that the basis of the reduced growth was a shortening of the elongation zone. In begonia leaves there was no change in wall thickness, optical refraction (an indicator of microfibril orientation), or turgor pressure between elongating and mature epidermal cells, suggesting that changes in wall extensibility controlled cell expansion (Serpe and Matthews, 1994). This is consistent with reported correlations in growing maize leaves between cell expansion and xyloglucan endotransferase activity, thought to facilitate wall loosening by cleaving and re-forming cell wall cross-linking carbohydrates (Palmer and Davies, 1996). Stress analysis also suggested that the epidermis was not the sole source of resistance to turgor-driven organ expansion (Serpe and Matthews, 1994).

Leaf shape mutants provide evidence that cell expansion is regulated by at least two independent polar processes. The *angustifolia* and *rotundifolia* mutants of *Arabidopsis* have leaves that are shortened or narrowed, respectively (Tsuge *et al.*, 1996). Both have the same number of cells as wild type but less expanded, specifically in the longitudinal or lateral dimension.

C. Hormone Targets

The potential role of the epidermis in controlling organ expansion raises the question of whether the epidermis is the target of hormone action. Attempts to address this question have produced controversial results. The majority of reports are consistent with the epidermal layer being of primary importance in auxin-mediated growth. Removal of the outer epidermal cell layer inhibited auxin-induced elongation of maize and oat coleoptiles (Pope, 1982; Kutschera *et al.*, 1987). These reports are challenged by Cleland, who essentially repeated the same experiments and found peeled and intact coleoptiles equally responsive to exogenous auxin. Similarly, peeled and unpeeled pea epicotyl segments elongated to the same extent and with the same kinetics in response to auxin (Rayle *et al.*, 1991). However, additional studies support the role of the epidermis in auxin-mediated elongation. Cellulose microfibrils in the outer epidermal wall of maize coleoptiles were reported to reorient upon auxin-induced elongation (Bergfeld *et al.*, 1988). One of the cell wall components that shows auxin-induced hydrolysis is β-D-glucans. Antibodies raised against oat β-D-glucans inhibited auxin-induced elongation of maize coleoptiles but only when applied to coleoptiles with an abraded outer epidermis (Hoson *et al.*, 1992). Coleoptiles with an abraded inner epidermis were not inhibited, nor were nonabraded coleoptiles, presumably because the antibodies could not penetrate the cuticle. Genes for endo-1,4-β-D-glucanase and xyloglucan endotransglycosylase, two of the enzymes that catalyze auxin-induced cell wall changes, are both expressed predominantly in the epidermis of tomato hypocotyls, although exogenous auxin application induced expression of both in the subepidermal cortical tissue (Catalá *et al.*, 1997). Thus the bulk of the literature supports a prominent, but perhaps not exclusive, role for the epidermis in regulating hormone-induced organ elongation.

D. Relation to Plant Morphogenesis

Several models of plant morphogenesis place a prominent role on the epidermis for determining the overall pattern of expansion and therefore the final form of plant organs. Single-celled algae exist with elaborate fronds that resemble fern leaves, indicating that multicellularity is not required for the production of complex forms (Kaplan and Hagemann, 1991) and the pattern of growth must therefore be regulated at the outer cell wall. This wall would correspond to the outer epidermal wall of higher plants. In the shoot apical meristem, correlations have been observed between the patterns of cellulose microfibril orientation in the L1 and organ initiation (Green, 1984). This led to the hypothesis that morphogenesis is controlled through patterns of cellulose reinforcement and weakness in the epidermis, with turgor pressure of the internal tissues providing the driving force. Weaker areas would be more prone to forming a bulge which would lead to organ initiation. A similar view is taken with respect to leaf growth, where expansion is thought to be controlled by epidermal properties (Dale, 1988).

It is clear that the epidermis does not function independently in controlling morphogenesis. Many studies have shown unequivocally that internal tissues influence developmental events in the epidermis (reviewed in Szymkowiak and Sussex, 1996). Furthermore, mutants such as *crinkly4* of maize (Becraft *et al.*, 1996) and *pale cress* of *Arabidopsis* (Reiter *et al.*, 1994) have severely abnormal epidermal cell shape. The *cr4* mutants have thinner outer epidermal walls. Both mutants show a normal basic overall leaf shape, indicating that perturbing epidermal cell shape and cell wall characteristics does not necessarily lead to altered leaf shape. Furthermore, in the *angustifolia* and *rotundifolia* mutants with altered leaf shape due to altered cell elongation, the epidermis appeared less affected than internal cells (Tsuge *et al.*, 1996), indicating that internal cell expansion contributes to organ expansion.

VI. Interactions with Internal Tissues

A. Correlations among Structures

Although the cell lineages of the epidermis and internal tissues are separate since early in embryogenesis, events in epidermal development are clearly tied to mesophyll development. This is obvious from the correlation of anatomical features in the two tissues. Stomates are associated with intercellular air spaces in the mesophyll, epidermal fibers overlay vasculature in many species, and so on. These interrelationships are particularly evident in monocots where anatomical features are arranged in linear arrays (Freeling and Lane, 1994). For instance, stomatal files occur between vascular traces, rarely over them. The question is how development of the two tissues is coordinated. The temporal sequence of development suggests there is likely to be two-way communication. The differentiation of vascular traces in maize precedes the morphological differentiation of associated epidermal cells or stomates (Sharman, 1942), suggesting that the internal vascular tissues probably regulate epidermal differentiation. Conversely, guard cells in *Peperomia* differentiate prior to the appearance of interstitial air spaces, suggesting that the epidermis signals the underlying cells to form a cavity (Sachs and Novoplansky, 1993). Neither stomates nor epidermal glands showed any association with internal anatomical features at earlier stages. Thus descriptive studies suggest that signaling occurs in both directions to coordinate the development of the epidermis and internal tissues.

B. Genetic Mosaics

Considerable insight on tissue interactions can be gained from the study of genetic mosaics. Analysis of plants with genetically different epidermal and internal tissues has revealed that most epidermal characteristics such as trichome morphology and density appear to be genetically controlled by the epidermal cells (Stewart *et al.*, 1972; Kaddoura and Mantell, 1991; Szymkowiak and Sussex, 1992; Szymkowiak and Sussex, 1996). On the other hand, organ and epidermal growth is for the most part regulated by the internal cells. Periclinal chimeras were generated by intergeneric grafting of *Solanum laciniatum,* which has glabrous, lobed leaves, and *Nicotiana tabacum*, which has hairy, entire leaves (Kaddoura and Mantell, 1991). Size and shape of the leaves were determined by the makeup of internal tissues, in particular L2-derived cells, whereas the degree of hairiness was determined by epidermal genotype. Similar conclusions were drawn about the control of leaf growth in an interspecific *Camellia* chimera (Stewart *et al.*, 1972).

The influence of internal tissues on the formation of epidermal structures was clearly demonstrated for maize ligule formation. As mentioned, the ligule initiates by a series of precisely controlled cell divisions in the epidermis and loss-of-function mutations in the *liguleless1* gene block ligule initiation (Becraft *et al.*, 1990; Sylvester *et al.*, 1990). Genetic mosaic analysis indicated that normal *Liguleless1+* functions in both epidermis and internal cells (Becraft *et al.*, 1990). *Lg1+* appeared to act cell autonomously in the lateral dimension; that is *Lg1+* tissue was not able to rescue the mutant phenotype of neighboring *lg1* mutant tissue. However, wild type internal cells could induce a rudimentary ligule in mutant epidermis, indicating that epidermal development was influenced by directionally transmitted information from internal tissues.

Internal cells also influence epidermal development in abscission zone formation. The *jointless* mutant of tomato fails to form an abscission layer in fruit pedicels but wild type internal cells can induce the formation of an abscission layer in mutant epidermis (Szymkowiak and Sussex, 1989). Other examples of internal cells controlling epidermal development include several dominant developmental mutants. The control of epidermal identity by *Kn1* and related mutants expressed internally has already been discussed (Hake and Freeling, 1986; Sinha and Hake, 1990; Fowler *et al.*, 1996); ectopic expression of these homeobox genes, normally restricted to the shoot apical meristem, in internal tissues of the blade causes the epidermis to assume a sheath identity. Also previously discussed were the *Teopod1* (*Tp1*) and *Tp2* mutants. Both mutations prolong the juvenile phase of maize development, thereby affecting epidermal characteristics. Genetic mosaic analysis showed that both mutants act nonautonomously (Dudley and Poethig, 1993). It was not possible to assign sites of action to any particular tissues because wild type sectors of any constitution showed a mutant phenotype due to the influence of neighboring mutant tissues.

C. Cytoplasmic Connections

Cytoplasmic continuity between cells is established by plasmodesmata, regulated channels that span the cell wall between cells (Robards and Lucas, 1990). Plasmodesmatal connections allow cell communication necessary for normal development. *Arabidopsis* embryos are completely dye coupled (McLean *et al.*, 1997) but after germination there is the establishment of symplastic domains (Duckett *et al.*, 1994). Cotyledon and hypocotyl epidermis is dye coupled but dye movement from the epidermis to internal cells was not observed. There is also a barrier to dye movement at the junction between the hypocotyl and root epidermis. Elimination of plasmodesmata or creation of unregulated cytoplasmic connections has severe developmental consequences (reviewed in McLean *et al.*, 1997). Primary plasmodesmata occur during cell wall formation at cytokinesis whereas secondary plasmodesmata are formed in existing cell walls. Because the epidermal cell lineage is largely separate from the internal lineages, the intervening cell wall is derived from expansion of an existing wall and we can infer that plasmodesmata between the epidermis and internal cells are mainly secondary.

Dye-coupling experiments have generally shown cytoplasmic continuity between epidermal and mesophyll cells in mature leaves (Erwee et al., 1985). It is likely that this is required for coordinated development and is clearly required for protein trafficking which occurs between L2 and L1 layers of the meristem and presumably between vascular cells and the epidermis in *Knotted1* mutant maize leaves (Jackson et al., 1994; Lucas et al., 1995). It is possible that similar trafficking of unknown proteins is important for normal leaf development. Epidermal cells are also dye coupled with one another with the exception of guard cells, which become cytoplasmically isolated (Erwee et al., 1985; Palevitz and Hepler, 1985). Cytoplasmic isolation of guard cells is important for regulating the ion fluxes essential for controlling stomate aperture.

VII. Pattern Formation and Cellular Differentiation

The leaf epidermis provides an attractive system for studying the process of cell differentiation. It is readily accessible and replica techniques allow the sequential viewing of the same cells during development. Some cells such as trichomes are dispensable, allowing convenient genetic analysis. The various cell types each have their own characteristic spacing pattern and ontogeny. Epidermal cell spacing patterns have attracted the attention of theoreticians because the accessibility and two-dimensional nature of the epidermis simplify the analysis and because the spacing of stomates in particular is relevant to plant productivity. The two major models for how spacing patterns are established in two dimensions are based on cell lineage or lateral inhibition. There is evidence for both mechanisms in different situations.

A. Stomates

Stomates are gated pores that allow the gas exchange necessary for photosynthesis. The critical physiological role of stomates has generated much interest in their regulation and developmental control. A stomatal complex consists of a pair of guard cells and their associated subsidiary cells. Turgor changes in the guard cells, regulated by ionic flux from subsidiary cells, lead to stomatal opening and closing. Optimal density and spacing of trichomes allow sufficient CO_2 uptake for photosynthesis while minimizing water loss from the plant. Stomatal density is controlled by environmental and genetic factors (Schoch et al., 1984; Yang and Sack, 1995; Serna and Fenoll, 1997). A striking feature of stomatal spacing is that neighboring stomates rarely are in direct contact. The major models for how this spacing pattern is achieved include cell lineage and lateral inhibition.

The cell lineage model states that stomatal precursor cells undergo a defined series of cell divisions that produce a pair of guard cells surrounded by epidermal

1. Development of the Leaf Epidermis 19

cells derived from the same cell divisions (Fig. 3) (Bünning and Sagromsky, 1948; Sachs, 1978). Thus the pattern of cell division leading to formation of a stomatal complex serves to preclude direct contact between neighboring stomates. By the lateral inhibition model, a signal is sent by developing stomates to inhibit stomate formation in neighboring cells (Bünning and Sagromsky, 1948; Korn, 1981). Both models appear to have some validity, with the primary mechanism varying among species. Other models such as early morphogenetic prepatterning have also been proposed (Meinhardt, 1982) although there is little evidence to support such mechanisms.

Stomate development in most species begins with an asymmetric cell division (reviewed in Larkin *et al.*, 1997). This first division has considerable influence on stomatal spacing in both dicots and monocots but beyond that, development is quite different in the two groups. In *Arabidopsis*, the smaller of the two cells, called a meristemoid, undergoes several more asymmetric divisions, culminating in a guard mother cell (GMC) which then divides equally to form a pair of guard cells (Fig. 3; Bünning, 1953). The term meristemoid highlights the fact that the divisions leading to stomate formation occur after most other proliferative cell division in the epidermis has ceased. The pattern of cell division is such that the guard cells form in the center of a group of cells derived from the same precursor cell. Sometimes a satellite meristemoid forms from an asymmetric division in the sister of the GMC. This division is regulated such that the satellite meristemoid forms opposite the primary meristemoid, thus leaving an intervening cell. The satellite may either divide immediately into a guard cell pair or undergo several other divisions before forming a GMC. In *too many mouths* (*tmm*) mutants, there is an increased number of stomata and many are clustered and in direct contact with one another (Yang and Sack, 1995; Larkin *et al.*, 1997). This appears to result from the overproduction of satellite meristemoids and their incorrect orientation of the preceding asymmetric division leading to their placement adjacent to the primary stomate (Larkin *et al.*, 1997). Thus *TMM* represents a critical controlling element in stomatal patterning.

In monocots the contribution of ordered cell divisions to stomate spacing is as follows: stomatal initials are formed in longitudinal cell files. Each initial is derived from an asymmetric division that produces a meristemoid and an interstomatal cell. The smaller meristemoid cell always occurs toward the distal pole with respect to the leaf axis and generally divides only once more to form guard cells. The consequence is that every stomate in a file is separated by at least one interstomatal cell (Tomlinson, 1974; Charlton, 1990; Croxdale *et al.*, 1992).

Stomatal spacing is also regulated at three additional levels: the selection of meristemoid mother cells from among protodermal cells, the regulation of whether stomate initials mature, and the control of the cell identity within a stomatal complex. In *Arabidopsis* the selection of mersitemoid mother cells appears random (Kagan *et al.*, 1992; Sachs and Novoplansky, 1993; Larkin *et al.*, 1997) whereas in other species the spacing pattern is nonrandom (Korn, 1981; Charlton, 1990;

Croxdale *et al.*, 1992). Nonrandom distributions suggest either interactions among meristemoids (lateral inhibition) or interactions with some other component of the developing leaf. Korn has developed computer models based on inhibitory influences of developing meristemoids on neighboring cells that simulate stomatal patterning of several species (Korn, 1981, 1993) demonstrating such patterns can be achieved by this mechanism. However, these models have been criticized because they are derived from mature stomatal patterns rather than observed developmental sequences (Sachs, 1994), leaving this issue unresolved.

Monocot leaves grow by a basal meristem. As cells move out of the meristematic region by intercalary growth of cells below them, they pass through different stages of development and maturation (Freeling, 1992). Distal to the zone of proliferative cell division is a zone of formative cell division where the asymmetric divisions leading to stomate formation occur. Zones of differentiation and maturation follow. The zones of proliferation and formative divisions are separated by a zone of little cell division; it is in this zone that stomatal patterning is thought to occur (Charlton, 1990; Croxdale *et al.*, 1992; Chin *et al.*, 1995). In *Tradescantia*, stomates tend to occur in linear clusters or strings (Charlton, 1990; Croxdale *et al.*, 1992). Within a string, stomates are separated by single epidermal cells, while strings are separated from one another in a file by a variable but nonrandom number of epidermal cells. Charlton (1990) proposed that cells which entered the zone of stomatal patterning in a particular stage of the cell cycle were competent to become stomates and that strings represent small clones of cells which were in roughly the same stage of the cell cycle because they were all derived from the same recent cell division. The observation that stomatal initials within a string divide synchronously is consistent with this hypothesis (Chin *et al.*, 1995). It has also been noted that because protodermal cells are displaced distally by intercalary growth, cells of a string will not pass a given distance from the leaf base simultaneously and therefore not enter the zone of stomatal specification in the same phase of the cell cycle (Charlton, 1990). This problem is explained by the observation that the zones of stomatal differentiation do not occur at fixed distances from the leaf base and appear to occur independently in each cell file (Chin *et al.*, 1995). Thus strings appear to be determined as a group, independently in different files and not necessarily at a fixed position.

Not all stomates that are initiated complete differentiation and stomatal arrest contributes significantly to the mature pattern of stomates in several species (Kagan and Sachs, 1991; Croxdale *et al.*, 1992; Kagan *et al.*, 1992; Boetsch *et al.*, 1995). The control of stomatal maturation is influenced by several factors, including proximity of neighboring stomates. Statistical analysis of patterns of stomatal arrest and maturation indicates that initials in proximity to a neighboring stomate are less likely to complete maturation than those separated by a distance (Kagan and Sachs, 1991; Boetsch *et al.*, 1995; Chin *et al.*, 1995). In *Pisum* the inhibition of neighboring stomate maturation appears to follow strict rules as stomates in direct contact are not observed in the mature epidermis (Kagan *et al.*, 1992). In *San-*

1. Development of the Leaf Epidermis

sevieria there do not appear to be strict rules but rather statistical influences because although the probability of neighboring stomates completing maturation is low, it does occur (Kagan and Sachs, 1991). In *Tradescantia* maturation was not influenced by proximity of cells within the same file but rather to cells in adjacent files (Boetsch *et al.*, 1995). Laser ablation of stomate initials did not release neighboring initials from inhibition, suggesting that the decision to arrest was made early (Croxdale *et al.*, 1992). Maturation may also be regulated by cumulative influences of stomatal density over a broad area as the number of stomata in samples areas of *Sansevieria* epidermis was relatively constant (Kagan and Sachs, 1991).

The final influence over stomate spacing in some species, including *Arabidopsis*, is in the fate decisions for cells produced during the formation of a stomatal complex. The normal sequence is that a meristemoid undergoes three asymmetric divisions to produce four cells, one of which becomes the GMC and the other three accessory cells. The importance of this cell fate control is apparent when normal development is perturbed either by mutation (Yang and Sack, 1995; Larkin *et al.*, 1997) or by manipulating environmental conditions (Serna and Fenoll, 1997). In *four lips* (*flp*) mutants the GMC is not properly specified. Following the symmetric cell division, instead of two guard cells forming, two GMCs form, each of which produces a pair of guard cells (Yang and Sack, 1995; Larkin *et al.*, 1997).

Phenocopies of *tmm* and *flp* can be induced by growing *Arabidopsis* plants in a secluded atmosphere, which limits the potential for gas exchange (Serna and Fenoll, 1997). Because most other epidermal cell division has ceased at the time stomatal cell divisions occur, it is possible to recognize prestomatal cells by the expression of cell cycle genes such as *CDC2a* and *CYC1a*. Promoters for these and the *RHA1* gene fused to the GUS reporter provide convenient markers for developing stomata. Meristemoids express *CDC2a* and *CYC1a*, GMCs express *CDC2a*, *CYC1a*, and *RHA1*, while mature guard cells, subsidiary cells, or pavement cells do not express any of these (Terryn *et al.*, 1993; Serna and Fenoll, 1997). Using these markers to follow development of stomatal clusters induced by atmospheric seclusion, Serna and Fenoll (1997) found that meristemoids were not clustered; rather, subsidiary cells assumed GMC fate. Furthermore, in a developing cluster, only one pair of guard cells expressed the markers, indicating that the stomata within a cluster were not synchronous. Thus environmental conditions altered the fate of subsidiary cells, resulting in markedly altered stomatal patterning.

Differentiation of stomatal complexes involves dramatic examples of cellular induction and asymmetric divisions. Subsidiary cells form in cells adjacent to the GMCs. In monocots, this involves cells in neighboring cell files of separate lineage, indicating that the GMC induces the formation of subsidiary cells (Fig. 3B). The unusual cell division leading to the formation of the subsidiary cells is highly asymmetric, producing a small lens-shaped cell on the side of the subsidiary mother cell (SMC) in contact with the GMC (Wick, 1991). An unusual aspect of this division is that both ends of the new cell wall form junctions with the same side wall of the mother cell instead of bisecting the cell as in a typical division. The ability to predict the future cell division by proximity to a GMC allows the convenient study of cytoskeletal behavior in this unusual cytokinesis. Prior to division actin filaments accumulate on the lateral walls of the GMC and of the SMC adjacent to the GMC, marking the site of the future spindle pole (Cho and Wick, 1990; Wick, 1991; Cleary and Mathesius, 1996). Subsequently, the SMC nucleus migrates and becomes appressed to the wall adjacent to the GMC. Microtubules form a structure called the preprophase band, which predicts the plane of nearly all plant cell divisions, including subsidiary cells (Cho and Wick, 1989). In subsidiary cell division the preprophase band forms an arc which intersects the SMC wall adjacent to the GMC. Actin filaments are necessary for the proper positioning of the SMC nucleus and preprophase band as cytochalasin B treatments disrupt the normal arrangement (Cho and Wick, 1990). Following mitosis, cytokinesis proceeds through the formation of a new arc-shaped cell wall in the prior location of the preprophase band. Thus GMCs induce an unusual cell division requiring the intricate interplay of various cytoskeletal components coordinated between neighboring cells.

B. Trichomes

Trichomes are epidermal hairs which, although performing a protective role on the leaf surface, are dispensable. The genetic control of trichome differentiation has been elegantly dissected in *Arabidopsis* (reviewed in Marks, 1997). Trichomes of *Arabidopsis* leaves are relatively large, single-celled structures containing 2–5 branches. Mutants have been identified that eliminate trichomes, alter trichome spacing, or alter trichome morphology. The *glabrous1* (*gl1*) and *transparent testa glabra* (*ttg*) mutants cause nearly complete elimination of trichomes from the leaf surface. Whereas *gl1* is specific to trichome development on aerial surfaces (Marks, 1997), *ttg* has pleiotropic effects, including increased root hair formation and loss of pigmentation in the seed coat (Korneef, 1981). *GL1* encodes an MYB domain protein and is expressed in leaf primordia and at later stages of development, specifically in trichomes (Oppenheimer *et al.*, 1991; Larkin *et al.*, 1993). *GL1*, under the control of the constitutive 35S promoter in transgenic plants, was not able to rescue *ttg* mutants, indicating that *GL1* is not downstream of *TTG* in a simple hierarchy. The *ttg* mutant can be complemented by the maize *R* gene, which is similarly unable to rescue the *gl1* mutant phenotype. Constitutive expression of both *GL1* and *R* resulted in nearly every epidermal cell being converted to a trichome. Thus trichome initiation requires the functions of both *GL1* and *TTG* genes which act to determine trichome cell identity (Larkin *et al.*, 1994).

Trichomes do not normally form in adjacent cells. Statistical analysis indicated that spacing is not random but that some developmental mechanism must be controlling spacing (Larkin *et al.*, 1996). Trichome spacing appears to be controlled more by lateral inhibition than by cell lineage mechanisms. Clonal analysis indicated that trichome complexes (hair plus multicellular base) could be variously bisected by sector boundaries (Larkin *et al.*, 1996). Thus trichome complexes are derived from variable cell lineages, which would not be expected if a set pattern of cell division produced the observed spacing pattern. In *ttg1*/+ plants expressing *GL1* under 35S control, trichome clusters were commonly observed (Larkin *et al.*, 1994). This was interpreted as being due to a decreased level of an inhibitory factor resulting from decreased *TTG* dose. This implies that *TTG* regulates a factor that helps determine trichome spacing by inhibiting the formation of trichomes in proximity to an existing one. Mutations in *TRYPTOCHON* (*TRY*) produce similar trichome clusters, suggesting a relationship to the putative inhibitory factor regulated by *TTG*.

Interestingly, constitutive expression of *GL1* alone caused a reduction in trichome number on adaxial leaf surfaces (Larkin *et al.*, 1994). The *try* mutation allows increased trichome formation and clustering (Fig. 3) in response to *GL1* overexpression but most epidermal cells still do not form trichomes, indicating that additional factors regulate trichome development (Szymanski *et al.*, 1998b). A genetic screen of mutagenized *35S::GL1* lines identified a mutation, *cot1*, which leads to increased abundance of trichomes in *GL1*-overexpressing lines but which has a normal phenotype alone (Szymanski *et al.*, 1998b). The *cot* and *try* mutants interact synergistically to allow a dramatic overabundance of clustered trichomes in response to *GL1* overexpression.

Time is also an important controlling factor in trichome spacing. Trichome initiation occurs over an extended period of leaf development during which new trichomes initiate between existing ones. A genetic locus that influences trichome spacing was identified by a polymorphism in trichome density between two *Arabidopsis* ecotypes. This variation was conditioned primarily by a single locus, *REDUCED TRICHOME NUMBER* (*RTN*) (Larkin *et al.*, 1996). The basis for the phenotype was that in leaves of *RTN* plants the developmental window during which trichome initiation occurred was abbreviated, causing less intercalary trichome initiation and lower final density.

Trichomes present an attractive system for studying the process of cellular morphogenesis. A variety of mutants have been identified that affect various aspects of trichome expansion and branching patterns (Hülskamp *et al.*, 1994; Marks, 1997). *GLABRA2* (*GL2*) encodes a homeodomain protein required for an early step in trichome morphogenesis (Rerie *et al.*, 1994). The *gl2* mutant inhibits trichome expansion outward from the leaf surface and enhances lateral expansion. Similar to *GL1*, *GL2* is expressed throughout leaf primordia, becoming restricted to trichome cells later in development (Szymanski *et al.*, 1998a). In early leaf development, expression of a GUS reporter under the regulation of the *GL2* promoter is equivalent in trichomes and nontrichome cells. The *GL2::GUS* reporter expressed in leaf primordia of *gl1* and *ttg* mutants, but not in the epidermis, indicating that specification of trichome cell fate is required for late *GL2* expression. In *gl1 ttg* double mutants primordial *GL2* expression was also reduced. These results are consistent with epistatic relationships which place *GL2* downstream of *GL1* and *TTG* (Marks, 1997) and suggest that *GL2* could be regulated directly by *GL1* (Szymanski *et al.*, 1998a).

After the initial outgrowth, *Arabidopsis* trichomes branch. Genetic analysis indicates that branching is controlled at several levels. There is a correlation between cell size, the amount of endoreduplication, and the degree of branching; mutants such as *glabra3* that cause smaller cells reduce the number of branches whereas those such as *kaktus* increase cell size and branching (Hülskamp *et al.*, 1994; Folkers *et al.*, 1997). There is also a specific control of branching pattern. Branching can be separated into primary and secondary branching events (Hülskamp *et al.*, 1994; Folkers *et al.*, 1997). The three-branched trichome is produced by two sequential branching events. Following the first branching, two unequal branches are produced, one of which (the one oriented distally relative to the leaf axis) produces a secondary branching. Overall branching is controlled by two genes, *STICHEL* and *NOECK*. Mutants of *stichel* have unbranched trichomes whereas *noeck* mutants are more highly branched; neither mutation affects trichome cell size (Hülskamp *et al.*, 1994; Folkers *et al.*, 1997). Specific control of each branching event is demonstrated by *stachel* mutants, which skip primary branching but are normal for secondary, whereas *zwichel* (*zwi*) mutants perform primary branching but not secondary (Hülskamp *et al.*, 1994; Folkers *et al.*, 1997). *ZWI* encodes a kinesin-like protein with a myosin domain, motor domain, and calmodulin binding domain (Oppenheimer *et al.*, 1997). This suggests that the *ZWI* gene product is involved in transporting cellular components along microtubules in a Ca^{2+}-regulated manner, tying together the cytoskeleton and Ca^{2+} in the regulation of branching patterns.

Little is known about trichome development in monocots; however, like stomates, the mechanisms regulating spacing and distribution are likely to be quite different from those of dicots. Maize leaves contain three distinctly different types of trichomes: macrohairs, prickle hairs, and bicellular hairs (Freeling and Lane, 1994) and there are several subtypes of macrohair. Each hair type shows a distinct pattern of distribution on the leaf. Macrohairs occur on the leaf margin and in bulliform cell files wherein they appear regularly spaced (Fig. 1). Prickle hairs occur predominantly in files neighboring bulliform rows and are more closely spaced than macrohairs. Thus the distribution of these trichomes is coupled to other epidermal cell types and occurs in linear files typical of the monocot epidermis. Bicellular hairs seem to be randomly distributed although an analysis of their distribution has not been reported. It is possible that their spacing might be achieved by mechanisms similar to those for *Arabidopsis* trichomes.

VIII. Determined State of the Epidermis

A. Determination of Epidermal Fate

During embryogenesis the protoderm is established as one of the elements of radial pattern (Mayer *et al.*, 1991). The *knolle* and *keule* mutations were identified as disrupted in radial pattern formation in a genetic screen to identify genes controlling embryonic pattern formation (Jürgens et al., 1991; Mayer et al., 1991). Both *knolle* and *keule* mutants cause incomplete cytokinesis (Assaad et al., 1996; Lukowitz et al., 1996) and *KNOLLE* encodes a syntaxin-like protein thought to be involved in vesicle fusion (Lukowitz et al., 1996). The *keule* defect appears less severe as mature embryos eventually form complete cross walls and are able to establish a radial pattern as evidenced by provascular differentiation and positionally correct expression of the protodermal marker, *AtlTP1* (Vroemen et al., 1996). Mutants of *knolle* do not form complete cross walls during cytokinesis and do not establish any elements of radial pattern (Lukowitz et al., 1996; Vroemen et al., 1996).

The effect of incomplete cytokinesis on pattern formation has several possible explanations. One possibility is that normally, cytoplasmic determinants are localized to specific regions of the embryo and specify cell fate accordingly. A scenario similar to that for *Drosophila* embryos where *bicoid* mRNA is localized and the translated protein freely diffuses (reviewed in St. Johnson and Nüsslein-Volhard, 1992) could be envisaged. Perhaps a message is localized in the outer cytoplasmic region and following the protodermal cell divisions is translated. Incomplete cytokinesis would not allow sequestration of the translation product, resulting in improper pattern formation. It is also possible that determinants might be localized to the outer cell wall (Vroemen et al., 1996) as demonstrated for *Fucus* embryos (Berger et al., 1994), and again subsequent cytoplasmic sequestration would be required to maintain the differentially specified cell fates between surface and internal cells. The presence of a cuticle on *Citrus* zygotes has led to the suggestion that zygotes possess epidermal identity and that internal cell types differentiate as they are isolated from the surface (Bruck and Walker, 1985a). The observed expression of the *Atl TP1* protodermal marker in internal cells of *knolle* mutant embryos is consistent with this model (Vroemen et al., 1996).

In postembryonic plant growth, cells of the shoot apical meristem are generally considered to be undifferentiated and not until the formation of organ primordia is the surface cell layer considered protoderm. However, the presence of a cuticle and the restriction of cell division to the anticlinal plane have led to the suggestion that epidermal identity is already determined in the L1, which would more aptly be considered as protoderm (Bruck and Walker, 1985a). L1-specific regulation of gene expression (Jackson et al., 1994; Lu et al., 1996) supports the idea that these cells have a specific identity and are not completely undifferentiated. Specification of epidermal identity appears to occur only at distinct times during plant development—during embryogenesis and in the formation of adventitious shoot meristems (Bruck and Walker, 1985b). Wounded tissue, even in cultured embryos, forms a periderm but does not redifferentiate an epidermis. Thus mere contact with the atmosphere or lack of contact with neighboring cells is insufficient to induce epidermal differentiation. Specific developmental cues are required and these only occur at discrete times in plant development.

B. Epidermal Potency

A key function of the epidermis is to maintain the individual identity of adjacent organs during development by preventing organ fusions. One feature of the epidermis related to this function is its stably determined state; it is difficult to induce the epidermis dedifferentiate and adapt alternative developmental pathways. Graft unions will not form unless the epidermis is removed and epidermal cells rarely participate in the wound healing associated with graft formation (Moore, 1984; Walker and Bruck, 1985; Bruck et al., 1988). The inability to graft may be a property of the epidermal differentiation state rather than an effect of an impermeable cuticle preventing cell interactions. Removal of the cuticle by gentle abrasion did not permit graft formation between leaf epidermises (Bruck et al., 1988). This treatment allowed diffusion of calcofluor white into the outer epidermal cell wall, indicating that molecular signals were free to pass between the cells in contact. Similarly, in wound healing the epidermis does not participate in the formation of periderm. It was concluded that the inability to graft was a property of epidermal cells per se and that the epidermis is terminally differentiated (Bruck et al., 1988).

Several recent reports contradict the notion that epidermal cells are irreversibly differentiated and indicate that at least some epidermal cells remain totipotent. Cultured leaf explants of an interspecific *Helianthus* line showed exceptionally high frequencies of somatic embryogenesis and it was found that the embryos were derived from epidermal cells (Fambrini et al., 1996). Explants formed embryos on medium lacking hormones, a condition viewed as permissive rather than inductive, leading to the conclusion that embryogenic potential was predetermined in the epidermal cells of this line. Somatic embryos also occurred in cultured epidermal peels of *Gaillardia picta* (Pillai et al., 1992). Guard cells did not participate in the formation of these embryos, which was interpreted as the guard cells being more highly differentiated than other epidermal cells. However, embryogenic callus produced from cultured leaf epidermal peels of sugar beet was derived exclusively from guard cells (Hall et al., 1996). It is difficult to draw generalizations from these few reports but it appears clear that at least some species retain totipotency in some cells of the leaf epidermis.

Epidermal dedifferentiation and organ fusion occur at very specific points in plant development, most commonly in the fusion of floral organs. A clear example is in postgenital carpel fusion in *Catharanthus roseus*. Two carpels are initiated separately and as they grow together, epidermal cells in the area of contact dedifferentiate and adapt an altered fate of internal parenchyma (Walker, 1975a, 1975b, 1975c). Complete fusion of the two organs is evident in the cytoplasmic continuity established by the formation of plasmodesmata between the contacting epidermises (van der Schoot et al., 1995). Dedifferentiation was blocked when impermeable barriers were placed between the carpels but proceeded when permeable barriers separated the developing organs, showing that a diffusible factor induced the response (Verbeke and Walker, 1986; Siegel and Verbeke, 1989). These agar-impregnated barriers became loaded with the factor and, when placed on a noncontacting carpel surface, were able to induce ectopic dedifferentiation. The loaded agar was unable to induce dedifferentiation in anther epidermis, demonstrating that responsiveness is specific to carpel epidermis.

Although the leaf epidermis is not normally competent to dedifferentiate or fuse, there are several mutations that cause epidermal fusion. *Arabidopsis fiddlehead1* (*fdh1*) mutants display epidermal fusions among lateral organs of the shoot, including floral organs and leaves (Lolle et al., 1992). Fusions occur postgenitally in regions of contact between developing organs but unlike carpel fusion, the epidermal layers remain distinct. An electron-dense material accumulates between epidermises in the region of adherence, apparently acting to glue intact epidermises together. Stomata can be recognized in these regions, indicating the epidermal cell types remain differentiated. Pollen germinates on *fdh1* leaf surfaces, suggesting that the organ fusion is part of a syndrome involving ectopic expression of a developmental program normally restricted to the gynoecium (Lolle and Cheung, 1993). The cell wall and cuticle of *fdh1* mutants are more permeable than normal and wall proteins more prone to leaching by aqueous buffer washes (Lolle et al., 1997). Several specific bands on protein gels appeared particularly enhanced in mutants. In addition, lipid composition of the cuticle showed an altered profile. Thus organ fusion in *fdh1* could be mediated by altered epidermal cell wall and cuticular properties. A maize mutant, *adherent1* (*ad1*), shows similar, but distinct, phenotypic characteristics to *fdh1*; *ad1* mutant leaves adhere but epidermal cells have normal morphology and identity. Mutant leaves have altered epicuticular wax structure and cell wall characteristics, including increased thickness, higher abundance of unesterified pectins, reduced levels of some arabinogalactans, and the occurrence of callose (Sinha and Lynch, 1998). Unlike *fdh1*, there is no evidence that the *ad1* mutant phenotype involves ectopic expression of a gynoecial program.

The *crinkly4* (*cr4*) mutation of maize also causes adherent leaves and floral organs (Becraft et al., 1996) but the *cr4* phenotype contrasts significantly with *fdh1* and *ad1*. Unlike *fdh1* and *ad1*, *cr4* mutants have striking effects on epidermal morphology (Fig. 4). Mutant epidermal cells are abnormally large, thin walled, and irregularly shaped and all epidermal cell types are affected. Tissues other than the epidermis are also affected but the epidermal defects are most pronounced (Ping Jin and P. W. Becraft, unpublished). Regions of adherence form graftlike unions with apparently complete fusion between adjoining epidermises and it is often difficult to determine the organ of origin for the cells involved. Occasionally, a piece of leaf will tear from its donor organ and remain attached via the region of adherence to the recipient. In such cases, the tissue remains alive, indicating that nutrients are transferred across the fused areas (P. Becraft, unpublished). Cell junctions within an epidermal layer are less tightly formed in the *cr4* mutant than wild type. Thus, epidermal cell associations that are normally very strong form weakly in *cr4* mutants and cell associations that are normally prevented are able to occur. The

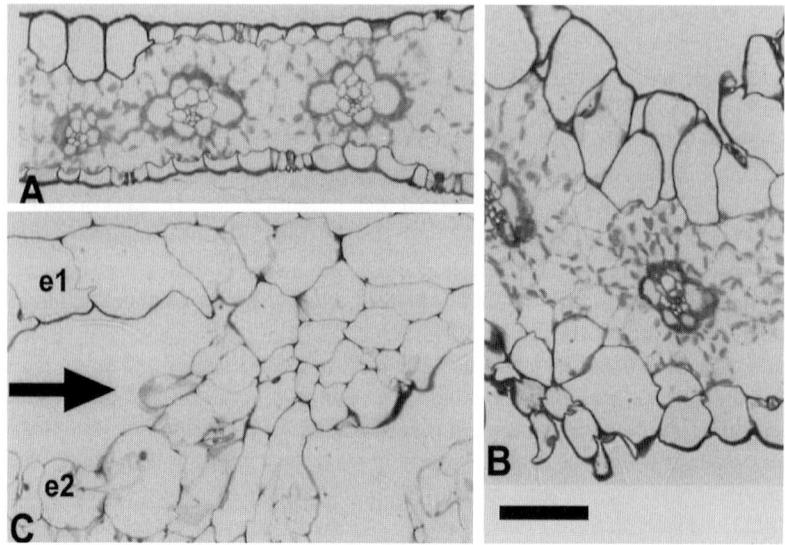

Fig. 4 Phenotype of *cr4* mutant leaves. (A) Transverse section through a wild type maize leaf. (B) Transverse section through a *cr4* mutant leaf at the same magnification. The internal cells have relatively normal proportions but the epidermal cells are grossly enlarged, misshapen, and multilayered. (C) Region of adherence (arrow) between two leaves. The abaxial epidermis of one leaf is designated e1 and the adaxial epidermis of the second leaf is designated e2. The size bar represents 100 μM.

phenotype indicates that the *cr4* gene product is necessary for proper epidermal differentiation (Becraft *et al.*, 1996).

The *cr4* gene codes for a putative receptor kinase (Becraft *et al.*, 1996). This implies that an extracellular signal induces proper epidermal differentiation and raises the question of what cells are the source of the signal. One possibility is that internal cells signal to the epidermis to induce differentiation. A second model is that the CR4 receptor functions in a self-recognition mechanism whereby epidermal cells signal one another, leading to the formation of a unified layer of tightly associated cells. Formation of such a cohesive layer could then preclude formation of further unions between epidermises during development. Identification of the signal ligand for the CR4 receptor will reveal much about how plant tissues, in particular the epidermis, are organized during development.

Several additional mutants cause epidermal fusions. Two *Arabidopsis* mutants designated *wax* have glossy leaf surfaces due to deficiencies of epicuticular wax (Jenks *et al.*, 1996a). Both mutants show suturelike epidermal fusions similar to *fdh1*. In a genetic screen for organ fusion mutants, 29 mutants were identified which fall into 9 complementation groups, including *fdh* (Lolle *et al.*, 1998). Some caused only floral organ fusion, but *conehead*, *deadhead*, and *thunderhead* showed leaf fusions as well. Pollen hydrated on the leaf surfaces of these three mutants, similar to *fdh1*, suggesting that these mutants also express aspects of a gynoecial developmental program in the leaves. These three mutants also showed cuticular defects as evidenced by the increased rate of chlorophyll extraction from leaf tissues and, for *deadhead*, the eceriferum phenotype.

IX. The Cuticle

The cuticle is a waxy layer that forms a protective covering on the surface of the outer epidermal wall. It reduces transpirational water loss and protects against insects, pathogens, UV light, and frost. As mentioned in the previous section, cuticular characteristics also appear to be important to normal development. Cuticle biosynthesis has been comprehensively reviewed elsewhere (Kolattukudy, 1981; Lemieux, 1996; Post-Beittenmiller, 1996) and will only be highlighted here.

The cuticle is not a uniform layer but changes structurally and compositionally from inside to out (Martin and Juniper, 1970). The cuticle proper consists of the inner secondary cuticle and the outer primary cuticle. The secondary cuticle often contains reticulate fibrillae that intermesh with the outermost portion of the cell wall whereas the primary cuticle generally appears amorphous. The surface consists of the epicuticular wax layer, which either can be smooth or can form microcrystalline arrays.

The main components of the cuticle are cutin and wax, with proteins forming a minor constituent. Cutin consists of a matrix of fatty acid polymers. Monomers of fatty acid derivatives such as hydroxy fatty acids, dihydroxy fatty acids, diols, and dicarboxylic acids are joined by ester linkages and the resultant polymers are cross esterified (Kolattukudy, 1981). Cutin is embedded in cuticular waxes comprised of complex mixtures of primarily very long chain ($>C18$) fatty acids, hydrocarbons, alcohols, aldehydes, ketones, esters, triterpenes, sterols, and flavonoids (Post-Beittenmiller, 1996). The most abundant cuticular protein in broccoli is a lipid transfer protein, thought to be involved in transport of cuticular wax precursors (Pyee *et al.*, 1994; Pyee and Kolattukudy, 1995).

The structure and composition of the cuticle vary significantly among species and are genetically and developmentally regulated. For example, maize leaf wax contains mainly primary alcohols, aldehydes, and wax esters but very little alkanes (Bianchi *et al.*, 1985). *Arabidopsis* leaves contain mostly alkanes with less primary alcohols and very little aldehydes or wax esters (Jenks *et al.*, 1995). Developmental regulation is seen in the 10-fold higher wax load per unit area in *Arabidopsis* stems compared to leaves. In addition, leaf wax had longer average chain lengths than stem and a higher proportion of alkanes and 1-alcohols with a large reduction in secondary alcohols, ketones, and esters (Jenks *et al.*, 1995). The levels of free fatty acids in *Arabidopsis* leaf wax decreased 4.5-fold between 7 and 15 days after germination whereas the wax composition of *Arabidopsis* stems changed little (Jenks *et al.*, 1996b). Similarly, the epicuticular wax of maize seedling leaves contains a microcrystalline structure whereas on adult leaves it is smooth (Poethig, 1990).

The biosynthesis of cuticles is complex and poorly understood. Biochemical studies have identified many of the enzymes involved in wax biosynthesis (Evenson and Post-Beittenmiller, 1995; Liu and Post-Beittenmiller, 1995; Post-Beittenmiller, 1996). Molecular genetic analysis of epicuticular wax mutants provides a complementary approach to identify components of the regulatory and biosynthetic machinery (Lemieux, 1996). Microcrystalline epicuticular wax, called wax bloom, causes a dull, blue-green luster (glaucousness) and readily sheds water (Schnable *et al.*, 1994). Mutants that disrupt wax biosynthesis cause a bright green, glossy luster and will bead water when sprayed with a mist. In maize such mutants are called *glossy* (*gl*) and in *Arabidopsis* they are called *eceriform* (*cer*), although in *Arabidopsis* only the stem contains wax bloom and not leaves. Over 20 mutants are known to affect wax deposition in both maize (Fig. 5) and *Arabidopsis* (Schnable *et al.*, 1994; Jenks *et al.*, 1995; Jenks *et al.*, 1996a). Analysis of wax composition of the various mutants provides clues as to the function of the respective genes and allows them to be placed in a relative order (Jenks *et al.*, 1995). The mutants identify genes encoding enzymes in wax biosynthesis, regulators of wax biosynthesis, or genes that affect cuticular wax indirectly. An example of the latter is the maize *gl15* gene, discussed previously, which controls juvenile epidermal cell identity (Evans *et al.*, 1994; Moose and Sisco, 1994, 1996).

Molecular cloning of several genes involved in wax production have recently been reported. *CER1* of *Arabidopsis* and *gl1* of maize encode similar proteins (Hansen *et al.*, 1997). *CER1* was predicted to encode an aldehyde decarbonylase based on biochemical analysis of mutant waxes (McNevin *et al.*, 1993; Aarts *et al.*, 1995); however, others proposed a transport function (Jenks *et al.*, 1995). GL1 was predicted to be integral membrane protein with five or seven membrane-spanning regions and similarity to a class of mammalian receptors including β-chemokine receptors which are involved in membrane fusions (Hansen *et al.*, 1997). This similarity suggests that CER1 and GL1 could be involved in transport vesicle fusion with the plasma membrane, releasing cuticular wax components (Hansen *et al.*, 1997). *CER2* and *gl2* also encode similar proteins which do not show similarity to other proteins in databases (Tacke *et al.*, 1995; Negruk *et al.*, 1996; Xia *et al.*, 1996). Biochemical analysis indicates that *cer2* and *gl2* mutations affect the terminal elongation step in very long chain fatty acid biosynthesis (Hannoufa *et al.*, 1993; Jenks *et al.*, 1995), leading to the proposal that these genes encode an elongase specific for C26 substrates (Xia *et al.*, 1996). A regulatory function for *CER2* has alternatively been proposed because the *cer2* mutation only affects wax composition of the stem and not leaves (Jenks *et al.*, 1995). *CER6*, on the other hand, affects the same classes of fatty acids as *CER2* but in both stems and leaves, suggesting an enzymatic function in terminal elongation (Jenks *et al.*, 1995). *CER3* encodes a protein that does not match known protein sequences (Hannoufa *et al.*, 1996). A predicted nuclear localization signal sug-

1. Development of the Leaf Epidermis

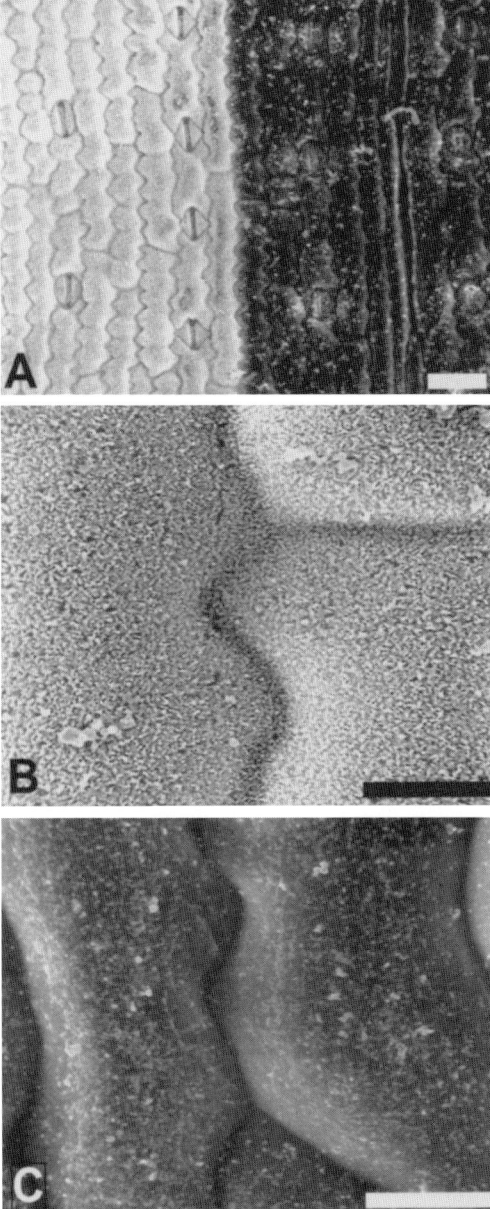

Fig. 5 (A) SEM of a revertant sector on a juvenile *gl1* mutant maize leaf. (B and C) High-magnification SEM of epicuticular waxes on cells of wild type and *gl1* mutant juvenile maize leaves, respectively. Courtesy of Patrick S. Schnable and Tsui-Jung Wen.

gests that *CER3* might have a regulatory function. Thus far the molecular analysis of these genes has produced few obvious answers to their molecular function. Further analysis is required to resolve how cuticular wax production is genetically controlled.

X. Summary

The leaf epidermis is essential to plant survival not only because of its protective role at the interface with the plant's environment but also because of crucial developmental functions. The protoderm is set aside early in embryogenesis, possibly in the zygote. Epidermal identity is determined by the interactions of a complex set of factors, including developmental phase of the plant, regional identity within the leaf, and axiality. For the most part, these characteristics appear to be specified by internal tissues. On the other hand, the epidermis has a key role in regulating organ growth and expansion; thus interactions between the epidermis and internal tissues regulate the overall leaf architecture. Overlying this is the specification of different cell types within the epidermis. Some aspects of this appear to involve interactions with internal tissues but the patterning of many epidermal cell types seems to occur within the two-dimensional field of the epidermis itself and to require both cell signaling and cell lineage dependent mechanisms.

Genetic analyses have provided much of the insight into the underlying principles that regulate epidermal development and a number of molecules important for various aspects of the process have been identified. Yet, for the most part, our understanding of the molecular basis for each component of epidermal development is still rudimentary and we have not yet scratched the surface of understanding how these pieces are integrated. The emerging technologies of functional genomics will provide powerful tools for solving these problems and the near future is likely to produce rapid progress.

Acknowledgments

The author thanks David Marks, Patrick Schnable, and Tsui-Jung Wen for providing figures, David Marks, Susan Lolle, and Neelima Sinha for providing preprints, and Diane Luth, Ping Jin, and Yvonne Asuncion-Crabb for critical reading of the manuscript. Research in the author's laboratory is supported by the National Science Foundation.

References

Aarts, M. G. M., Keijzer, C. J., Stiekema, W. J., and Pereira, A. (1995). Molecular characterization of the *CER1* gene of *Arabidopsis* involved in epicuticular wax biosynthesis and pollen fertility. *Plant Cell* **7,** 2115–2127.

1. Development of the Leaf Epidermis

Assaad, F. F., Mayer, U., Wanner, G., and Jurgens, G. (1996). The *KEULE* gene is involved in cytokinesis in *Arabidopsis*. *Mol. Gen. Genet.* **253**, 267–277.

Becraft, P. W., and Freeling, M. (1989). Use of the scanning electron microscope to ascribe leaf regional identities even when normal anatomy is disrupted. *Maize Genet. Coop. Newsl.* **63**, 37.

Becraft, P. W., and Freeling, M. (1994). Genetic analysis of *Rough sheath1* developmental mutants of maize. *Genetics* **136**, 295–311.

Becraft, P. W., Bongard-Pierce, D. K., Sylvester, A. W., Poethig, R. S., and Freeling, M. (1990). The *liguleless-1* gene acts tissue specifically in maize leaf development. *Dev. Biol.* **141**, 220–232.

Becraft, P. W., Stinard, P. S., and McCarty, D. R. (1996). CRINKLY4: A TNFR-like receptor kinase involved in maize epidermal differentiation. *Science* **273**, 1406–1409.

Berger, F., Taylor, A., and Brownlee, C. (1994). Cell fate determination by the cell wall in early *Fucus* development. *Science* **263**, 1421–1423.

Bergfeld, R., Speth, V., and Schopfer, P. (1988). Reorientation of microfibrils and microtubules at the outer epidermal wall of maize coleoptiles during auxin-mediated growth. *Bot. Acta* **101**, 31–41.

Bianchi, G., Avato, P., and Salamini, F. (1985). Biosynthetic pathways of epicuticular wax of maize as assessed by mutation, light, plant age and inhibitor studies. *Maydica* **30**, 179–198.

Boetsch, J., Chin, J., and Croxdale, J. (1995). Arrest of stomatal initials in *Tradescantia* is linked to the proximity of neighboring stomata and results in the arrested initials acquiring properties of epidermal cells. *Dev. Biol.* **168**, 28–38.

Bossinger, G., Maddaloni, M., Motto, M., and Salamini, F. (1992). Formation and cell lineage patterns of the shoot apex in maize. *Plant J.* **2**, 311–320.

Bruck, D. K., and Walker, D. B. (1985a). Cell determination during embryogenesis in *Citrus jambhiri*. I. Ontogeny of the epidermis. *Bot. Gaz.* **146**, 188–195.

Bruck, D. K., and Walker, D. B. (1985b). Cell determination during embryogenesis in *Citrus jambhiri*. II. Epidermal differentiation as a one-time event. *Am. J. Bot.* **72**, 1602–1609.

Bruck, D. K., Alvarez, R. J., and Walker, D. B. (1988). Leaf grafting and its prevention by the intact and abraded epidermis. *Can. J. Bot.* **67**, 303–312.

Bünning, E. (1953). "Entwicklungs- und Bewegungsphysiologie der Pflanze." Springer-Verlag, Berlin.

Bünning, E., and Sagromsky, H. (1948). Die Bildung des Spaltoffnungsmusters in der Blattepidermis. *Z. Naturforsch B* **3**, 203–216.

Catalá, C., Rose, J. K. C., and Bennet, A. B. (1997). Auxin regulation and spatial localization of an endo-1,4-β-D-glucanase and a xyloglucan endotransglycosylase in expanding tomato hypocotyls. *Plant J.* **12**, 417–426.

Cerioli, S., Marocco, A., Maddaloni, M., Motto, M., and Salamini, F. (1994). Early event in maize leaf epidermis formation as revealed by cell lineage studies. *Development* **120**, 2113–2120.

Charlton, W. A. (1990). Differentiation in leaf epidermis of *Chlorophytum comosum* Baker. *Ann. Bot.* **66**, 567–578.

Chien, J. C., and Sussex, I. M. (1996). Differential regulation of trichome formation on the adaxial and abaxial leaf surfaces by gibberellins and photoperiod in *Arabidopsis thaliana* (L.) Heynh. *Plant Physiol.* **111**, 1321–1328.

Chin, J., Wan, Y., Smith, J., and Croxdale, J. (1995). Linear aggregations of stomata and epidermal cells in *Tradescantia* leaves: Evidence for their group patterning as a function of the cell cycle. *Dev. Biol.* **168**, 39–46.

Cho, S., and Wick, S. M. (1989). Microtubule orientation during stomatal differentiation in grasses. *J. Cell Sci.* **92**, 581–594.

Cho, S., and Wick, S. M. (1990). Distribution and function of actin in the developing stomatal complex of winter rye (*Secale cereale* cv. Puma). *Protoplasma* **157**, 154–164.

Cleary, A. L., and Mathesius, U. (1996). Rearrangements of F-actin during stomatogenesis visualised by confocal microscopy in fixed and permeabilised *Tradescantia* leaf epidermis. *Bot. Acta* **109**, 15–24.

Cosgrove, D. J. (1997). Relaxation in a high-stress environment: The molecular bases of extensible cell walls and cell enlargement. *Plant Cell* **9**, 1031–1041.

Croxdale, J., Smith, J., Yandell, B., and Johnson, J. B. (1992). Stomatal patterning in *Tradescantia*: An evaluation of the cell lineage theory. *Dev. Biol.* **149**, 158–167.

Dale, J. E. (1988). The control of leaf expansion. *Annu. Rev. Plant Physiol. Plant Mol. Biol.* **39**, 267–295.

Dawe, R. K., and Freeling, M. (1991). Cell lineage and its consequences in higher plants. *Plant J.* **1**, 3–8.

Duckett, C. M., Oparka, K. J., Prior, D. A. M., Dolan, L., and Roberts, K. (1994). Dye-coupling in the root epidermis of *Arabidopsis* is progressively reduced during development. *Development* **120**, 3247–3255.

Dudley, M., and Poethig, R. S. (1993). The heterochronic *Teopod1* and *Teopod2* mutations of maize are expressed non-cell-autonomously. *Genetics* **133**, 389–399.

Erwee, M. B., Goodwin, P. B., and van Bel, A. J. E. (1985). Cell–cell communication in the leaves of *Commelina cyanea* and other plants. *Plant Cell Environ.* **8**, 173–178.

Esau, K. (1977). "Anatomy of Seed Plants." Wiley, New York.

Evans, M. M., and Poethig, R. S. (1995). Gibberellins promote vegetative phase change and reproductive maturity in maize. *Plant Physiol.* **108**, 475–487.

Evans, M. M., Passas, H. J., and Poethig, R. S. (1994). Heterochronic effects of *glossy15* mutations on epidermal cell identity in maize. *Development* **120**, 1971–1981.

Evenson, K. J., and Post-Beittenmiller, D. (1995). Fatty acid-elongating activity in rapidly expanding leek epidermis. *Plant Physiol.* **109**, 707–716.

Fahn, A. (1990). "Plant Anatomy." Pergamon, Elmsford, NY.

Fambrini, M., Cionini, C., and Pugliesi, C. (1996). Development of somatic embryos from morphogenetic cells of the interspecific hybrid *Helianthus annuus X Helianthus tuberosus*. *Plant Sci.* **114**, 205–214.

Flanders, D. J., Rawlins, D. J., Shaw, P. J., and Lloyd, C. W. (1990). Nucleus-associated microtubules help determine the division plane of plant epidermal cells: Avoidance of four-way junctions and the role of cell geometry. *J. Cell Biol.* **110**, 1111–1122.

Folkers, U., Berger, J., and Hülskamp, M. (1997). Cell morphogenesis of trichomes in *Arabidopsis*: Differential control of primary and secondary branching by branch initiation regulators and cell growth. *Development* **124**, 3779–3786.

Fowler, J. E., and Freeling, M. (1996). Genetic analysis of mutations that alter cell fates in maize leaves: Dominant *Liguleless* mutations. *Dev. Genet.* **18**, 198–222.

Fowler, J. E., Muehlbauer, G. J., and Freeling, M. (1996). Mosaic analysis of the Liguleless3 mutant phenotype in maize by coordinate suppression of *Mutator*-insertion alleles. *Genetics* **143**, 489–503.

Freeling, M. (1992). A conceptual framework for maize leaf development. *Dev. Biol.* **153**, 44–58.

Freeling, M., and Lane, B. (1994). The maize leaf. *In* "The Maize Handbook" (M. Freeling and V. Walbot, Eds.), pp. 17–28. Springer-Verlag, New York.

Giles, K. L., and Shehata, A. I. (1984). Some observations on the relationship between cell division and cell determination in the epidermis of the developing leaf of corn (Zea mays). *Bot. Gaz.* **145**, 60–65.

Goodbody, K. C., Venverloo, C. J., and Lloyd, C. W. (1991). Laser microsurgery demonstrates that cytoplasmic strands anchoring the nucleus across the vacuole of premitotic plant cells are under tension. Implications for division plane alignment. *Development* **113**, 931–939.

Green, P. B. (1984). Shifts in plant cell axiality: Histogenetic influences on cellulose orientation in the succulent, *Graptopetalum*. *Dev. Biol.* **103**, 18–27.

Hake, S., and Freeling, M. (1986). Analysis of genetic mosaics shows that the extra epidermal cell divisions in *Knotted* mutant maize plants are induced by adjacent mesophyll cells. *Nature* **320**, 621–623.

Hall, R. D., Riksen-Bruinsma, T., and Weyens, G. (1996). Stomatal guard cells are totipotent. *Plant Physiol.* **112,** 889–892.

Hannoufa, A., McNevin, J., and Lemieux, B. (1993). Epicuticular wax of *Arabidopsis thaliana* eceriform (*cer*) mutants. *Phytochemistry* **33,** 851–855.

Hannoufa, A., Negruk, V., Eisner, G., and Lemieux, B. (1996). The *CER3* gene of *Arabidopsis thaliana* is expressed in leaves, stems, roots, flowers and apical meristems. *Plant J.* **10,** 459–467.

Hansen, J. D., Pyee, J., Xia, Y., Wen, T. J., Robertson, D. S., Kolattukudy, P. E., Nikolau, B. J., and Schnable, P. S. (1997). The *glossy1* locus of maize and an epidermis-specific cDNA from *Kleinia odora* define a class of receptor-like proteins required for the normal accumulation of cuticular waxes. *Plant Physiol.* **113,** 1091–1100.

Harper, L., and Freeling, M. (1996). Interactions of *liguleless1* and *liguleless2* function during ligule induction in maize. *Genetics* **144,** 1871–1882.

Hoson, T., Masuda, Y., and Nevins, D. J. (1992). Comparison of the outer and inner epidermis. Inhibition of auxin-induced elongation of maize coleoptiles by glucan antibodies. *Plant Physiol.* **98,** 1298–1303.

Hülskamp, M., Miséra, S., and Jürgens, G. (1994). Genetic dissection of trichome cell development in *Arabidopsis*. *Cell* **76,** 555–566.

Jackson, D., Veit, B., and Hake, S. (1994). Expression of maize *KNOTTED1* related homeobox genes in the shoot apical meristem predicts patterns of morphogenesis in the vegetative shoot. *Development* **120,** 405–413.

Jenks, M. A., Tuttle, H. A., Eigenbrode, S. D., and Feldmann, K. A. (1995). Leaf epicuticular waxes of the *eceriferum* mutants in *Arabidopsis*. *Plant Physiol.* **108,** 369–377.

Jenks, M. A., Rashotte, A. M., Tuttle, H. A., and Feldmann, K. A. (1996a). Mutants in *Arabidopsis thaliana* altered in epicuticular wax and leaf morphology. *Plant Physiol.* **110,** 377–385.

Jenks, M. A., Tuttle, H. A., and Feldmann, K. A. (1996b). Changes in epicuticular waxes on wildtype and *eceriferum* mutants in *Arabidopsis* during development. *Phytochemistry* **42,** 29–34.

Johnson, H. B. (1975). Plant pubescence: An ecological perspective. *Bot. Rev.* **41,** 233–258.

Jürgens, G., Mayer, U., Torres Ruiz, R. A., Berleth, T., and Miséra, S. (1991). Genetic analysis of pattern formation in the *Arabidopsis* embryo. *Development Suppl.* **1,** 27–38.

Kaddoura, R. L., and Mantell, S. H. (1991). Synthesis and characterization of *Nicotiana-Solanum* graft chimeras. *Ann. Bot.* **68,** 547–556.

Kagan, M. L., and Sachs, T. (1991). Development of immature stomata: Evidence for epigenetic selection of a spacing pattern. *Dev. Biol.* **146,** 100–105.

Kagan, M. L., Novoplansky, N., and Sachs, T. (1992). Variable cell lineages form the functional pea epidermis. *Ann. Bot.* **69,** 303–312.

Kaplan, D. R., and Hagemann, W. (1991). The relationship of cell and organism in vascular plants. *Bioscience* **41,** 693–703.

Karabourniotis, G., Kotsabassidis, D., and Manetas, Y. (1995). Trichome density and its protective potential against ultraviolet-B radiation damage during leaf development. *Can. J. Bot.* **73,** 376–383.

Kiesselbach, T. A. (1949). The structure and reproduction of corn. *Nebr. Agric. Exp. Stn., Res. Bull.* **161,** 1–96.

Kolattukudy, P. E. (1981). Structure, biosynthesis, and biodegradation of cutin and suberin. *Annu. Rev. Plant Physiol.* **32,** 539–567.

Korn, R. W. (1981). A neighboring-inhibition model for stomate patterning. *Dev. Biol.* **88,** 115–120.

Korn, R. W. (1993). Evidence in dicots for stomatal patterning by inhibition. *Int. J. Plant Sci.* **154,** 367–377.

Korneef, M. (1981). The complex syndrome of *ttg* mutants in *Arabidopsis*. *Arabid. Inform. Serv.* **18,** 45–51.

Kutschera, U., Bergfeld, R., and Schopfer, P. (1987). Cooperation of epidermis and inner tissues in auxin-mediated growth of maize coleoptiles. *Planta* **170,** 168–180.

Larkin, J. C., Oppenheimer, D. G., Pollock, S., and Marks, M. D. (1993). *Arabidopsis GLABROUS1* gene requires downstream sequences for function. *Plant Cell* **5,** 1739–1748.

Larkin, J. C., Oppenheimer, D. G., Lloyd, A. M., Paparozzi, E. T., and Marks, D. M. (1994). Roles of the *GLABROUS1* and *TRANSPARENT TESTA GLABRA* genes in *Arabidopsis* trichome development. *Plant Cell* **6,** 1065–1076.

Larkin, J. C., Young, N., Prigge, M., and Marks, M. D. (1996). The control of trichome spacing and number in *Arabidopsis*. *Development* **122,** 997–1005.

Larkin, J. C., Marks, M. D., Nadeau, J., and Sack, F. (1997). Epidermal cell fate and patterning in leaves. *Plant Cell* **9,** 1109–1120.

Laux, T., and Jürgens, G. (1997). Embryogenesis: A new start in life. *Plant Cell* **9,** 989–1000.

Lawson, E. J. R., and Poethig, R. S. (1995). Shoot development in plants: Time for a change. *Trends Genet.* **11,** 263–268.

Lemieux, B. (1996). Molecular genetics of epicuticular wax biosynthesis. *Trends Plant Sci.* **1,** 312–318.

Liu, D., and Post-Beittenmiller, D. (1995). Discovery of an epidermal stearoyl-acyl carrier protein thioesterase: Its potential role in wax biosynthesis. *J. Biol. Chem.* **270,** 16962–16969.

Lloyd, A., Schena, M., Walbot, V., and Davis, R. (1994). Epidermal cell fate determination in *Arabidopsis*: Patterns defined by a steroid-inducible regulator. *Science* **266,** 436–439.

Lloyd, C. W. (1991). How does the cytoskeleton read the laws of geometry in aligning the division plane of plant cells? *Development Suppl.* **1,** 55–65.

Lolle, S. J., and Cheung, A. Y. (1993). Promiscuous germination and growth of wildtype pollen from *Arabidopsis* and related species on the shoot of the *Arabidopsis* mutant *fiddlehead*. *Dev. Biol.* **155,** 250–258.

Lolle, S. J., Cheung, A. Y., and Sussex, I. M. (1992). *Fiddlehead*: An *Arabidopsis* mutant constitutively expressing an organ fusion program that involves interactions between epidermal cells. *Dev. Biol.* **152,** 383–392.

Lolle, S. J., Berlyn, G. P., Engstrom, E. M., Krolikowski, K. A., Reiter, W.-D., and Pruitt, R. E. (1997). Developmental regulation of cell interactions in the *Arabidopsis fiddlehead-1* mutant: A role for the epidermal cell wall and cuticle. *Dev. Biol.* **189,** 311–321.

Lolle, S. J., Hsu, W., and Pruitt, R. E. (1998). Genetic analysis of organ fusion in *Arabidopsis thaliana*. *Genetics* **149,** 607–619.

Lu, P., Porat, R., Nadeau, J. A., and O'Neill, S. D. (1996). Identification of a meristem L1 layer-specific gene in *Arabidopsis* that is expressed during embryonic pattern formation and defines a new class of homeobox genes. *Plant Cell* **8,** 2155–2168.

Lucas, W. J., Bouché-Pillon, S., Jackson, D. P., Nguyen, L., Barker, L., Ding, B., and Hake, S. (1995). Selective trafficking of KNOTTED1 homeodomain protein and its mRNA through plasmodesmata. *Science* **270,** 1980–1983.

Lukowitz, W., Mayer, U., and Jürgens, G. (1996). Cytokinesis in the *Arabidopsis* embryo involves the syntaxin-related KNOTTED gene product. *Cell* **84,** 61-71.

Mansfield, S. G., and Briatry, L. G. (1991). Early embryogenesis in *Arabidopsis thaliana*. II. The developing embryo. *Can. J. Bot.* **69,** 461–476.

Marks, M. D. (1997). Molecular genetic analysis of trichome development in *Arabidopsis*. *Annu. Rev. Plant Physiol. Mol. Biol.* **48,** 137–163.

Martin, J. T., and Juniper, B. E. (1970). "The Cuticles of Plants." Edward Arnold, London.

Mayer, U., Torres Ruiz, R. A., Berleth, T., Miséra, S., and Jürgens, S. (1991). Mutations affecting body organization in the *Arabidopsis* embryo. *Nature* **353,** 402–407.

McDaniel, C. N., and Poethig, R. S. (1988). Cell-lineage patterns in the shoot apical meristem of the germinating maize embryo. *Planta* **175,** 13–22.

McLean, B. G., Hempel, F. D., and Zambryski, P. C. (1997). Plant intercellular communication via plasmodesmata. *Plant Cell* **9,** 1043–1054.

McNevin, J. P., Woodward, W., Hannoufa, A., Feldmann, K. A., and Lemieux, B. (1993). Isolation

and characterization of *eceriform* (*cer*) mutants induced by T-DNA insertions in *Arabidopsis thaliana*. *Genome* **36**, 610–618.
Meinhardt, H. (1982). "Models of Biological Pattern Formation." Academic Press, New York.
Moore, R. (1984). Cellular interactions during the formation of approach grafts in *Sedum telephoides* (Crassulaceae). *Can. J. Bot.* **62**, 2476–2484.
Moose, S. P., and Sisco, P. H. (1994). *Glossy15* controls the epidermal juvenile-to-adult phase transition in maize. *Plant Cell* **6**, 1343–1355.
Moose, S. P., and Sisco, P. H. (1996). *Glossy15*, an *APETALA2*-like gene from maize that regulates leaf epidermal cell identity. *Genes Dev.* **10**, 3018–3027.
Negruk, V., Yang, P., Subramanian, M., McNevin, J. P., and Lemieux, B. (1996). Molecular cloning and characterization of the *CER2* gene of *Arabidopsis thaliana*. *Plant J.* **9**, 137–145.
Oppenheimer, D. G., Herman, P. L., Sivakumaran, S., Esch, J., and Marks, M. D. (1991). A Myb gene required for leaf trichome differentiation in *Arabidopsis* is expressed in stipules. *Cell* **67**, 483–493.
Oppenheimer, D. G., Pollock, M. A., Vacik, J., Szymanski, D. B., Ericson, B., Feldmann, K., and Marks, M. D. (1997). Essential role of a kinesin-like protein in *Arabidopsis* trichome morphogenesis. *Proc. Natl. Acad. Sci. USA* **94**, 6261–6266.
Palevitz, B. A., and Hepler, P. K. (1985). Changes in dye coupling of stomatal cells of *Allium* and *Commelina* demonstrated by microinjection of Lucifer yellow. *Planta* **164**, 473–479.
Palmer, S. J., and Davies, W. J. (1996). An analysis of relative elemental growth rate, epidermal cell size and xyloglucan endotransglycosylase activity through the growing zone of ageing maize leaves. *J. Exp. Bot.* **47**, 339–347.
Paolillo, D. J., Jr. (1995). The net orientation of wall microfibrils in the outer periclinal epidermal walls of seedling leaves of wheat. *Ann. Bot.* **76**, 589–596.
Pillai, K. G., Rao, I. U., Rao, I. V. R., and Ram, H. Y. M. (1992). Induction of division and differentiation of somatic embryos in the leaf epidermis of *Gaillardia picta*. *Plant Cell Rep.* **10**, 599–603.
Poethig, R. S. (1984). Cellular parameters of leaf morphogenesis in maize and tobacco. *In* "Contemporary Problems in Plant Anatomy" (R. A. White and W. C. Dickson, Eds.), pp. 235–259. Academic Press, New York.
Poethig, R. S. (1987). Clonal analysis of cell lineage patterns in plant development. *Am. J. Bot.* **74**, 581–594.
Poethig, R. S. (1990). Phase change and the regulation of shoot morphogenesis in plants. *Science* **250**, 923–930.
Poethig, R. S., and Sussex, I. M. (1985). The cellular parameters of leaf development in tobacco: A clonal analysis. *Planta* **165**, 170–184.
Pope, D. G. (1982). Effect of peeling on IAA-induced growth in *Avena* coleoptiles. *Ann. Bot.* **49**, 493–501.
Post-Beittenmiller, D. (1996). Biochemistry and molecular biology of wax production in plants. *Annu. Rev. Plant Physiol. Plant Mol. Biol.* **47**, 405–430.
Pyee, J., and Kolattukudy, P. E. (1995). The gene for the major cuticular wax-associated protein and three homologous genes from broccoli (*Brassica oleracea*) and their expression patterns. *Plant J.* **7**, 49–59.
Pyee, J., Yu, H., and Kolattukudy, P. E. (1994). Identification of a lipid transfer protein as the major protein in the surface wax of broccoli (*Brassica oleracea*) leaves. *Arch. Biochem. Biophys.* **311**, 460–468.
Randolph, L. F. (1936). Developmental morphology of the caryopsis in maize. *J. Agric. Res.* **53**, 881–916.
Rayle, D. L., Nowbar, S., and Cleland, R. E. (1991). The epidermis of the pea epicotyl is not a unique target tissue for auxin-induced growth. *Plant Physiol.* **97**, 449–451.
Reiter, R. S., Coomber, S. A., Bourett, T. M., Bartley, G. E., and Scolnik, P. A. (1994). Control of leaf and chloroplast development by the *Arabidopsis* gene *pale cress*. *Plant Cell* **6**, 1253–1264.

Rerie, W. G., Feldman, K. A., and Marks, M. D. (1994). The *GLABRA2* gene encodes a homeo domain protein required for normal trichome development. *Genes Dev.* **8,** 1388–1399.
Robards, A. W., and Lucas, W. J. (1990). Plasmodesmata. *Annu. Rev. Plant Physiol. Plant Mol. Biol.* **41,** 369–419.
Sachs, T. (1978). The development of the spacing pattern in the leaf epidermis. *In* "The Clonal Basis of Development" (S. Subtelny and I. M. Sussex, Eds.), pp. 161–183. Academic Press, New York.
Sachs, T. (1994). Both cell lineages and cell interactions contribute to stomatal patterning. *Int. J. Plant Sci.* **155,** 245–247.
Sachs, T., and Novoplansky, N. (1993). The development and patterning of stomata and glands in the epidermis of *Peperomia*. *New Phytol.* **123,** 567–574.
Satina, S., Blakeslee, A. F., and Avery, A. G. (1940). Demonstration of the three germ layers in the shoot apex by means of induced polyploidy in periclinal chimeras. *Am. J. Bot.* **44,** 311–317.
Schnable, P. S., Stinard, P. S., Wen, T. J., Heinen, S., Weber, D., Zhang, L., Hansen, J. D., and Nikolau, B. J. (1994). The genetics of cuticular wax biosynthesis. *Maydica* **39,** 279–287.
Schnittger, A., Grini, P. E., Folkers, U., and Hulskamp, M. (1996). Epidermal fate map of the *Arabidopsis* shoot meristem. *Dev. Biol.* **175,** 248–255.
Schnyder, H., Seo, S., Rademacher, I. F., and Kühbach, W. (1990). Spatial distribution of growth rates and of epidermal cell lengths in the elongation zone during leaf development in *Lolium perenne* L. *Planta* **181,** 423–431.
Schoch, P. G., Jacques, R., Lecharny, A., and Sibi, M. (1984). Dependence of the stomatal index on environmental factors during stomatal differentiation in leaves of *Vigna sinensis* L. *J. Exp. Bot.* **35,** 1405–1409.
Serna, L., and Fenoll, C. (1997). Tracing the ontogeny of stomatal clusters in *Arabidopsis* with molecular markers. *Plant J.* **12,** 747–756.
Serpe, M. D., and Matthews, M. A. (1994). Growth, pressure, and wall stress in epidermal cells of *Begonia argenteo-guttata* L. leaves during development. *Int. J. Plant Sci.* **155,** 291–301.
Sharman, B. C. (1942). Developmental anatomy of the shoot of *Zea mays* L. *Ann. Bot.* **6,** 245–284.
Siegel, B. A., and Verbeke, J. A. (1989). Diffusible factors essential for epidermal cell redifferentiation in *Catharanthus roseus*. *Science* **244,** 580–582.
Sinha, N., and Hake, S. (1990). Mutant characters of *Knotted* maize leaves are determined in the innermost tissue layers. *Dev. Biol.* **141,** 203–210.
Sinha, N., and Hake, S. (1994). The *Knotted* leaf blade is a mosaic of blade, sheath, and auricle identities. *Dev. Genet.* **15,** 401–414.
Sinha, N., and Lynch, M. (1998). Fused organs in the *adherent1* mutation in maize show altered epidermal walls with no perturbations in tissue identities. *Planta* **206,** 184–195.
Sinnott, E. W., and Bloch, R. (1941). The relative position of cell walls in developing plant tissues. *Am. J. Bot.* **28,** 607–617.
Smith, L. G., Greene, B., Veit, B., and Hake, S. (1992). A dominant mutation in the maize homeobox gene *Knotted-1* causes its ectopic expression in leaf cells with altered fates. *Development* **116,** 21–30.
Smith, L. G., Hake, S., and Sylvester, A. W. (1996). The *tangled-1* mutation alters cell division orientations throughout maize leaf development without altering leaf shape. *Development* **122,** 481–489.
Stebbins, G. L., and Shah, S. S. (1960). Developmental studies of cell differentiation in the epidermis of monocotyledons. II. Cytological features of stomatal development in the Gramineae. *Dev. Biol.* **2,** 477–500.
Steeves, T. A., and Sussex, I. M. (1989). *"Patterns in Plant Development."* Cambridge Univ. Press, New York.
Stewart, R. N. (1978). Ontogeny of the primary body in chimeral forms of higher plants. *In* "The Clonal Basis of Development" (S. Subtelny and I. Sussex, Eds.), pp. 131–159. Academic Press, New York.

1. Development of the Leaf Epidermis

Stewart, R. N., and Dermen, H. (1975). Flexibility in ontogeny as shown by the contribution of the shoot apical layers to leaves of periclinal chimeras. *Am. J. Bot.* **62,** 935–947.

Stewart, R. N., and Dermen, H. (1979). Ontogeny in monocotyledons as revealed by studies of the developmental anatomy of periclinal chloroplast chimeras. *Am. J. Bot.* **66,** 47–58.

Stewart, R. N., Meyer, F. G., and Dermen, H. (1972). Camellia + "Daisy Eagleson," a graft chimera of *Camellia sasanqua* and *C. japonica*. *Am. J. Bot.* **59,** 515–524.

Stewart, R. N., Semeniuk, P., and Dermen, H. (1974). Competition and accommodation between apical layers and their derivatives in the ontogeny of chimeral shoots of *Pelargonium X hortorum*. *Am. J. Bot.* **61,** 54–67.

St. Johnson, D., and Nüsslein-Volhard, C. (1992). The origin of pattern and polarity in the *Drosophila* embryo. *Cell* **68,** 201–219.

Sussex, I. M. (1955). Morphogenesis in *Solanum tuberosum* L: Experimental investigation of leaf dorsiventrality and orientation in the juvenile shoot. *Phytomorphology* **5,** 286–300.

Sylvester, A. W., Cande, W. Z., and Freeling, M. (1990). Division and differentiation during normal and *liguleless-1* maize leaf development. *Development* **110,** 985–1100.

Sylvester, A. W., Smith, L., and Freeling, M. (1996). Acquisition of identity in the developing leaf. *Annu. Rev. Cell Dev. Biol.* **12,** 257–304.

Szymanski, D. B., Jilk, R. A., Pollock, S. M., and Marks, M. D. (1998a). Control of *GL2* expression in *Arabidopsis* leaves and trichomes. *Development* **125,** 1161–1171.

Szymanski, D. B., Klis, D. A., Larkin, J. C., and Marks, M. D. (1998b). *cot1*: A regulator of *Arabidopsis* trichome initiation. *Genetics* **149,** 565–577.

Szymkowiak, E. J., and Sussex, I. M. (1989). Chimeric analysis of cell layer interactions during development of the flower pedicel abscission zone. *In* "Cell Separation in Plants" (D. J. Osborne and M. B. Jackson, Eds.), pp. 363–368. Springer-Verlag, Berlin.

Szymkowiak, E. J., and Sussex, I. M. (1992). The internal meristem layer (L3) determines floral meristem size and carpel number in tomato periclinal chimeras. *Plant Cell* **4,** 1089–1100.

Szymkowiak, E. J., and Sussex, I. M. (1996). What chimeras can tell us about plant development. *Annu. Rev. Plant Physiol. Plant Mol. Biol.* **47,** 351–376.

Tacke, E., Korfhage, C., Michel, D., Maddaloni, M., Motto, M., Lanzini, S., Salamini, F., and Doring, H. P. (1995). Transposon tagging of the maize *Glossy2* locus with the transposable element *En/Spm*. *Plant J.* **8,** 907–917.

Telfer, A., and Poethig, R. S. (1994). Leaf development in *Arabidopsis*. *In* "Arabidopsis" (E. M. Meyerowitz and C. R. Somerville, Eds.), pp. 379–401. Cold Spring Harbor Laboratory Press, Cold Spring Harbor, NY.

Telfer, A., Bollman, K. M., and Poethig, R. S. (1997). Phase change and the regulation of trichome distribution in *Arabidopsis thaliana*. *Development* **124,** 645–654.

Terryn, N. Brito Arias, M., Engler, G., Tiré, C., Villarroel, R., Van Montagu, M., and Inzé, D. (1993). *rha1*, a gene encoding a small GTP binding protein from *Arabidopsis*, is expressed primarily in developing guard cells. *Plant Cell* **5,** 1761–1769.

Tomlinson, P. B. (1974). Development of the stomatal complex as a taxonomic character in the monocotyledons. *Taxon* **23,** 109–128.

Tsuge, T., Tsukaya, H., and Uchimiya, H. (1996). Two independent and polarized processes of cell elongation regulate leaf blade expansion in *Arabidopsis thaliana* (L.) Heynh. *Development* **122,** 1589–1600.

van der Schoot, C., Dietrich, M. A., Storms, M., Verbeke, J. A., and Lucas, W. J. (1995). Establishment of a cell-to-cell communication pathway between separate carpels during gynoecium development. *Planta* **195,** 450–455.

Verbeke, J. A., and Walker, D. B. (1986). Morphogenetic factors controlling differentiation and dedifferentiation of epidermal cells in the gynoecium of *Catharanthus roseus*. II. Diffusible morphogens. *Planta* **168,** 43–49.

Vroemen, C. W., Langeveld, S., Mayer, U., Ripper, G., Jürgens, G., van Kammen, A., and De Vries,

S. C. (1996). Pattern formation in the *Arabidopsis* embryo revealed by position-specific lipid transfer protein gene expression. *Plant Cell* **8,** 783–791.
Waites, R., and Hudson, A. (1995). *phantastica*: A gene required for dorsoventrality of leaves in *Antirrhinum majus*. *Development* **121,** 2143–2154.
Waites, R., Selvadurai, H. R. N., Oliver, I. R., and Hudson, A. (1998). The *PHANTASTICA* gene encodes a MYB transcription factor involved in growth and dorsoventrality of lateral origans in *Antirrhinum*. *Cell* **93,** 779–789.
Walker, D. B. (1975a). Postgenital carpel fusion in *Catharanthus roseus* (Apocyanaceae). I. Light and scanning electron microscopic study of gynoecial ontogeny. *Am. J. Bot.* **62,** 457–467.
Walker, D. B. (1975b). Postgenital carpel fusion in *Catharanthus roseus*. II. Fine structure of the epidermis before fusion. *Protoplasma* **86,** 29–41.
Walker, D. B. (1975c). Postgenital carpel fusion in *Catharanthus roseus*. III. Fine structure of the epidermis during and after fusion. *Protoplasma* **86,** 43–63.
Walker, D. B., and Bruck, D. K. (1985). Incompetence of stem epidermal cells to dedifferentiate and graft. *Can. J. Bot.* **63,** 2129–2131.
Wick, S. M. (1991). Spatial aspects of cytokinesis in plant cells. *Curr. Opin. Cell Biol.* **3,** 253–260.
Williams, S. E., and Pickard, B. G. (1972). Receptor potentials and action potentials in *Drosera* tentacles. *Planta* **103,** 193–221.
Xia, Y., Nikolau, B. J., and Schnable, P. S. (1996). Cloning and characterization of *CER2*, an *Arabidopsis* gene that affects cuticular wax accumulation. *Plant Cell* **8,** 1291–1304.
Yang, M., and Sack, F. D. (1995). The *too many mouths* and *four lips* mutations affect stomatal production in *Arabidopsis*. *Plant Cell* **7,** 2227–2239.
Zhang, J., Specht, J. E., Graef, G. L., and Johnson, B. E. (1992). Pubescence density effects on soybean seed yield and other agronomic traits. *Crop Sci.* **32,** 641–648.

2
Genes and Their Products in Sea Urchin Development

Giovanni Giudice
Department of Cellular and Developmental Biology
University of Palermo
90128 Palermo, Italy

    I. Introduction
   II. Histones
      A. CS Histones
      B. H2A and H2B Histones
      C. H1 Histone Genes
      D. H4 Histone Genes
      E. H3 Histone Genes
  III. Cytoskeletal Actin Genes
      A. CyIIIa Gene
      B. CyI Gene
      C. CyIIIb Gene
      D. CyIIa Gene and S9 Gene
  IV. Endo16 and Other Endoderm-Specific Genes
   V. Ectoderm-Specific Genes
  VI. Genes of the Primary Mesenchyme or However Relevant to the Extracellular Matrix
 VII. Genes for Cortical Granule Content and Genes Related to the Wnt Pathway and Gastrulation
      A. Genes for Cortical Granule Content
      B. Genes Related to the Wnt Pathway
      C. Genes Related to Gastrulation
VIII. Homeobox-Containing Genes and Genes for Transcription Factors That Do Not Contain Homeobox
      A. Homeobox-Containing Genes
      B. Genes for Transcription Factors That Do Not Contain Homeobox
  IX. Tubulin Genes and Other Genes Related to Ciliogenesis and to the Mitotic Apparatus
   X. Genes for Cyclin and Other Cell Division Related Proteins
  XI. Genes for Kinases, a Fibroblast Growth Factor Receptor, and Metallothionein
      A. Genes for Kinases
      B. Gene for a Fibroblast Growth Factor Receptor
      C. Genes for Metallothionein
  XII. Genes for Heat Shock Proteins (hsps) and Ubiquitin Genes
      A. Genes for Heat Shock Proteins
      B. Ubiquitin Genes
XIII. Genes That Are Differentially Expressed along the Animal–Vegetal Axis

XIV. Genes for Arylsulfatase, Myogenic Factors, and Ribosomal Proteins, Genes with an EGF Domain, and Genes for Poly(A)-Binding Proteins
    A. Genes for Arylsulfatase
    B. Genes for Myogenic Factors
    C. Genes for Ribosomal Proteins
    D. Genes with an EGF Domain
    E. Genes for Poly(A)-Binding Proteins
XV. Genes for Jelly Proteins and the Egg Receptor for Sperm
    A. Genes for Jelly Proteins
    B. Genes for the Egg Receptor for Sperm
XVI. Genes Expressed by Coelomocytes
    A. Genes for Vitellogenin
    B. Genes for SpCoel1 (a Sea Urchin Profilin Gene)
    C. Other Genes
XVII. Miscellaneous Genes
    A. Genes for Fascin
    B. Genes for Calmodulin
    C. Genes Related to Mitochondria
    D. Genes for Distrophin-Related Proteins
    E. Genes for Bindin
    F. Genes for Other Sperm-Associated Proteins
    G. Genes for Lamin
    H. Genes for snRNAs
    I. Genes for Elongation Factors
XVIII. Proto-oncogenes and Retroposons
    A. Proto-oncogenes
    B. Retroposons
    References

I. Introduction

Sea urchin embryos were among the first eukaryotes in which regulation of gene expression was studied and still represent one of the main developing systems for which most of the literature has accumulated in the past and also in recent years. The interest in the subject has, however, shifted from the molecular biology field—i.e., from the study of the mechanisms of regulation of gene expression—to a field in which the role of genes and their products on development is being considered.

It is the aim of the present article to briefly review the literature of the eighties and to report in some deeper detail that of the nineties (see Giudice (1993, 1995) for previous reviews) to provide the reader with an ordered list of the sea urchin genes and their products, with special attention to the mechanisms of regulation of their temporal and spatial expression and, where possible, their function in development. An attempt has been made to make the present review comprehensive, but, being aware that the attempt was not completely successful, I apologize for those papers which for any reason have been omitted.

To understand the molecular biological interest of the material, it is important to recall here that the first eukaryotic mRNA was identified in sea urchins (Kedes and Gross, 1969; Nemer and Lindsay, 1969) and that the first genes were cloned from sea urchins (Kedes *et al.*, 1975a, 1975b); these were the genes for histones. The interest in this biological material does not rest merely on historical reasons, but on the suitability of the system to address important questions about the mechanisms of regulation of gene expression both at different developmental times and in different embryonic territories and to try to understand what the effect of gene action on early development and morphogenesis is. This suitability is due to a number of properties of these eggs and embryos: their transparency, which allows easy observation of morphogenesis; their permeability to labeled macromolecule precursors; the synchrony of their development, which allows the study of macromolecules extracted in high amounts from great numbers of embryos at the same developmental stage; the external fertilization; the possibility of injecting constructs which become stably and inheritably incorporated into nuclei; the invariant pattern of cleavage, which, coupled to cell interactions, originates five polyclonal territories by the sixth cleavage; the knowledge of the cell lineage (Davidson, 1989; Cameron and Davidson, 1991; Coffman and Davidson, 1992); and the possibility of studying cell interaction by microtransplantation experiments (Hörstadius, 1939, 1973) and also by dissociation–reaggregation studies, with the restitution of entire embryos from dissociated cells (Giudice, 1962, 1973, 1986). We will start our review with histones, because they represent the first genes whose expression regulation in various developmental stages was studied.

II. Histones

It is important to recall that histone proteins are divided into five species (H4, H2B, H3, H2A, and H1), each coded by one corresponding gene (see Thatcher and Gorovsky (1994) for a phylogenetic analysis), and that in sea urchins these genes, at least those expressed till the blastula stage, are clustered in that order within the same DNA strand in units tandemly repeated about 400 times. The histone proteins are divided into subtypes which are synthesized at different developmental stages. This has rendered histone genes a suitable model to study the mechanisms underlying the temporal regulation of gene expression during early development (see Giudice (1986, 1993, 1995), Davidson (1986), Spinelli and Albanese (1990), and Romano (1992) for reviews); namely, the subtype called CS (from cleavage stage) is synthesized in the oocyte, in the egg, and in the embryo up to the third cleavage, then the synthesis of the so-called early (or α) subtype occurs, till the blastula stage, and eventually the subtype late is expressed from the mesenchyme blastula on, continuing in the adult life. It should be recalled that the late genes are not clustered but rather scattered in the genome. Two sperm specific subtypes (SpH1 and SpH2) have to be added to the list, which are synthesized during spermiogenesis and incorporated into sperm chromatin. In addition to the use of different genes for the synthesis of these subtypes, posttranslational modifications of histones occur such as phosphorylation (Green et al., 1995; Hoshino et al., 1997), acetylation (Jasinskas et al., 1997), and ADP-ribosylation (Furuya et al., 1994).

A. CS Histones

The cDNAs of these histones have been isolated and identified by Birnstiel's group (Mandl et al., 1997) for all five subtypes with the help of partial peptide sequence information from eggs of *Psammechinus miliaris*. These genes, being independent from cell replication, contain introns, poly(A) addition signals, and long non-translated sequences. Their transcripts, already present in the oocytes, can be detected up to the early blastula stage. Their sequences are distantly related to those of the early, late, or sperm histones but at least that of the CS H1 shows a high homology with the corresponding maternal H1 of *Xenopus*, therefore suggesting that the CS histones are of ancient evolutionary origin and may perform similar functions in different species. This conclusion is reinforced by further amino acid sequence analysis carried out by Brandt et al. (1997).

B. H2A and H2B Histones

The first experiments on the mechanism of temporal regulation of such genes are due to Birnstiel and Busslinger, who described the fact that the sperm-specific H2B1 gene is no longer transcribed when the sperm penetrates the egg, because a new protein binding a sequence adjacent to the CAAT box of the promoter prevents the binding to the CAAT box of a sperm-specific transcription factor (Barberis et al., 1987). The same group showed that if the same HB1 gene was microinjected into eggs, but after substituting its promoter with that of the H2A early histone gene, then HB1 was expressed at the same time as the endogenous H2A (Vitelli et al., 1988). Analyses of the H2A promoter were reported by Grosschedl and Birnstiel (1980) and Grosschedl et al. (1983) for *P. miliaris* and by Palla et al. (1989, 1993, 1994, 1997) for *Paracentrotus lividus*. Birnstiel's group was the first to describe a sequence of the pre-H2A AT-rich spacer which modulates the expression of the gene and was therefore called modulator. Spinelli et al., through a series of experiments based on sequence analysis and microinjection into homologous oocytes of various constructs, found a similar sequence in *Paracentrotus* and demonstrated that it has the properties of an enhancer; they also described a modulator binding protein. The latter authors recently found that an enhancer blocking sequence is located near the 3' end of the H2A gene, which may explain why the described enhancer does not influence the transcription of the downstream histone genes, which, on the contrary, appear to be independently regulated. The termination of the transcripts of the early histone genes was also analyzed by Birnstiel's group, who described the importance of some sequences around the 3' terminus such as a palindrome and the sequence CAAGAAAGA (see Birnstiel et al. (1985) for a review) as well as the importance of the participation of the U7 RNA in processing. The H2B early gene was also studied by Birnstiel's group (Mous et al., 1985), who provided evidence for the regulation of their synthesis through transcription factors.

The importance of the octamer ATGCAAAT for the regulation of the H2B histone gene in the early stages was described by Bell et al. (1992). The same group of authors (Char et al., 1993, 1994) described a protein of S. purpuratus with a POU domain, produced by a gene called SpOct, which is able to bind the octamer; microinjection of antisense RNA blocks cleavage, in a manner which is reversed by injection of SpOct mRNA. The latter result contradicts the old finding of Gross and Cousineau (1963), who found that blockage of RNA synthesis by actinomycin starting from before fertilization permits cleavage till early blastula. One way to reconcile the two findings is, in my opinion, to assume that SpOct mRNA synthesis is especially resistant to actinomycin inhibition or that the antisense blocks also a maternal SpOct mRNA.

The expression of the early H2B gene was increased by its hypermethylation, experimentally achieved through microinjection of 5-methyl-dCTP, whereas the expression of the late H2B gene was not affected by such a treatment, which indicates that repression of the expression of these genes does not operate through their methylation (Chen et al., 1993).

Busslinger and Barberis (1985) and others (Maxson et al., 1983, 1988; Knowles and Childs, 1984; Lai and Childs, 1986; Ito et al., 1988) have dealt with the regulation of the late H2A and H2B histone genes on *Psammechinus* and *Strongylocentrotus*, respectively. Also for these the expression is regulated at a transcriptional level as demonstrated by the presence of their messenger RNAs only in the late stages, at least in substantial amounts. A transcription factor for the promoters of four subtypes of H2A and H2B late histone genes was found by Barberis et al. (1989) and by Maxson et al. (1988) for the late H2B promoter, as it was an enhancer after the 3' end of this gene (Zhao et al., 1990, 1991), which is able to bind a factor with a homeodomain of the antennapedia type. These late genes have the characteristics of the cell-cycle-regulated histone genes, because they contain a terminal palindrome at their 3' end instead of the polyadenylation site and do not contain introns (Birnstiel et al., 1985). MacIsaac et al. (1992) found that the mRNA for the subclass H2A.F/Z is present at equal levels in the cytoplasm of all cells of S. purpuratus mesenchyme blastulae, therefore suggesting that synthesis of this special subclass is not coordinated with that of DNA, a hypothesis supported by DNA synthesis inhibition experiments.

C. H1 Histone Genes

The group led by G. Childs has produced a great deal of work concerning the regulation of the synthesis of this histone class at both early and late stages (Lieber et *al.*, 1986; Knowles *et al.*, 1987; Lai and Childs, 1988; Lai *et al.*, 1988, 1989; Di Liberto *et al.*, 1989; Fei and Childs, 1993). Promoters, enhancers, and transcription factors have been described. It is interesting, among other things, that a sequence identical to the Sp1 binding site in mammals differs for only one nucleotide in early and late genes: one G of the former is replaced by a T in the latter. If late genes in which this T has been artificially changed into a G are microinjected into eggs, they will now be expressed at early stages. As to transcription factors, De Falco and Childs (1996) characterized a new activation domain of the so-called SSAP (stage-specific activator protein), and found that two sequences rich in glycine/glutamine and serine/threonine, respectively, constitute an extremely powerful activation domain, which *in vitro* interacts with many basal transcription factors. Del Gaudio *et al.* (1996) reported a 97.6% identity between the sequence of an H1 *P. lividus* histone gene with the homologous gene of the polychaete *C. variopedatus*, which led the authors to hypothesize a horizontal gene transfer. Without wanting to draw conclusions, it may be pertinent to recall here that Britten *et al.* (1995) have described retrotransposons of the class Gypsy/Ty3 in the DNA of marine invertebrates, including sea urchins (see also Nisson *et al.* (1988) and Springer *et al.* (1995).

D. H4 Histone Genes

The group led by E. Weinberg worked on these genes, mainly assaying their expression in nuclear extracts from both early and late stages (Tung *et al.*, 1989, 1990; Lee *et al.*, 1991). Regulatory sequences of the promoter and transcription factors were analyzed.

The observation of I. Albanese (personal communication, submitted for publication) that about 230 nucleotides of the 3' terminal part of one H4 histone mRNA of *P. lividus* are complementary to a part of the 26S sea urchin ribosomal RNA is quite intriguing. The meaning of this unexpected finding is not known, but Mauro and Edelman (1997), after a database search, published that rRNA-like sequences occur in mRNAs of many eukaryotes and hypothesized for them a role in the control of gene expression.

E. H3 Histone Genes

The problem of the coordinated regulation of the synthesis of the entire set of histones has not yet received a solution. Spinelli's group, however, has reported that the H3 and H2A genes share at least one site for the binding of transcription factors, and probably the transcription factor itself, which binds to the already-described enhancer (Shinelli *et al.*, 1991; Palla *et al.*, 1993, 1994). A recent analysis of the H3 late histone protein is due to Fucci *et al.* (1997).

2. Genes and Their Products in Sea Urchin Development

In addition to the well-demonstrated regulation of histone synthesis at the transcriptional level, other levels of regulation occur, as, for example, the storage within the female pronucleus of the mRNA synthesized during oogenesis (Venezky *et al.*, 1981; Showman *et al.*, 1982) and the delayed recruitment of this RNA on polysomes (Wells *et al.*, 1981).

III. Cytoskeletal Actin Genes

The study of the synthesis of histones has served the purpose of trying to understand the mechanisms underlying the temporal regulation of the expression of different classes of genes at different times of development but has provided no information about the spatial regulation of this expression in different embryonic territories, because the ubiquitous histone genes are expressed in all the embryonic cells (Angerer *et al.*, 1985). On the other hand, with the study of genes such as those coding for cytactins, which are expressed in different embryonic territories, a new era was started in which, from one side, the mechanisms of territorial activation of such genes and, from the other, their possible roles in the differentiation of the embryonic territories or in the establishment of embryonic axes was investigated. Also the development of the whole-mount *in situ* hybridization to detect expression of reporter genes has played a central role for the investigation in the detection of the territorial gene expression, as it will in the near future for the further application of the green fluorescent protein technique to sea urchins (Arnone *et al.*, 1997).

Five types of cytoskeletal actins have been described in sea urchins, named CyI, CyIIa, CyIIb, CyIIIa, and CyIIIb, plus a muscular actin named M, as partially reviewed by Davidson (1986, 1989). Their quantitative expression during development and in the various embryonic territories has been thoroughly studied (Cameron *et al.*, 1989; Flytzanis *et al.*, 1989; Lee *et al.*, 1992). CyI and CyIIb are two examples in which in early stages the expression is in all embryonic territories and then becomes restricted to the oral ectoderm and to the intestine (Cox *et al.*, 1986).

A. CyIIIa Gene

The structure of CyIIIa was first described by Akhurst *et al.* (1987). Its transcripts appear during the late cleavage and only in the precursor cells of the aboral ectoderm, where they become localized at the pluteus stage. The properties of a regulatory region of 2300 nucleotides at the 5′ terminus have been thoroughly studied by the group led by E. H. Davidson by a variety of techniques including the microinjection of constructs made by this region followed by a reporter gene. Hough-Evans *et al.* (1988) demonstrated that these constructs become stably incorporated in the genome in an inheritable manner in about 25% of cases and then randomly distributed among the daughter nuclei. Livant *et al.* (1991) then showed that repeated injections into the same egg may sometime yield a stable incorporation into over 95% of nuclei, which is a result even superior to that reported by Franks *et al.* (1988b), who microinjected the constructs directly into egg nuclei. This gene offered the first possibility to titrate *in vivo* the amount of transcription factors of the gene present in the embryo.

The strategy was the following: First, various amounts of this regulatory region were injected into the egg to see at which point the injected fragments had competed off all the transcription factors, as revealed by the inhibition of the expression in the embryo of a fusion gene made by the CyIIIa regulatory sequence followed by a reporter gene. This gave the figure that 105 (± 40) copies of injected construct are needed to half-saturate the transcription factors (Livant *et al.*, 1988), although the same authors (Livant *et al.*, 1991) warn that the endogenous CyIIIa synthesis is little affected when the synthesis of the product of a reporter gene preceded by the CyIIIa regulatory region has been 85% competed off. In further work, the regulatory region was subfractionated into sequences assumed to bind transcription factors, these were microinjected in various amounts, and the expression of a CAT reporter gene preceded by the entire regulatory region was assayed to test the importance of single transcription factors and to titrate them (Calzone *et al.*, 1988; Thèzè *et al.*, 1990; Cutting *et al.*, 1990). It was found that six sequences, called P1, P2, P3B, P4, P7I, and P8, were relevant for CyIIIa transcription and that these were able to bind about 10 proteins at 20 binding sites. Calzone *et al.* (1988) measured the amount of various proteins able to bind these regulatory regions at developmental times at which the gene is expressed or not, finding a parallelism for some of them but not for others. This search became better addressed when, thanks to an automated method developed by Coffman *et al.* (1992), these DNA-binding proteins started to be purified in bulk (see also Harrington *et al.*, 1992). The cDNAs of all but two of these proteins were then cloned by Coffman and Davidson (1994). A complete map of the CyIIIa regulatory region with a schematic representation of the cloned transcription factors binding to them was reported by Coffman *et al.* (1996), with a further subdivision of the previously described regions as represented in Fig. 1, showing its separation into three main modules—proximal, middle, and distal—whose functions will be described later. A new transcription factor binding the region 7I was also described, which was called SpRunt-1 because it contains a highly conserved "runt" domain similar to those described in transcription factors of *Drosophila*, mouse, and humans. Since its concentration in the embryo dramatically increases when the CyIIIa gene expression increases, i.e., after the blastula stage, **SpRunt-1** is a possible candidate for the upregulation of CyIIIa expression during development. Calzone *et al.* (1997), by means of a new and more efficient method than those previously employed, found that at least eight different transcription factors able to actively interact with the reg-

2. Genes and Their Products in Sea Urchin Development

ulatory region of CyIIIa were already present in the cytoplasm of *S. purpuratus* eggs. Also, a new high-sensitivity capillary electrophoresis method for DNA–protein binding assay has facilitated such research (Xian *et al.*, 1996). One of the hypotheses which had been raised to explain the mechanisms which regulate the territorial expression of CyIIIa was that two of the sequences of its regulatory region bound some repressors in the cells of the territories not expressing the gene. The idea stemmed from the following observations: A construct made by this regulatory region from *S. purpuratus* followed by the CAT reporter gene, when injected into the egg of another species, *Lytechinus variegatus*, was expressed at the correct developmental time but also in other territories besides the expected ones (Franks *et al.*, 1988a, 1990). This was supposedly due to the lack of appropriate repressors in the other species, which is in agreement with the lack of conservation between these two genera of the regulation of the expression of the actin gene family described by Fang and Brandhorst (1996) after a thorough analysis of this family in *Lytechinus pictus*. Later, after injecting an excess of seven protein binding sequences of the CyIIIa regulatory region of *S. purpuratus* into eggs of the same species to compete off transcription factors, Hough-Evans *et al.* (1990) found that the injection of two of them caused the ectopic expression of the gene. This was also tentatively explained by the fact that two repressors binding the two injected sequences had been competed off. What are these repressing factors? Calzone *et al.* (1991) and Höög *et al.* (1991) purified by *S. purpuratus* two factors called **SpP3A1** and **SpP3A2** which interact with sequences containing the element (T/C)N(T/C)GCGC(A/T). Zeller *et al.* (1995a,b,c) measured by immunological methods the amounts of these two proteins in the nuclei and cytoplasms of embryos at different developmental stages and found that both are present and enter the nuclei in early stages, but only SpP3A2 remains in the nuclei at later stages in which the territorial expression is observed, which made the authors suggest that SpP3A2 may be the territorial repressor. Further work of the Caltech group (Wang *et al.*, 1995a, 1995b) identified another element of the regulatory region of the CyIIIa gene which binds a protein with 12 Zn fingers, called **SpZ12-1**. If this element, characterized by inverted repeats (Arnone *et al.,* 1994), is experimentally mutated in this region and then a construct made by such a mutated regulatory region followed by the CAT reporter gene is injected into the egg, ectopic expression of CAT, i.e., also in the mesenchyme cells, is observed at the gastrula stage, therefore indicating that SpZ12-1, which is the only nuclear protein that binds this element *in vitro*, is the trans factor mediating the repression of CyIIIa expression in mesenchyme cells. Coffman *et al.* (1997) further demonstrated that another protein with a myb domain, **SpMyb**, specifically binds the middle module of the CyIIIa gene at the P7II site, thus contributing to repression of ectopic expression of CyIIIa. The organization of the CyIIIa regulatory region in modules as indicated in Fig. 1 is not fortuitous but stems from a hypothetical model that the Caltech group elaborated first for this gene and then for other genes and other species (Kirchhamer

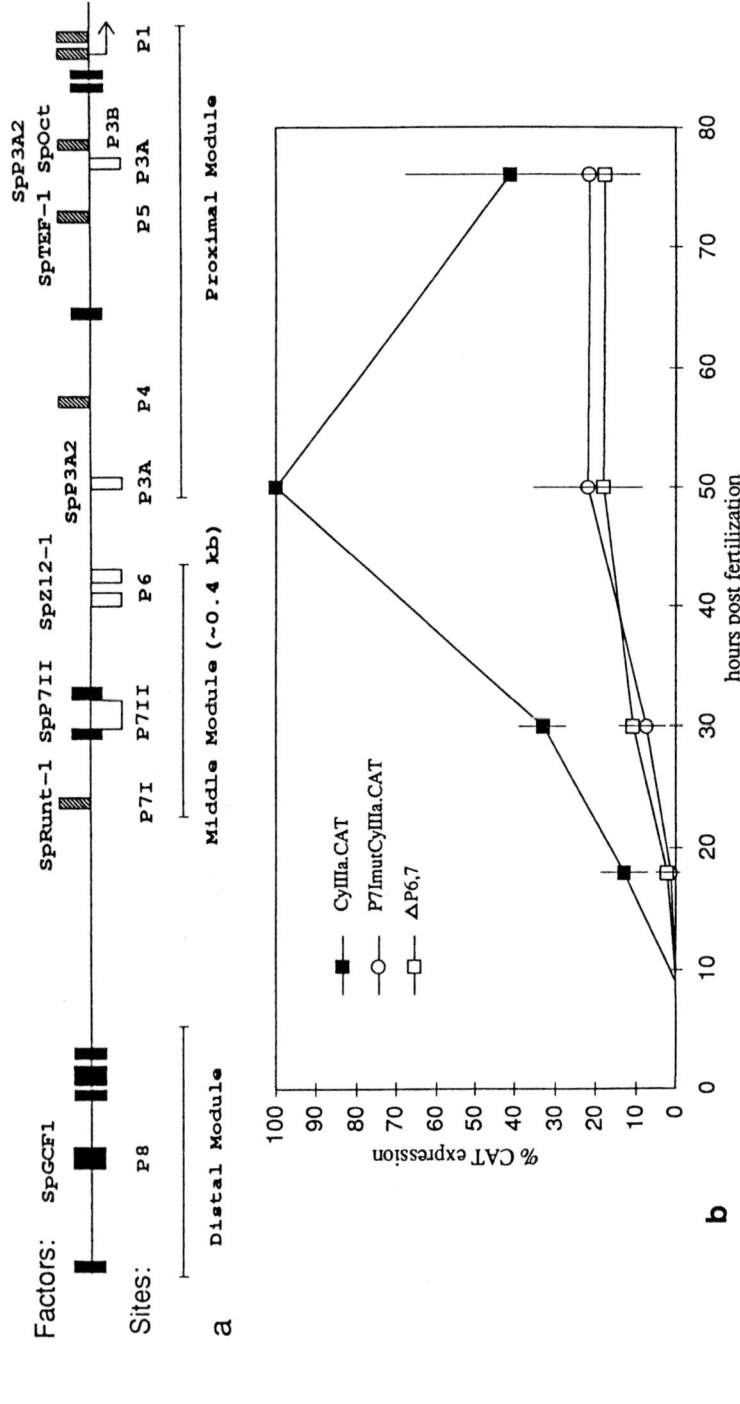

2. Genes and Their Products in Sea Urchin Development

and Davidson, 1996; Kirchhamer *et al.*, 1996a; Arnone and Davidson, 1997). Each module is defined as one part of a regulatory region which, through the attachment of specific transcription factors, regulates either temporally or spatially the expression of the gene, as can be experimentally demonstrated by the injection of a construct made by the module followed by a reporter gene into a particular cell (for example, derived from a particular embryonic territory or from a particular developmental stage) where it executes a regulatory function that is a subfraction of the overall regulatory function accomplished by the complete system. Details of the functions of the three modules of the CyIIIa regulatory region and of the functional interactions among them have been described by Kirchhamer and Davidson (1996). In summary, the proximal module (with respect to the transcription initiation site) has the function of establishing CyIIIa expression in the precursors of the aboral ectoderm as soon as this territory is specified by the sixth cleavage and of activating the gene late in cleavage. The middle module has the function of highly enhancing CyIIIa expression during gastrulation and of controlling the correct territorial expression later on. The distal module has no function in the control of territorial expression but has a general positive effect on CyIIIa expression, and, most important, interaction of this module with the other two is required for the normal level of function of both of them. An example of physical interaction of the modules is illustrated in Fig. 2, in which *in vitro* looping of the CyIIIa regulatory region is observed after experimental addition of one of the transcription factors, **SpGCF1**, able to bind various sites of all three modules. This factor is interesting in itself for the following features: it binds certain target sequences containing a CCCN core (where N is usually T); it is present in five forms with molecular masses from 37 to 55 kDa, with increasing extents of the N-terminal proline-rich domain, which are translated from the same mRNA by alternative utilization of five in-

---

Fig. 1 The P7I target site binds a positive regulator of CyIIIa expression. (a) Schematic representation of the CyIIIa 5' regulatory domain. Specific cis-regulatory target sites (not to scale), designated P1–P8, are shown below the line, while cloned trans-acting factors known to interact with these elements are indicated above the line. Sites for the ubiquitous factor SpGCF1 (Zeller *et al.*, 1995a, 1995b) are shown as black boxes; hatched boxes above the line represent sites known to have a positive effect on transcription (Franks *et al.*, 1990; Kirchhamer and Davidson, 1996), while open boxes below the line represent sites known to have a negative effect on transcription (Hough-Evans *et al.*, 1990; Wang *et al.*, 1995a; Kirchhamer and Davidson, 1996). The P7I site is located within the middle module, which also includes sites for the transcription factors SpGCF1, SpP7II, and SpZ12-1. The arrow indicates the transcriptional start site. (b) Effect of deleting the entire middle module (ΔP6,7, open box) or mutating five bases in the P7I element (P7ImutCyIIIaCAT, open circle) on reporter gene expression, compared to wild-type CyIIIaCAT expression (black box). To facilitate comparison, data points of each separate experiment were normalized to the peak of expression of CyIIIa•CAT at 50 h, which was set to 100%, and then averaged ($n = 4$). Vertical lines through the data points represent standard deviations and were also calculated after normalization of each value obtained. While the initial data points in this experiment were taken at 16 h, the curves start at 8 h postfertilization, which is the time at which endogenous CyIIIa is activated in development (Lee *et al.*, 1992). (From Coffman *et al.*, 1996.)

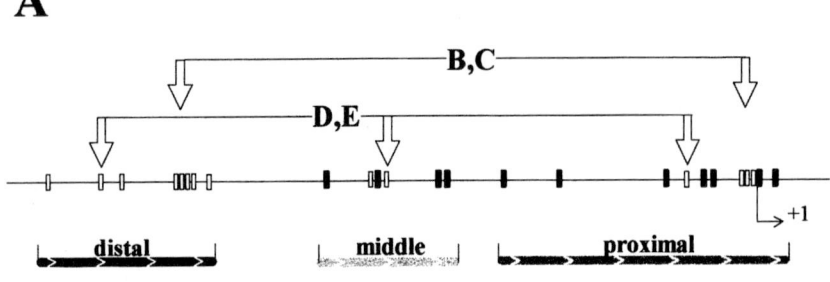
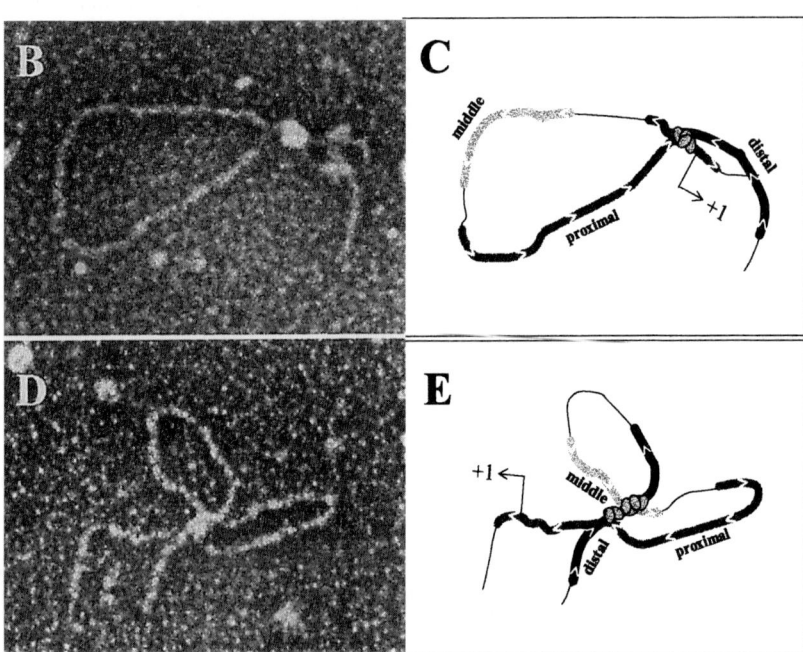

Fig. 2 Looping by SpGCF1 *in vitro* and interpretations. (A) Location of SpGCF1 sites in the CyIIIa cis-regulatory domain. (B and D) DNA loops visualized by electron microscopy, formed by complexes between recombinant SpGCF1 protein and CyIIIa cis-regulatory DNA fragment. The protein/DNA complex is the white aggregation at the crossover point of the loop. (B) The loop joins elements of the distal and proximal modules. (D) The distal SpGCF1 site cluster donates SpGCF1 molecules to a complex that includes elements of both proximal and distal modules. Interpretations are shown in C and E, respectively; ovals represent multimeric complexes of SpGCF1 molecules (other factors not shown). (From Kirchhamer *et al.*, 1996. Copyright 1996 National Academy of Sciences, U.S.A.)

2. Genes and Their Products in Sea Urchin Development

frame ATG codons; it interacts with target sites of genes that are expressed in at least three different embryonic territories, as CyIIIa (expressed in the aboral ectoderm), Endo16 (expressed in the vegetal plate and in the midgut), and SM30 (expressed in the skeletogenic territories); it is able to multimerize with itself and, at least *in vitro*, to loop DNA (Zeller *et al.*, 1995b, 1995c).

B. CyI Gene

Although the studies of the mechanism of regulation of the expression of such a gene are less advanced than those for CyIIIa, some useful information is available: Collura and Katula (1992), for example, found a correlation between cell division and CyI gene expression and, by injecting into *S. purpuratus* eggs constructs made by the regulatory region of such a gene followed by the β-gal reporter gene, located two sequences required for the correct spatial expression, one within the first 195 nucleotides and the second one within the first intron. Wang *et al.* (1994) described a homologous gene in *Tripneustes gratilla*, which, in spite of extensive sequence homology, has a different temporal and territorial expression.

C. CyIIIb Gene

Niemeyer and Flytzanis (1993), continuing preliminary work by Flytzanis *et al.* (1989), studied the regulatory region of this gene by microinjecting into *S. purpuratus* eggs constructs containing its upstream region, as such or variously deleted, followed by the CAT reporter gene. Their results indicated that three regions upstream of the promoter, called C1L, C1R, and E1, relevant to the quantitative expression were able to interact with nuclear proteins and that mutations within one such region, E1, brought about a decrease of CAT expression at pluteus but not at blastula. In further work carried out in three species (*S. purpuratus, L. pictus, and L. variegatus*) the same group (Xu *et al.*, 1996) concluded that E1 is involved in the temporal regulation of the expression and that all three elements are necessary and sufficient to confer aboral ectoderm specificity to the proximal promoter. This territorial restriction is due to the fact that in tissues different from the aboral ectoderm the CyIIIb promoter is repressed by the cooperative interaction of these cis-acting elements. Actually, the binding of the proteins **SpCoup-TF** (present only in the ciliated band according to Vlahou *et al.* (1996) and **SpSHR2** (present in all pluteus cells) to the C1R element (Kontrogianni-Konstantopoulos *et al.*, 1996) may be necessary and sufficient for the repression of CyIIIb expression in territories different from the ectoderm, whereas the additional presence of both C1L and E1 is required for restriction of this expression to the aboral ectoderm. These results are schematically summarized in Fig. 3.

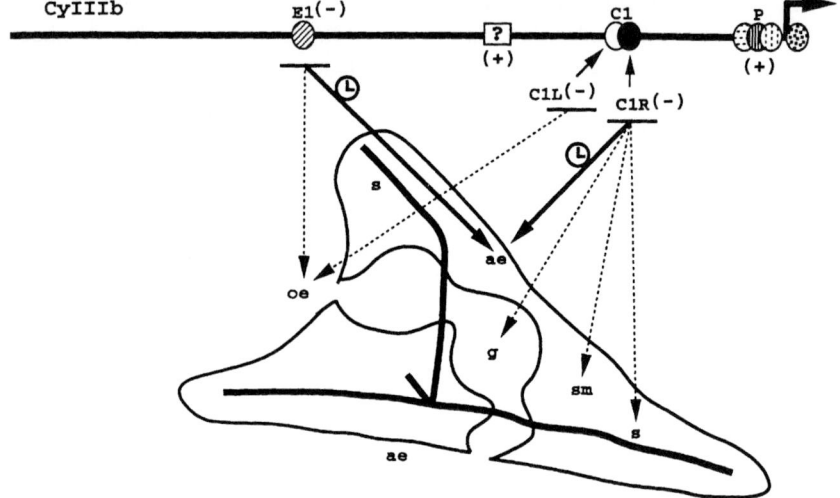

Fig. 3 Graphic summary of the proposed function of the CyIIIb upstream regulatory elements at pluteus stage. Arrows with dashed lines indicate repression, whereas the solid lines depict the temporal regulation (marked with a clock) exhibited by the combination of E1 and C1R within the aboral ectoderm. The box with the question mark indicates the possible presence of a still unidentified positive element. ae, aboral ectoderm; oe, oral ectoderm; g, gut; s, spicules; sm, secondary mesenchyme cells. Reprinted from *Mech. Dev.* **60**; N. Xu, C. C. Niemeyer, M. G. Gonzalez-Rimbau, E. A. Bogosian, and C. N. Flytzanis; Distal *cis*-acting elements restrict expression of the CyIIIb actin gene in the aboral ectoderm of the sea urchin embryo, 151–162. Copyright 1996, with permission from Elsevier Science.

D. CyIIa Gene and S9 Gene

Miller et al. (1996) studied by whole mounts the expression of these two genes, whose transcripts appear in the vegetal plate of *S. purpuratus* late blastulae, but in different territories within this field. Thus cells expressing S9 are broadly distributed in the vegetal plate except for a central zone and later on migrate as secondary mesenchyme to the dorsal wall of the embryo: cells expressing CyIIa are first located in the ventral side of the blastula as a group of 10–15 cells and later on are found as primary and secondary mesenchyme and as part of the gut. A ventralizing treatment, as with $NiCl_2$, causes extension of the expression of CyIIa to a higher number of cells in a radially symmetrical pattern. This shows that these gene products can be used to distinguish the ventral from the dorsal side of the vegetal plate, which therefore are diversified at a transcriptional level already before gastrulation. As to the regulation of the territorial expression of CyIIa in *S. purpuratus*, Arnone et al. (1998) found that it is completely reproduced by a construct made of a short upstream sequence of less than 450 bp followed by a green fluorescent protein (GFP) reporter gene. No evidence for a negative regulator was found; moreover, if the CyIIa gene of *S. purpuratus* is introduced into *L. pictus*, there is also here no ectopic expression. The evolution of the actin gene family of some sea urchins has been described by Kissinger *et al.* (1997).

IV. Endo16 and Other Endoderm-Specific Genes

Another gene for which the modular composition of the cis-regulatory region has been studied is Endo16 (Yuh and Davidson, 1996; Yuh *et al.*, 1994, 1996): In earlier studies McClay *et al.* (1983), Angerer and Davidson (1984), Wessel and McClay (1985), Wessel *et al.* (1989), and Nocente-McGrath *et al.* (1989) described the endodermal localization of the transcripts of this gene in *S. purpuratus* and of its equivalent LvN1.2 (and its protein Endo1) in *L. variegatus*. Ransick *et al.* (1993) found that its RNA appears in *S. purpuratus* 18 h postfertilization in the blastomeres deriving from the veg2 ring of the sixth cleavage and can be seen later in their descendants, while they are part of the vegetal plate, then in the archenteron of gastrulae, and eventually only in the midgut of the postgastrular stages. Its synthesis is switched off in the secondary mesenchyme cells as they delaminate from the tip of the archenteron. It is localized also in the endoderm of Li-induced exogastrulae, and it is expressed also in embryos whose gastrulation has been prevented by the collagen inhibitor β-aminoproprionitrile. Godin *et al.* (1996) found that Endo16 produces two transcripts of 8.5 kb in addition to one of 6.6 kb by alternative splicing.

The signal to the macromeres to initiate differentiation of the vegetal plate is provided by micromeres (Ransick and Davidson, 1995). Interestingly, if the embryos are dissociated into cells at the stage of 4–8 cells and then reaggregated, they continue to express Endo16, but if not reaggregated, they fail to express it (Godin *et al.*, 1997): The situation recalls that described long ago by Giudice and Pfohl for the role of cell interaction in the expression of another intestinal marker, i.e., the alkaline phosphatase (Pfohl and Giudice, 1967; Giudice, 1986).

The details of the location of the Endo16 protein, described by Soltysik-Espanola *et al.* (1994), show that it is found on the cell surface as in the extracellular matrix, which indicates its role in cell interactions during gastrulation, as also suggested by a Ca-binding domain in its sequence.

How is such a complicated spatial and territorial regulation brought about? The Caltech group has shown that the 2.3 kb cis-regulatory region consists of at least six modules. Three of them, called A, B, and G, synergistically activate transcription in the endodermal lineage; the other three, DC, E, and F, are necessary to repress transcription in the nonendodermal cell, and precisely, DC represses it in the primary mesenchyme, whereas E or F suffice to repress it in the ectoderm cells, that is respectively in the lower and upper ring of cells with respect to those originating endoderm. Kirchhamer *et al.* (1996b) experimentally produced chimeric cis-regulatory systems by combining modules of this gene with regulatory elements of another gene, that is, SM50, which is expressed in a different cell lineage (the skeletogenic primary mesenchyme), to investigate, through the expression of the CAT reporter gene, if predictable combinations of spatial expressions could be generated. The interesting result was that the positive Endo16 regulatory modules plus the positive regulatory sequences of SM50 (which does not have negative ones) cause CAT expression in both the Endo16 and SM50 territories but that the addition of the repressory modules of Endo16 to the positive regulatory regions of SM50 does not repress the CAT expression in the territory of SM50, therefore showing that to exert their negative effect on Endo16 ectopic expression, these modules have to interact with the other Endo modules, which cannot be replaced by the SM50 regulatory sequences; also the combination of the positive and negative modules of Endo16 plus the regulatory sequences of SM50 does not prevent CAT expression in the endoderm, suggesting that the Endo DC repressor region is not a dominant silencer of transcription (see Table I for a very schematic representation of these results, derived from Table I of Kirchhamer *et al.* (1996b)).

A mathematical calculation of the quantitative functional interactions of the various modules of the Endo16 gene has been reported by Yuh *et al.* (1996, 1998), who elaborated a computer program in which the regulatory functions of the Endo16 promoter can be predicted and experimentally tested. It is demonstrated that the activating effect of module A is switched on by the binding of SpOtx to one of its regulatory sequences, that if A interacts with B there is a further increase of Endo16 expression by a factor of 4.2, and that the modules DC, E, and F need to interact with A to repress ectopic Endo16 expression.

V. Ectoderm-Specific Genes

These have been studied mainly in two species, that is, *S. purpuratus* and *L. pictus*. The first ones have been named Spec1, Spec2a, Spec2b, Spec2c, and Spec2d; they belong to the family of the calmodulin–troponin C-myosin light chain and

Table I   CAT Expression in Three Different Embryonic Tissues, as Regulated by Different Modules in the Chimeric Regulatory Region

	CAT expression		
Chimeric regulatory region	In PMC	In gut	In ectoderm
SM50 + CAT	+	−	−
GBA (positive for Endo16) + CAT	−	+	−
GBA (positive for Endo16) + SM50 + CAT	+	+	−
DC (negative for Endo16) + SM50 + CAT	+	−	−
GBA (positive for Endo16) + DC (negative for Endo16) + SM50 + CAT	+	+	−
GBA (positve for Endo16) + DC (negative for Endo16) + CAT	−	+	−

are expressed in the aboral ectoderm cells and in their precursors (Hardin *et al.*, 1988). To these the gene Spec3 has to be added, whose expression is related to ciliogenesis (Eldon *et al.*, 1987), as immunologically detailed by Eldon *et al.* (1990). Microinjections of constructs made by the 5′ regulatory regions of the Spec1, -2a, or -2c genes followed by the CAT reporter gene have clearly shown the importance of a conserved region between the nucleotides -1565 and -697 for the correct territorial expression of at least Spec2a (Gan *et al.*, 1990a, 1990b). Within this region an enhancer was found between nucleotides -631 and -443; this sequence, called RSR (for "repeat–spacer–repeat," to indicate a structure with two direct repeats separated by a spacer), proved necessary to confer maximal activity and partial territorial specificity to a LacZ reporter gene: the reporter is expressed in the ectoderm, but not exclusively in the aboral one. In further work, Gan and Klein (1993) focused their attention on this enhancer in Spec2a and found that the RSR contains an AT-rich palindromic sequence that binds a transcription factor called A/T binding protein (A/TBP) whose specificity of binding suggests a relation to bicoid. As to the level of regulation of the territorial expression, Gagnon *et al.* (1992), using probes against introns, made the interesting observation that the levels of mRNA precursors of Spec1 and CyIIIa are comparable at the gastrula stage in tissues that produce and do not produce their proteins, whereas it becomes different at pluteus, suggesting that a regulation at the level of splicing is operating at gastrula, whereas a regulation at the level of transcription is operating at pluteus. Experiments of run on with isolated nuclei confirm this observation and extend it to Spec2c and **SpEGF1** (*S. purpuratus* epidermal growth factor 1).

The equivalent of the Spec1 gene of *S. purpuratus* is called LpS1 in *L. pictus*; this starts to be expressed at blastula, peaks at gastrula, and then declines during the larval stages (Tomlinson and Klein, 1990; Tomlinson *et al.*, 1990). Looking at hybrids between *Strongylocentrotus* and *Lytechinus*, Brandhorst *et al.* (1991) observed that Spec1 is expressed preferentially with respect to LpS1 both *in vivo* and in experiments of run on with isolated nuclei, although there is no loss or rearrangement of the LpS1 gene in the hybrid genome. *In situ* hybridizations (Nisson *et al.*, 1992; Brandhorst and Klein, 1992) showed that at the pluteus stage only few embryos, about 1%, expressed only LpS1, another 1% expressed mutually exclusive patches of LpS1 and Spec1, and the rest expressed only Spec1. At the gastrula stage, on the other hand, about 50% of embryos expressed both genes. Whatever the mechanism of predominance of Spec1 expression over LpS1, it must become operative only after the gastrula stage.

Looking at the promoter of LpS1, Xiang *et al.* (1991) found that a G string of six guanines between nucleotides -75 and -70 is important for the correct territorial expression. Two transcription factors bind to this sequence: one is typical of the ectodermal nuclei and the other is found also in the endoderm–mesoderm nuclei. More recently, Seid *et al.* (1996) stressed the importance for LpS1 correct territorial expression of an upstream stimulatory transcription factor (USF), which binds to the core sequence 5′-CACGTG-3′ of the LpS1 promoter, because reporter genes within constructs with mutations in the USF binding site are expressed also in undued territories at both gastrula and pluteus. This USF is therefore a good candidate for being a repressor of LpS1 ectopic expression. The hypothesis that ADP-ribosylation of nuclear proteins contributes to ectoderm cell differentiation was considered by the group led by Yasuamsu (Kamata *et al.*, 1993a, 1993b). They found that the **ADP-ribosyltransferase** activity peaks at the morula and gastrula stages in *Hemicentrotus pulcherrimus* and that inhibitors of this enzyme affect differentiation of ectodermal structures more than that of endodermal structures.

Differentiation in sea urchins in many instances is due to cell interactions or cell–matrix interactions. Also LpS1 expression obeys such a rule. Ramachandran *et al.* (1993) indeed found that if collagen deposition in the extracellular matrix (ECM) is inhibited by β-aminoproprionitrile in *Lytechinus*, LpS1 expression is inhibited together with gastrulation and spiculogenesis; if, however, platelet-derived growth factor β is added, gastrulation, spiculogenesis, and LpS1 expression resume. Govindarajan *et al.* (1995) found that antibodies against human **TGFβ** or **TGFα** immunoprecipitated specific proteins isolated from *Lytechinus* embryos and inhibited gastrulation and spiculogenesis; they also found that peptides representing the heparan sulfate proteoglycan binding site of human PDGF-β inhibited gastrulation and caused the formation of multiple radially arranged spicules, therefore reinforcing the author's hypothesis of the existence of a growth factor mediated signal transduction pathway. We shall come back to the subject of signal transduction pathways in sea urchins, which have recently started to be investigated.

Seid *et al.* (1997) found that a region of 128 nucleotides (from -108 to $+7$) of the LpS α or β gene is probably the ECM responsive element (ECM RE) as suggested by the fact that, if inserted into a metallothionein gene, it confers to this ECM dependence and that mutations of a G string within this ECM RE make LpS1 independent from ECM.

Wikramanayake *et al.* (1995) also investigated the dependence of the expression of Spec1 from cell interactions. Isolated mesomeres of *S. purpuratus* embryos differentiate a squamous epidermis expressing Spec1 at the same time as the entire embryos and specifically in this epidermis but not in the rest of the embryoids, therefore suggesting that an oral–aboral polarity was developed independently from cell interactions. However, the protein **EctoV**, a marker of the oral ectoderm, was detected throughout the embryos; moreover, the partial independence from cell interactions found in *S. purpuratus* was not shared by *L. pictus*. In the latter in fact specific aboral ectoderm genes were not expressed in animal explants, although the overall morphology of these was similar in the two species, and this expression required either recombination with vegetal blastomeres or treatment of the animal explants with lithium. Wikramanayake and Klein (1997b) also demonstrated that LiCl treatment of isolated *L. pictus* mesomeres at dosages that do not induce endoderm, if started 7 h postfertilization, caused complete oral–aboral polarization but if started 16 h postfertilization caused polarized expression of the aboral-specific LpS1 protein, but ubiquitous expression of EctoV and no induction of stomodeum or ciliary band. At least in this case, a role of the endoderm in complete oral–aboral polarization is therefore suggested; however, in the first case no endoderm induction is needed.

Arylsulfatase (see also p. 81) is another gene whose products accumulate in the ectoderm. Actually, its RNA begins to accumulate in all cells at blastula, but then, at gastrula, it starts to be restricted to the aboral ectoderm (Yamada *et al.*, 1989; Yang *et al.*, 1989a; Akasaka *et al.*, 1990). The relevant sequences for the proper temporal expression have been investigated by Yamada *et al.* (1992, 1993) by microinjecting into *H. pulcherrimus* eggs constructs containing the CAT gene: At least two positive regulatory elements and one negative regulatory element were identified in the promoter and an enhancer in which a G string and the sequence GATCTCCC are sites of interaction with nuclear proteins, whose concentration changes during development.

Yang *et al.* (1993) by immunoblot with polyclonal antibodies against an arylsulfatase fusion protein identified in *S. purpuratus* several peptides of 65–70 kDa and showed by immunohistochemistry that this enzyme is at first uniformly distributed along all presumptive ectoderm and at pluteus accumulates along the apical surface of the aboral ectoderm and to a lesser extent in part of the oral ectoderm. The authors identified a second arylsulfatase gene similar to that of *H. pulcherrimus* and called it SpARII.

Other genes are also expressed exclusively in the ectoderm: We shall recall here the already mentioned CyIIIa, and we shall see some other examples, which because of other properties more relevant to understanding gene role in development than the simple ectodermal location, will be described later.

VI. Genes of the Primary Mesenchyme or However Relevant to the Extracellular Matrix

Proteins characteristic of the primary mesenchyme cells (PMC) have been described (Harkey and Whiteley, 1983; McClay *et al.*, 1983; Wessel and McClay, 1985; DeSimone and Spiegel, 1985, 1986; Benson *et al.*, 1987; Farach *et al.*, 1987; Leaf *et al.*, 1987; Sucov *et al.*, 1987; Decker *et al.*, 1988; Matsuda *et al.*, 1988; Shimizu *et al.*, 1988; Tamboline and Burke, 1989; George *et al.*, 1991; Ettensohn and Ruffins, 1993; Harkey *et al.*, 1995; Killian and Wilt, 1996).

Wessel and McClay (1985) were the first to describe **Meso1** as a protein which appears in the PMC as they delaminate from the blastula wall. A thorough analysis of the integral matrix of the *S. purpuratus* embryonic spicules was carried out by Killian and Wilt (1996): these authors after *in vivo* labeling the embryos with ^{35}S-methionine detected 12 intensely labeled matrix proteins and about three dozen less labeled matrix proteins, most with an acidic p*I*; the proteins encoded by the genes **SM50** and **SM30** were identified by Western blots; a portion of those encoded by SM30 has an amino acid sequence with a domain similar to the carbohydrate recognition domain of the C-type lectin proteins.

As to the relative genes, Benson et al. (1987) and Sucov et al. (1987) were the first to isolate SM50 from *S. purpuratus* and to determine its structure. Leaf et al. (1987) identified again in *S. purpuratus* a cDNA clone encoding a surface protein specific of the PMC, called msp130.

Katoh-Fukui et al. (1991) established a more correct sequence of the already mentioned SM50 of *S. purpuratus*, whereas Livingston et al. (1991) cloned the homologous cDNA in *L. pictus* and named it pLSM1. The expression of both genes is restricted to the skeletogenic micromere lineage.

Katoh-Fukui et al. (1993) then isolated a homologous cDNA clone from *H. pulcherrimus* and called it pHPSMC. The product of the latter is exclusively localized in the PMC of gastrulae and appears in the isolated micromeres only after 48 h of culture. Another cDNA encoding for an acidic protein found in the matrix of the spicules was isolated by George et al. (1991) from *S. purpuratus* and called SM30. Frudakis and Wilt (1995) then investigated the regulatory regions which permit a correct territorial expression of SM30 and by microinjecting a construct made by the CAT reporter gene preceded by a regulatory region upstream 5' found that the region between nucleotides -1628 and -300 is especially important for the correct spatial expression. This region contains the highly repetitive sequence motifs (G/A/C)CCCCT and (T/C)(T/A/C)CTTTT(T/A/C); these elements interact with nuclear proteins, and coinjection of an excess of one such element with the above-described constructs causes their ectopical expression, suggesting that the element binds a negative regulator, which is competed off in the coinjection experiment. Both sequences, however, are necessary for correct spatial expression of SM30, as shown by experiments in which a construct containing only one or both of these elements was fused to a misregulated reporter gene: only in the second case was misexpression suppressed.

In 1995 Harkey et al. (1995) published the results of a long work in which they described the isolation of cDNA and genomic clones encoding **PM27**, i.e., another protein targeted to the spicules, which is peculiar of the PMC. Since it is known (Harkey et al., 1988) that during skeletogenesis at least five transcripts accumulate coordinately in the PMC, Raman et al. (1993), looking for a mechanism for coordinating the expression of the transcripts for skeleton matrix proteins, analyzed three genes coding for them in *S. purpuratus*, i.e., SM50, PM27, and msp130, and found that all three contain a cluster of four types of conserved sequence elements, where nuclear proteins which appear after PMC ingression, can bind. Two of the binding sites are conserved in all three genes and show reciprocal competition for binding proteins. These observations, although promising, do not yet solve the problem of coordinated expression.

Part of the same group (Klueg et al., 1997) then addressed the interesting question of how the expression of msp130 is shifted to later developmental stages in a species, *Heliocidaris erythrogramma*, which, having evolved to a direct development, makes a much reduced larval skeleton and expresses this gene later in development. Experiments of injection of constructs into eggs of species at indirect or direct development show that the promoter of the latter species still contains elements able to direct correct spatial expression in the host at indirect development and indicate that the difference should be looked for in trans-acting factors. The authors also found that an alternate processing of this gene is a characteristic of the indirect-development species and not of the adult skeletogenesis.

Following earlier work of Sucov *et al.* (1988), a more detailed analysis of the SM50 regulatory region from 440 bp upstream to 120 bp downstream of the transcription starting site was published by Makabe *et al.* (1995). Three regulatory elements (not yet called modules in this paper) were identified and called, starting from 5', C, D, and A. A series of microinjection experiments of constructs containing the CAT reporter gene preceded by these elements in various combinations and variously mutated established that C is the element which is primarily responsible for the correct spatial expression and was therefore named the "locator element," since it requires the action of the other two regions for stimulating transcription.

A major protein involved in skeletogenesis is collagen. The first reports on **collagen** synthesis in sea urchin embryos and adults were due to Pucci-Minafra *et al.* (1972, 1975, 1978). An important series of papers followed, which were, however, addressed to—more than the problem of regulation of gene expression—the analysis of the sequences of the various collagen genes, to the types of collagens synthesized, to their relation to spiculogenesis, and, more recently, to their spatial distribution. Gambino *et al.* (1997), for example, observed that the mRNA for collagen type I is already present in the unfertilized egg of *P. lividus* and restricted to one side of the egg, which, judging from its subsequent localization during development, seems to be the oral side. Suzuki *et al.* (1997) found that RNAs for fibrillar collagens $\alpha 1$ and $\alpha 2$ appeared in primary and secondary mesenchyme cells of *S. purpuratus* at late gastrula and then, at pluteus, were restricted to the cells associated with spicules and intestine; transcripts of the basement membrane $\alpha 3$ appeared already in the PMC of blastulae.

We shall only briefly recall here some of the works on collagen genes: Data on sequence analysis and time of developmental expression have been reported by Venkatesan *et al.* (1986), Butler *et al.* (1987), Angerer *et al.* (1988), Nemer and Harlow (1988), Saitta *et al.* (1989), D'Alessio *et al.* (1989, 1990), Benson *et al.* (1990), Wessel *et al.* (1991), Exposito et al. (1992a, 1992b, 1993, 1994), and Thurmond and Trotter (1994). Some relevant conclusions are that at least seven collagen types are present in sea urchin embryos; that four cDNAs coding for collagens have been isolated and sequenced, two for fibrillar genes, called COLL1α and COLL2α, and two for nonfibrillar collagens, called COLLP3α and COLLP4α; that the collagen fibrils of at least some echinoderms are bipolar; and that collagen is important for spiculogenesis and for spicular mineralization but is also present in other embryonic structures such as the hyaline layer (Spiegel and Spiegel,

1979). A function in spiculogenesis for the globular part of the N-propeptide of the 2a fibrillar collagen was proposed by Exposito et al. (1997).

Other macromolecules characteristic of the PMC have also been studied. In 1983 Lennarz (see also Grant et al., 1985) described in S. purpuratus a cell surface glycoprotein which is necessary for spicule formation and gastrulation and is recognized by the monoclonal antibody mAb1223, which inhibits mineralization and spiculogenesis (Farach-Carson et al., 1989) as well as glycoprotein processing (Kabakoff and Lennarz, 1990). In later work, the same group (Kabakoff et al., 1992) identified two glycoproteins which, besides the already mentioned msp130, contain the epitope to **mAb1223** and established that all three proteins are sulfated and contain GPI anchors.

Some studies have been dedicated to understanding the mechanism of interaction of the matrix proteins of the spicules with crystalline $CaCo_3$ (Berman et al., 1988, 1990; Decker and Lennarz, 1988; Cho et al., 1996). The conclusion is that these proteins are distributed within the spicules in intimate relationship with the crystals (Drager et al., 1989) and that probably the same set of genes is expressed during calcification in embryos and adults and in different species.

Other proteins that appear relevant to $CaCO_3$ deposition in the spicules are the enzymes H^+,K^+-ATPase, **Cl^-,HCO_3^--ATPase**, and **Na^+,K^+-ATPase** (Mitsunaga et al., 1987, 1989). Mitsunaga-Nakatsubo et al. (1992), however, having found that the levels of the α subunit of the last enzyme reach a maximum between hatching and gastrula and that they are higher in animalized than in vegetalized embryos, proposed that this enzyme might be a marker for ectoderm cell differentiation. The genes for the α subunit of the Na^+,K^+-ATPase have been studied in *H. pulcherrimus* by Yamazaki et al. (1997a, 1997b). In the process of spiculogenesis a role is also played by metalloproteinases, as suggested by the adverse effect of their inhibitors (Ingersoll and Wilt, 1998).

It has long been known that isolated micromeres can form spicules if induced by a treatment with horse serum (Okazaki, 1975), which may be thought to mimic an *in vivo* signal in which the extracellular matrix may be involved. Page and Benson (1992) found that isolated micromeres lose their competence to respond *in vitro* to horse serum after 50 h of development in *S. purpuratus*, but not the ability to synthesize the SM50 mRNA or the msp130 protein, therefore indicating that the synthesis of these two molecules is not a sufficient condition for spiculogenesis. Kiyomoto and Tsukahara (1991) found that the blatocoelic fluid stimulates spicule formation by isolated micromeres. An interesting review on the skeleton morphogenesis was published by Ettensohn et al. (1997) and related considerations were reviewed by Wilt (1997).

As to the question of cell interactions and cell–matrix interactions, an important role is that of fibronectin-like proteins. Iwata and Nakano (1981, 1983) and DeSimone et al. (1985) were the first to purify **fibronectin** from sea urchin gonads and embryos, respectively. A fibronectin-binding acid polysaccharide was described by Iwata and Nakano (1985) and a collagen-binding protein was reported by the same authors (Nakano and Iwata, 1982). A comparative analysis of sea urchin fibronectins was carried out on Japanese and Mediterranean species by Matranga *et al.* (1995). The same group (Zito *et al.*, 1998) by means of Fabs of monoclonal antibodies against a *P. lividus* nectin (**Pl nectin**) showed that these inhibited skeletogenesis if externally added to early blastulas but not if injected into the blastocoel, thus reinforcing the idea of the importance of appropriate ectoderm cell interaction for skeletogenesis, as elegantly shown by Armstrong *et al.* (1993). Also, Guss and Ettensohn (1997) demonstrated the importance of the ectoderm influence for the expression of three genes of the primary mesenchyme cells of *L. varigatus* (the homologues of SM50, SM30, and msp130) and for skeletal rod growth. It should be recalled, however, that the specific pattern of skeletal morphology is dictated by the primary mesenchyme cells, as demonstrated by Armstrong and McClay (1994) by injecting into embryos of one species the PMCs of a different species: skeletal morphology depends entirely upon the injected PMCs.

Alliegro and McClay (1988) characterized a series of proteins which are part of the hyaline layer and share a common carbohydrate epitope. They are secreted after fertilization from a class of vesicles different from the cortical granules. Some of them may belong to the group described as **apical lamina proteins**.

Alliegro *et al.* (1988) called **echinonectin** a fibronectin-like molecule, which they found in the hyaline layer of *L. variegatus* embryos. Its precise localization was detailed by immunoelectron microscopy by Fuhrman *et al.* (1992), who found it in intracellular membrane-bound vesicles. It was shown that this component of the hyaline layer is responsible for cell adhesion (Alliegro *et al.*, 1990) and that the adhesive properties of embryonic cells to this molecule *in vitro* change during development in a way that reflects their morphogenetic movements *in vivo* (Burdsal *et al.*, 1991). This is in agreement with previous work of Fink and McClay (1985) on the adhesion of PMC to the protein hyalin whose ultrastructure, made by several **hyalin** molecules associated with a high molecular weight globular core, was described by Adelson *et al.* (1992). The hyalin structure is stabilized by calcium, a old notion that was investigated at the molecular level by Robinson *et al.* (1988). Coffman and McClay (1990) characterized in *L. variegatus* a structural component of the hyaline layer, previously called EctoV antigen, which is a large filamentous glycoprotein, associated with the protein hyalin, concentrated around the tip of microvilli. A maternal form of 350 kDa is present in the egg, and a much larger one is synthesized at late gastrula, before being processed, becoming then localized in the oral ectoderm and in the foregut. Wessel *et al.* (1998) isolated the mRNA for hyalin and found its progressive confinement to the aboral ectoderm for the part of it which reappears at mesenchyme blastula. The same authors also identified a repeated region of the protein which is important for cell adhesion both in *S. purpuratus* and in *L. variegatus*. Matese *et al.* (1997) immunologically detailed the storage compartments where the hyaline layer components are in the unfertilized egg and the timing of their release following fertilization. The first compartment to release its content is that of the cortical granules, accompanied by that of the so-called basal lamina vesicles. The content of the apical vesicles is released about 5 min later, then the echinonectin vesicles release their content into the inner side of the hyaline layer, and finally about 30 min later maternal cadherins are deposited from a fifth compartment. It is relevant in this context that Wessel and Chen (1993) found that PMC, during their ingression into the blastocoel transiently accumulate mRNA for **α-spectrin**, therefore suggesting a role for microfilaments in their movement inside the blastocoel.

Other lectin-like proteins have been purified from sea urchins: A mucin-like lectin with an RDG sequence, named **echinoidin**, was isolated from the coelomic fluid of *Anthocidaris crassispina* (Giga *et al.*, 1987), with adhesive properties *in vitro* (Ozeki *et al.*, 1991). Another protein from the same species called **SUEL** (Sasaki and Aketa, 1981) was found by immunofluorescence and by immunoelectron microscopy to be stored in small electron-dense granules of the unfertilized eggs. This protein moves to the cortex within 10 min after fertilization, then become localized in the cortical layer of the cleaving blastomeres, and at least at midgastrula is deposited into the hyaline layer (Ozeki *et al.*, 1995). Matranga *et al.* (1992) found another extracellular matrix protein with adhesive properties different from echinonectin. This protein is uniformly distributed through the cytoplasm before fertilization and then moves to the cortex and stays in the apical surface of the ectoderm and endoderm. Fishkind *et al.* (1990) by means of immunofluorescence microscopy identified spectrin in the oocytes of several sea urchin species and found that this protein also moves to the egg cortex following fertilization and then redistributes during cleavage mostly along the blastomere plasma membranes and at the cell surfaces of the blastulas. Also, the **moesin-like** molecule (characterized by Bachman and McClay, 1997) was found to move to the egg cortex following fertilization of *L. variegatus* eggs. This movement depends on an actin-based cytoskeleton, as does moesin localization at the apical ends of cells in the regions of intercellular junctions later in embryogenesis. Still another group of proteins responsible for adhesion in sea urchins and extractable with butanol from the cell surface, and therefore called **bep**, were found to move to the cortex (in this case at least, of the animal pole) after fertilization (Di Carlo, personal communication).

The story of the adhesion proteins extractable with butanol from embryos is a long one and starts with the observation of Noll *et al.* (1979, 1981) that an extraction with diluted butanol makes *P. lividus* embryonic cells unable to reaggregate, but readdition of the extract to the dissociated cells renders them again able to reaggregate and to differentiate into pluteus-like structures. The 22S proteic complex contained in the butanol extract was named **toposome** and was hypothesized to be responsible for the selective adhesion of cells from different parts of embryos to each other (Noll *et al.*, 1985; Matranga *et al.*, 1986, 1991; Gratwohl *et al.*, 1991). Actually, fluorescent monoclonal antibodies against different toposome subunits stain different embryonic territories, which is in agreement with this hypothesis, although not proving it.

2. Genes and Their Products in Sea Urchin Development 65

At least two **cadherins** have been isolated from *P. lividus* eggs and embryos (Ghersi and Vittorelli, 1990; Ghersi *et al.*, 1993, 1996). One of them, with a mass of 140 kDa, is already present in the unfertilized egg and then becomes predominantly localized in the invaginating endoderm; the other, with a mass of 125 kDa, appears only later in development and is localized on the surface of most epithelia. Monovalent antibodies against these cadherins disturb development and cause cell dissociation.

Direct evidence for the existence of three **β-integrin** subunits in *S. purpuratus* embryos has been provided by Marsden and Burke (1997), who found that at least one of them has a 58% homology with vertebrate integrins; this is expressed as a transcript with a peak at gastrulation and mostly in the ingressing PMCs, in the secondary mesenchyme, in the gut, and in the pigmented cells. The expression of its protein is also high in the ingressing PMCs but is downregulated in the archenteron through gastrulation. A fibronectin receptor immunologically related to β-integrin was isolated by Katow *et al.* (1997) from *Clypeaster japonicus*.

Among the other proteins related to cell membrane, a **receptor for WGA** has been studied by Yoshigaki (1997), who observed that this kind of molecule accumulates in the cleavage furrow during division of the eggs of *H. pulcherrimus or H. depressus*, suggesting its involvement in the formation of the contractile ring; and Kuno *et al.* (1994) describe an insulin receptor.

Some metalloproteases have been identified in the *Arbacia punctulata* extracellular matrix and assumed to play a role in tissue and matrix remodeling during development (Quigley *et al.*, 1993). It was known that inhibitors of metalloendoproteases block spiculogenesis (Roe *et al.*, 1989).

Additional information about extracellular matrix proteins will be provided under the paragraph dealing with their possible role in gastrulation.

VII. Genes for Cortical Granule Content and Genes Related to the Wnt Pathway and Gastrulation

A. Genes for Cortical Granule Content

Laidlaw and Wessel (1994) identified in *S. purpuratus* four mRNAs that encode proteins specifically targeted to cortical granules and showed that all of these are synthesized throughout oogenesis at comparable rates, which suggests a coordinated gene activation. Fujita *et al.* (1994) provided indirect evidence that a protein which binds the exogastrula inducing peptides is part of the hyaline layer. Following the earlier work of Vater and Jackson (1990), Brennan and Robinson (1994) characterized in *S. purpuratus* a protein of 32 kDa, named HLC-32, which was localized by immunofluorescence throughout the cytoplasm of the unfertilized egg, in the hyaline layer of the fertilized egg, and in the basal lamina of the blastula. Sequence analysis of the cDNA showed a 41% and 47% homology to the already mentioned bep1 and bep4 proteins of *P. lividus*. Berg and Wessel (1997) developed a method to obtain *in vitro* maturation of *L. variegatus* oocytes, which permitted them to analyze cortical granule movement and gene expression. Wessel *et al.* (1998) found that the mRNA for hyalin is present during oogenesis but disappears at oocyte maturation, to reappear at mesenchyme blastula and to become localized in the aboral ectoderm at gastrula and pluteus; the authors identified hyalin cDNAs from *S. purpuratus* and *L. variegatus* and analyzed the sequence of the corresponding fibrillar glycoprotein of about 330 kDa. Another protein, called pamlin, is present inside granules of the cortex smaller than the cortical granules and has been assumed to be involved in the ingression of PMCs into the blastocoel (Katow, 1995; Katow and Komazaki, 1996).

Haley and Wessel (1997) isolated from *S. purpuratus* a cDNA for a serine protease of 35 kDa, which is then processed to a 25-kDa form, which is located in the cortical granules and shown by means of a specific inhibitor to participate in blocking polyspermy. Brooks and Wessel (1997) also isolated a promoter specific for oogenesis and probably important for the coordinated activation of the synthesis of cortical granule proteins. The same group (Conner and Wessel, 1997; Conner *et al.*, 1997) provided evidence that the proteins **syntaxin**, **VAMP**, **tagmin**, and **rab3** are enriched in the regions of the embryo assumed to have a high secretory activity and also function in cortical granule exocytosis. These authors also found that microinjections of rab3 specifically inhibit cortical granule exocytosis.

Also, some metal-binding proteins are involved in the elevation of the fertilization membrane as indicated by Andacht and McClay (1997) by means of treatment with nickel chloride or various other metals which inhibit fertilization envelope elevation. These authors also isolated two nickel-binding proteins which are secreted at fertilization.

Bachman and McClay (1996) cloned from *S. purpuratus* a **β-1,3-glucanase** cDNA which is known to be part of the cortical granule content and is an enzyme of 499 amino acids. Its message is also present in early embryogenesis and in the gut.

B. Genes Related to the Wnt Pathway

The attention of many authors has recently focused on genes known to influence morphogenesis in other embryos more amenable to genetic analysis such as *Drosophila* to investigate whether similar cascade pathways can be found in a deuterostome as the sea urchin. Information about this comes from the analysis of the expression of some genes which are known to produce a signal cascade in *Drosophila* and again from some transplantation experiments in which either the autonomous or the induced cell potentialities are investigated.

Among the first type of experiments we shall mention here are those of Wikramanayake and Klein (1997a), who by injecting β-catenin mRNA in one blastomere of an animal half obtained results consistent with the idea that a Wnt-like signal from the micromeres may pattern the animal–vegetal axis. The search for such signals has proceeded in the hands of McClay's group by looking at the pattern of distribution of a **Notch** type receptor, by means of antibodies against what the authors called LvNotch. It was found (Sherwood and McClay, 1997) that at early blastula this antigen, absent from the vegetal pole, concentrated in the basolateral cell membranes of the animal side. Then at the mesenchyme blastula stage it appeared concentrated in a ring of cells around the vegetal plate and delimiting the boundary between the presumptive endoderm and secondary mesoderm cells, with a higher concentration in the aboral side of the vegetal plate. We also mention in this respect the work of Egaña *et al.* (1997), who, by RNAase protection, showed that many members of the sea urchin Wnt gene family (Suswnt-1, -4, -6, and -7) and a sea urchin **hedgehog** (Sushh) are expressed in early sea urchin development from cleavage (Suswnt-1 and -4) to early gastrula and that Suswnt-7 and Sushh are primarily expressed in the mesenchyme lineage. It is still too early to judge if these genes are part of a pathway that controls a signal cascade relevant to gut induction, but it is undoubtedly worth exploring what the role of the above gene is in sea urchin development.

Wikramanayake *et al.* (1998) provided evidence for the need for **β-catenin** in vegetal plate specification and endoderm formation, and proposed a similarity between the role of the wnt pathway in ventro-dorsal specification in vertebrates and vegetal–animal specification in sea urchins.

Miller and McClay (1997a, 1997b) found that β-catenin moves out of the adherens junctions of *L. variegatus* early blastulas in the primary and secondary mesenchyme cells as they differentiate from the epithelial ones and lose adhesion to each other; also, during gastrulation there is a decrease of the intercellular junction β-catenin in cells undergoing convergent–extension movements. These data together with the observation that also cadherin moves away from cell surfaces during PMC ingression but not during the convergent–extension of gastrulation speak for an integrated role of these two kinds of molecules in cell adhesion during these morphogenetic processes. These authors also found and analyzed for sequence a new cadherin with a molecular mass of 330 kDa, therefore named Goliath Lv cadherin.

C. Genes Related to Gastrulation

We have already mentioned some of the genes relevant to such a process. It may be of interest to quote here some more of them, with the hope of providing some answers to the morphogenetic problems of gastrulation. One of the facts that have been long ascertained through transplantation experiments is that micromeres are able to induce gut formation (Hörstadius, 1935; Ransick and Davidson, 1993). The question that has been asked is then: "Do micromeres in nature send an inductive signal to adjacent cells to make intestine and, if so, what is the nature of such a signal?" This seems to depend upon transfer of β-catenin to nuclei; in fact, such a transfer is experimentally prevented by overexpressing cadherin, micromeres lose their ability to induce intestine when transplanted on top of the animal pole of a 16 cell *L. variegatus* embryo (Logan *et al.,* 1999).

The archenteron is divided into three different pieces: foregut, midgut, and hindgut. McClay and Logan (1996) demonstrated that if any of these parts were experimentally removed, they were replaced by the other parts, restituting the normal archenteron morphology, and that this plasticity was conserved also long after each of the archenteron parts had expressed specific territorial markers. The authors proposed that this ability requires extensive and continuous short-range communication between cells of the archenteron. The same authors (Logan and McClay, 1997a) then detailed the descendance of the archenteron cell precursors by microinjecting into the 60 cell blastomers of *L. variegatus* the lipophilic dye DiI (C16) and found that both veg1 and veg2 originate archenteron cells and that, therefore, in contrast to previous results obtained with less refined techniques, the ectoderm and endoderm lineages are not yet segregated at the sixth cleavage. The segregation of veg1 and veg2 cells to ectoderm occurs relatively late in development and is unpredictable. This indicates that the future position of the veg1 descendant cell is more important in determining their ectodermal or endodermal destiny cell than the early cleavage pattern: one further score in favor of the short-range cell signals. Ransick and Davidson (1998), working on *S. purpuratus*, also with the aid of the expression of the gut marker Endo16, confirmed that Veg1 originates hindgut and half of midgut, whereas Veg2 originates the other half of midgut and the entire foregut, but in contrast to *Lytechinus*, in *Strongylocentrotus* the allocation of Veg1 descendants to the endoderm depends upon the position of the clone with respect to the oral–aboral axis. As to the information deriving from transplantation experiments, Logan and McClay (1997b) showed in *Lytechinus* that descendants of Veg1 and Veg2 form the gut structures independently from micromere signaling; they also showed that patterning of the endoderm requires cell–cell interactions between Veg1 and Veg2 populations. Benink *et al.* (1997), using chimeric embryos in which fluorescently labeled blastomeres were implanted into *Lytechinus* embryos, also found that implanted archenteron precursor cells autonomously differentiated into the various gut parts. Moreover, these cells, when ectopically implanted induced an additional skeleton made by the host cells. The results of the experiments of Benink *et al.* support the hypothesis of a sequential induction as proposed by Davidson (1989). Chen and Wessel (1996) asked the question if endoderm-specific genes such as **Endo1** and **LvN1.2** of *L. variegatus* can be switched on in the absence of cell interactions: the answer was that if cells of the vegetal plate destined to form endoderm were cultured *in vitro*, they expressed these genes also in the absence of cell interactions, if isolated after the late blastula, but not if dissociated before that stage, although cells destined to form mesenchyme or ectoderm differentiated in the same culture, also if isolated before late blastula, which again speaks in favor of early cell interactions for endoderm formation. Hardin and Armstrong (1997) carried out on *L. variegatus* a series of microsurgery and microtransplantation experiments detailing the information required for the ectoderm to form the oral field, leading to the indication that cell interactions before the gastrula stage are capable of inducing ectopic oral structures, that after the early gastrula stage the oral pattern is already specified, and that the oral structures do not require direct interactions between the mesentoderm of the archenteron and the oral endoderm. More information on cell lineage is beyond the scope of this review and therefore the reader is referred to the review of Ettensohn *et al.* (1996), to which we shall only add here the result of Ruffins and Ettensohn (1996) indicating that in *L. variegatus* the dorsal and ventral blastomeres of the four-cell stage do not contribute equally to the nonskeletogenic mesodermal precursors.

Two other functions of the extracellular matrix proteins have been studied and are worth mentioning here: one is their role in embryonic morphogenesis (see Ettensohn and Ingersoll (1992) and Ingersoll and Ettensohn (1994), and the second is their function in cell interactions. As to the first, we have already mentioned that experimental block of biosynthesis or processing of the extracellular matrix inhibits gastrulation; Berg *et al.* (1996) identified in *L. variegatus* another extracellular matrix protein important for gastrulation and called in **ECM18**. Antibodies against it inhibit PMC organization and endoderm morphogenesis; it accumulates in the basal lamina surrounding the archenteron and in other parts of the blastocoel wall. The fact that its mRNA is present throughout development but the protein is detectable only during gastrulation suggests a regulation at the translational level and probably, as indicated by the analysis of polysomal mRNA, at the level of polysome–messenger interaction. An ultrastructural description of the basal lamina was given by Amemiya (1989). It is interesting that the inhibition of protein tyrosine kinase following fertilization of *S. purpuratus* eggs inhibits a much later event such as gastrulation, although leaving unaffected mesoderm and endoderm formation (Livingston *et al.*, 1998). The artificial expression of a dominant/negative PDGF receptor in *L. pictus* embryos inhibits gastrulation and causes the ectopic expression of the oral ectoderm specific gene EctoV, whereas the aboral ectoderm specific genes are repressed (Ramachandran *et al.*, 1997).

Another extracellular matrix protein, called **ECM3** (Wessel and Berg, 1995), was hypothesized to be involved in targeting the archenteron to the blastocoel roof in *L. variegatus*, because it becomes localized during gastrulation in the basal lamina and adjacent blastocoelic region of the whole ectoderm except for those cells of the animal pole to which the invaginating archenteron will attach in this species. The process of target recognition by the archenteron in sea urchin embryos is a very specific one as demonstrated by the very elegant experiments of Ettensohn and McClay (1986) and Hardin and McClay (1990). One step of regulation of the expression of the ECM3 protein is that of its secretion from the vesicles within which it is stored, because it is specifically secreted into the blastocoel and not at the opposite side of the cells. Another step is probably at the level of transcription, because its RNA is distributed uniformly throughout the oocyte and the embryo till blastula but then is found exclusively where mentioned above. Evidence for a role of proteolysis of a basal laminar glycoprotein during gastrulation has been provided by Vafa *et al.* (1996). Luke *et al.* (1997) described a role for the **Spfkh1** gene in the formation of the gut in *S. purpuratus*. Actually, its expression is limited in time and space to the invaginating endoderm and is increased by LiCl or phorbol ester; its decline is prevented by extracellular matrix disruption. That phorbol ester alters cell fate during gastrulation, inducing ectodermal cells to differentiate as endodermal or mesodermal, had already been shown by Livingston and Wilt (1992), who attributed this to **PKC** stimulation (see also Livingston *et al.*, 1998).

The cellular aspects of sea urchin gastrulation have been carefully reviewed by Hardin (1996); we shall therefore add here only a few of the pertinent papers, such as those of Malinda *et al.* (1995) and Miller *et al.* (1995) on microscopic analysis of filopodia during gastrulation, that of Nakajima and Burke (1996) on the formation of bottle cells in gastrulation, that of Davidson *et al.* (1995b) on biomechanic models, that of Crawford and Burke (1994) on the stimulation of PMC impression by laminin, that of Lane *et al.* (1993) on the role of the apical ECM in epithelial invagination, and that of Yazaki *et al.* (1995) on the possible role for the protein **ES-1** in the attachment of the esophagus to the stomateum.

VIII. Homeobox-Containing Genes and Genes for Transcription Factors That Do Not Contain Homeobox

A. Homeobox-Containing Genes

The presence of such genes in an unsegmented invertebrate such as the sea urchin was looked for not long after their description in *Drosophila*: Dolecki *et al.* (1986, 1988) and Dolecki and Humphreys (1988) found that genes of the Antennapedia and Engrailed type exist in *Tripneustes gratilla* and that at least one of them is expressed starting at the blastula stage, with an increase at gastrula and a decrease at pluteus. Angerer *et al.* (1989) found that this gene is also expressed in *S. purpuratus* and that in both species its mRNA accumulates first in all the aboral ectoderm, restricting then to an area of it close to the vertex. It was also observed that these proteins are highly conserved in the two species also outside the homeodomain. A gene belonging also to the Antennapedia class was described by Pfeffer and von Holt (1991) in the South African *Parechinus angulosus*, which is expressed at mesenchyme blastula and also in adult tissues. Different classes of homeobox-containing genes were cloned by Di Bernardo *et al.* (1994, 1995) in *P. lividus*. One of them, named **PlHbox11**, is the homologue of the human and mouse HoxB3 and is expressed only in plutei and adults; another one, called **PlHbox12**, has a 55% homology with the paired class genes and is transiently expressed already at the 4-cell stage, with a maximum at the 64-cell stage, and only from one side of the macro- and mesomeres. Its expression depends on cell interactions. The early and localized expression makes it possible to hypothesize that it might play a role in the establishment of the oral–aboral axis, although crucial experiments to support such a hypothesis are still lacking. Bellomonte et al. (1998) found that another *P. lividus* homeobox gene (**PlHbox9**) is expressed beginning at late gastrula and that its RNA at pluteus is confined to the anal opening. Ishii et al. (1997) proposed that in *H. pulcherrimus* a homeobox-containing gene called **HpHox8** is involved in the irreversible establishment of the aboral ectoderm, because its protein is found in the nuclei of this region and its overexpression represses development of oral ectoderm, mesenchyme cells, and endoderm and the induction of oral ectoderm by $NiCl_2$.

Martinez et al. (1997) published the complete sequence of *S. purpuratus* Hox8 from a late gastrula cDNA and found that there is only one gene copy and only one Hox gene cluster per genome. Homeobox-containing genes of the orthodenticle class were described in *S. purpuratus* by a group of authors (Gan et al., 1995; Mao et al., 1996; Li et al., 1997). The two transcripts of one such gene, called **SpOtxα** and **SpOtxβ**, respectively, and generated by alternative RNA splicing, accumulated—one with a peak at blastula and the other with a peak at gastrula—first in all cells but then became gradually restricted to oral ectoderm and vegetal plate territories. They were already detectable in the egg cytoplasm and became transferred into nuclei between the 60- and 120-cell stage. It has been suggested that acetylcholinesterase might be involved in such a transfer from cytoplasm to nuclei, since it is inhibited by inhibitors of this enzyme (Falugi et al., 1997). It is of interest that the SpOtx protein binds with high specificity to elements of the control region of the already described Spec2a gene. In support of the theory of a regulatory function of SpOtx on Spec2a are the experiments of microinjection of high amounts of SpOtx mRNA into the cytoplasm of the egg, which caused abnormalities in embryonic development, turning the embryos into a ball of aboral ectoderm, as revealed by tissue marker transcripts. At least three distinct domains of the homeobox are required to produce the morphological abnormalities. Although this hypothesis is attractive and nicely supported by experiments, the Otx genes probably also have some other function, as hypothesized by the authors, since, for example, their proteins are found in the nuclei of all cells at both blastula and pluteus stages. Another gene expressed in the aboral ectoderm which might be regulated by the Otx protein is the gene for arylsulfatase of *H. pulcherrimus* (HpArs): Sakamoto et al. (1997) actually found that three types of Otx-related proteins interact with an enhancer of HpArs. These proteins appear and disappear at different times around the blastula stage. Mitsunaga-Nakatsubo et al. (1997) described two types of Otx RNAs in this species, one synthesized in early stages, predominantly in the presumptive endoderm, and the other in the micromere-derived cells, then in the vegetal plate, and, after PMC ingression, in the ectoderm. Microinjection of either RNA again causes the embryos to develop into epithelial spheres with no oral–aboral axis. Evidence that these two RNAs derive from a single gene by altering the transcription start site and by alternative splicing has been presented by Kiyama *et al.* (1998) upon analysis of the gene structure.

Martinez and Davidson (1997) also isolated from *S. purpuratus* a gene of the subclass of divergent homeobox regulators called H6 in humans and named here **SpHmx**. This codes for a transcription factor and is expressed only in pigment cells of prisms and plutei, although its transcripts are present in much lower amounts already in the unfertilized eggs and subsequent stages. These results suggest that SpHmx acts as a regulator of pigment cell differentiation.

Dobias *et al.* (1997) studied the sequence and expression of an Msx gene in *S. purpuratus* with a homeodomain very similar to that of vertebrates and called it SpMsx. This gene is expressed at blastula during gastrulation; its transcripts are localized in the invaginating archenteron, in the secondary mesenchyme, and in the oral ectoderm. Its expression depends on early cell interactions, because it is prevented by cell dissociation at the two-blastomere stage but remains in the same tissues after experimentally induced exogastrulation.

Czerny *et al.* (1997) isolated from *P. lividus* five Pax genes. Three of them bear no homology with those known in other animal phyla. One of these, named suPaxB, codes for the transcription factor TSAP, involved in the regulation of transcription of late histone genes. Morris *et al.* (1997) identified by PCR with degenerate primers seven Hox-type sequences from *Holopneustes purpurescens*. The sensitivity of hox genes to retinoic acid in vertebrates is well known. Also, sea urchin development is affected by retinoic acid, as demonstrated by Sconzo *et al.* (1996) and Sciarrino and Matranga (1995), which at least indicates that retinoic receptors are present in sea urchin eggs and embryos. A discussion on the changes in the roles of homeobox genes during echinoderm evolution was published by Lowe and Wray (1997).

B. Genes for Transcription Factors That Do Not Contain Homeobox

We have already mentioned some of them in our discussion of the regulation of several genes, and we recall here that many transcription factors are being continuously isolated by the group at Caltech, thanks to an automatic device (Coffman *et al.*, 1992). Harrington *et al.* (1997) analyzed by high-resolution two-dimensional electrophoresis nine transcription factors of *S. purpuratus* and tabulated the genes they control, their functions, and their structural characteristics. It was observed that these factors were covalently modified during development, usually through phosphorylation. We shall, however, add here some more information which may be relevant to the spatial control of gene expression.

Wang *et al.* (1996) isolated from *S. purpuratus* a transcription factor containing three Zn fingers and a leucine zipper, related to the Kruppel/Krox gene and called

SpKrox1. Its expression, low in the unfertilized egg, increased at blastula and decreased after gastrulation; interestingly, its RNA appeared in the macromeres of the 16-cell stage and then became restricted to a ring of cells of the vegetal plate around the blastopore. As we shall discuss later, it seems to be involved in the initial establishment of the vegetal plate.

Harada *et al.* (1996) identified in *H. pulcherrimus* an orthologue of the vertebrate forkhead (which is related to notochord formation) and called in **Hphnf3**. Its rare RNA is first detectable at the swimming blastula stage, accumulates at gastrula prism, and decreases at pluteus. It is again initially located in the vegetal plate, then in the cells surrounding the blastopore, and at prism in the entire gut; the same authors found that **forkhead** and **Brachyury** orthologues start to be expressed at blastula in the vegetal plate, and that the latter is expressed later in the secondary mesenchyme founder cells (see also Papaioannu and Silver (1998) for a recent review). A role for Spfkh1 in gut formation has already been described in this review. Vlahou *et al.* (1996) and Kontrogianni-Konstantopoulos *et al.* (1996) described the asymmetrical localization of the maternal transcripts for **SpCoup-TF**, encoding for the orphan steroid receptor, which are found only in one of the first two blastomeres both in *Strongylocentrotus* and in *Lytechinus*. The hypothesis that this may be related to the establishment of the dorsoventral axis is, however, not demonstrated yet (see Henry (1998) for a review on this point). A homolog of the human **BMP** has been characterized in *S. purpuratus* embryos by Hwang *et al.* (1994).

IX. Tubulin Genes and Other Genes Related to Ciliogenesis and to the Mitotic Apparatus

Transcripts for α- and β- tubulins as well as their proteins are already present in the unfertilized egg (for reviews, see Giudice (1986), Stephens (1995, 1997), and Suprenant and Foltz Daggett (1995)). Their amount, however, increases during ciliogenesis and their synthesis can be stimulated by deciliation or by the addition of an animalizing agent such as Zn, which stimulates ciliary overgrowth (Alexandraki and Ruderman, 1985; Gong and Brandhorst, 1987; Gianguzza *et al.*, 1989, 1992). Their transcripts are destabilized by the accumulation of unpolymerized tubulin, and microtubules can be severed by a katanin (MacNally and Vale, 1993). Gianguzza *et al.* (1995) found that six different genes for tubulins are expressed after the cleavage stage in *P. lividus*: transcripts of one, the α-tubulin gene Plα2, are first detectable at blastula and these locate first, together with Plβ3 transcripts, in the few cells from which neural territories will differentiate. The same authors then found (Casano *et al.*, 1996) that the other four types of embryonic mRNAs for α- and β-tubulins are localized in the oral ectoderm, in the ciliated band, and in the intestine in a way which suggests that the increase of such transcripts may be related to ciliogenesis or to gastrulation movements.

Another component of cilia is also in sea urchins, the protein **tektin**; following earlier work in which it was shown that this is synthesized during embryogenesis in a quantity which correlates with the amount of cilia, normally generated or experimentally induced, Norrander *et al.* (1995) investigated about the amount of tektin A mRNA. The data show that this is upregulated during ciliogenesis and correlates with the length of cilia in animalized embryos, whereas the pools of tektin B and C mRNAs are much larger and nonlimiting, therefore suggesting that it is the amount of tektin A mRNA that controls ciliary growth. A posttranslational control must also be present, because if theophylline is added to zinc-arrested embryos, a threefold ciliary length increase is induced and tektin mRNA is accumulated during zinc arrest but translated only after theophylline addition.

Kinesin and kinesin-like proteins (**KLPs**) have been found in sea urchin eggs, embryos, and coelomocytes (Scholey *et al.*, 1985, Wright *et al.*, 1991, 1993; Cole *et al.*, 1992, 1993; Henson *et al.*, 1992; Skoufias *et al.*, 1994). The results indicate that kinesin is bound to vesicles and ER of the mitotic aster and transports them toward the cell surface along microtubules. Whereas antibodies against kinesin permit mitosis, antibodies against some KLPs disrupt the mitotic spindle. The best characterized among the sea urchin KLPs is a plus-end-directed, heterotrimeric, kinesin-related microtubule motor protein called **KRP(85/95)**. Henson *et al.* (1995) by means of monoclonal antibodies and confocal microscopy demonstrated that this protein is concentrated in the pericentriolar region at prophase, in the kinetochore-to-pole microtubules at metaphase, and especially in the spindle interzone at anaphase. The results suggest that it may play a role in vesicle transport during embryonic cell division. Data obtained by the use of monoclonal antibodies against this KRP, however, show that these have again no effect on mitosis whereas they dramatically inhibit the formation of normal motile cilia at blastula and cause the lack in these of the central pair of microtubules, which suggests a role in the delivery of ciliary components for ciliogenesis (Morris and Scholey, 1997).

An important role in cell division has long been recognized for centrosomes in sea urchins. Kuriyama and Maekawa (1992) found that a 225-kDa **centrosomal protein** of *S. purpuratus* becomes phosphorylated during mitosis. **Protein kinases** have been purified from sea urchins (Peaucellier *et al.*, 1993; Moore and Kinsey, 1994; Rakow and Shen, 1994; Onodera *et al.*, 1997; see also later in this review) but whether some of these are involved in phosphorylation of mitotic-related proteins has not been demonstrated yet. There is indication for a transient tyrosine phosphorylation of the egg receptor for sperm (Abassi and Foltz, 1994). Another component of the sea urchin mitotic apparatus which is phosphorylated during cytokinesis has been described by Shuster *et al.* (1997), who called it **p62**. Ye and Sloboda (1997) isolated the cDNA encoding p62, which is required for mitosis progression and is bound to interphase chromatin; its mRNA is present from before fertilization till gastrula.

Hinchcliffe and Linck (1998) have isolated from *S. purpuratus* sperm flagella, besides tubulins and tektins, two new proteins, termed **Sp77** and **Sp83** after their molecular weights, which are found in microtubules and centrioles.

X. Genes for Cyclin and Other Cell Division Related Proteins

Recalling that cyclin was discovered in sea urchins (Evans *et al.*, 1983), we mention here that Kurokawa *et al.* (1997) cloned from *H. pulcherrimus* a cDNA encoding maternal cyclin E very similar to maternal cyclin E of some other eukaryote species. This gene is in a single copy; the level of its mRNA is high till morula and then declines. A splice variant of a gene for cyclin B has been described by Lozano *et al.* (1998) in *Sphaerechinus granularis* and found to produce a cell cycle delay and abnormal development if injected into the unfertilized egg. It is assumed to naturally play a role in keeping the oocytes in prophase and in slowing down the cell cycle in cells committed to differentiate during development. Meijer *et al.* (1997) demonstrated that roscovitine, a specific inhibitor of cyclin-dependent kinases, reversibly arrests sea urchin cell cycle in late prophase. Patel *et al.* (1997) demonstrated that caffein induces mitosis and cell division also in the presence of unreplicated DNA by stimulating tyrosine dephosphorylation of **p34cdc2** and switching on its protein kinase activity. Nemer and Stuebing (1996) isolated from *S. purpuratus* and *L. pictus* a cDNA encoding a protein with high similarity to the **Mik1** and **Wee1** protein tyrosine kinases. Its mRNA is low till the first five cell division cycles and then rises till midblastula; at later stages, it is absent from cells which do not divide such as those of the pluteus aboral ectoderm. Another gene involved in cell division is that of the **laminin binding protein**, which in *S. purpuratus* is expressed during oogenesis and then again from hatching; at prism both RNA and protein are detectable only in the oral epithelium, in the primary mesenchyme, and in the endoderm (Hung *et al.*, 1995).

It may be appropriate to add here a few other pieces of information regarding cell division: Yazaki *et al.* (1995) produced evidence suggesting that in *P. lividus* eggs the cytoskeleton plays a role in regulating calcium channel activity during the cell cycle. Yoshigaki (1997), as mentioned before, observed an accumulation of **WGA receptor** proteins in the region of the egg contractile ring. Bonder and Fishkind (1995) published a review on the role of actin in sea urchin membrane cytoskeletal dynamics. He and Dembo (1997) reconsidered the old question of whether polar relaxation or equatorial contraction is determinant in egg cleavage, concluding that a series of forces drive actin and myosin to the equatorial ring so that the poles relax in the absence of these molecules. Pudles *et al.* (1990) isolated from the so-called cytomatrix (a cortical material resistant to shear in nonionic detergents) of several Japanese eggs a giant protein with properties similar to those of **titin**. Sluder *et al.* (1994, 1995) investigated the mechanism of nuclear envelope breakdown and of sensing chromosome malorientation in all division.

Recent observations on the cell division inhibitory activity were reported for LiCl by Becchetti and Whitaker (1997) for the IMPase inhibitor L690,330 by Sconzo *et al.* (1998) and for nordidemnin by Pesando *et al.* (1995).

XI. Genes for Kinases, a Fibroblast Growth Factor Receptor, and Metallothionein

A. Genes for Kinases

Sakuma *et al.* (1997) identified by RT-PCR seven PTK sequences in *Anthocidaris crassispina*, of which five were not receptors, and were named NRTK1 to NRTK5, one (RTK1) recalled the EGF receptor family, and the last (RTK2) recalled the insulin receptor family. Only RTK1 was not maternal. Komatsu *et al.* (1997) identified and cloned from *H. pulcherrimus* a homologue of the mammalian **MAP-KAPK**, which is expressed throughout development and *in vitro* phosphorylates the light chain of myosin II. Other information about kinases has already been provided in the preceding section and in the section dealing with genes related to gastrulation.

B. Gene for a Fibroblast Growth Factor Receptor

McCoon *et al.* (1996) isolated from *S. purpuratus* a cDNA for a protein belonging to such a family. Two splicing variants of this gene accumulate in the embryos beginning at mesenchyme blastula and then become somewhat enriched in the oral ectoderm and endoderm.

C. Genes for Metallothionein

It was Nemer who first dedicated his attention to such genes (Nemer *et al.*, 1984, 1985, 1991a, 1995; Angerer *et al.*, 1986; Wilkinson and Nemer, 1987; Harlow *et al.*, 1990; Bai *et al.*, 1993). Two of them were described in *S. purpuratus* and called SpMTA and SpMTB; the latter was then subdivided into SpMTB1 and SpMTB2. The expression of SpMTA is restricted to the aboral ectoderm of gastrulae and plutei, and that of SpMTB1 to the gut and oral ectoderm, whereas another MT gene of *L. pictus* appears to be ubiquitously expressed. Microinjection experiments of the 5'-flanking regions of SpMTA, variously mutated, followed by different reporter genes identified three metal-responsive elements (MRE) in the promoter, which act in combination with intron 1, and established that the first 1.6 kb of the 5'-flanking region together with the 1.12 kb of the first intron are sufficient to con-

2. Genes and Their Products in Sea Urchin Development

fer territorial specificity to the expression. This specificity is largely suppressed by deletion of the first 405 bp of the first intron. Within this region there is a genomically repetitive segment of 295 bp, with characteristics of a transposon, which contains several motifs with high homology to regulatory elements of other sea urchin genes such as the binding sites for the already described P3A, P5, and P7II nuclear proteins of the CyIIIa promoter. The characterization of the *L. pictus* metallothionein (LpMT1) was reported by Cserjesi *et al.* (1997), who showed that it contains four transcribed exons and compared it to the closely related SpMT2. Its uninduced expression shows no spatial preference.

High levels of metallothionein were found by Scudiero *et al.* (1994) in eggs and embryos of *P. lividus* and *Sp. granularis* but not of *Arbacia lixula* or of three other Eastern species, whereas negligible amounts of metallothionein were found in the Antarctic species *Sterechinus neumayeri* by Scudiero *et al.* (1997), who found little homology between the primary structures of the metallothioneins of *Sphaerechinus* and *Sterechinus*. The mode of metal binding in sea urchin metallothionein has been studied with electronic absorption, chiroptical, NMR, and mass microscopic methods by Wang *et al.* (1995).

XII. Genes for Heat Shock Proteins (hsps) and Ubiquitin Genes

A. Genes for Heat Shock Proteins

The interest for these in sea urchin development is twofold: first, because there are times in which the hsps are synthesized following heat shock and times in which they are not, and therefore they represent a model in which the arisal of competence for activation of gene expression by external stimuli can be studied, and second, because of their general role as chaperones. Both of these aspects have been investigated for a long time at the University of Palermo (Roccheri *et al.*, 1981a). It was found (see Giudice (1989) for a review of the first data) that if *P. lividus* embryos are heated at 31°C at any stage following hatching, they undergo a decrease of the overall rate of protein synthesis, but they also start a wave of synthesis of hsps and then go back to the normal pattern of protein synthesis and continue to develop normally. If, on the other hand, they are heated at the same temperature at any stage from fertilization to hatching, they again undergo the inhibition of the overall protein synthesis but are unable to synthesize hsps in appreciable amounts and degenerate. Four genes coding for the major hsp, that of 70 kDa, have been cloned and sequenced (La Rosa *et al.*, 1990; Sconzo *et al.*, 1992). An interesting feature of one of them is that it contains introns, although it bears also the so-called heat shock elements in its promoter, which is indicative of heat inducibility. As to the mechanism of regulation of inducibility of the hsp synthesis only after hatching, it was found that transcription after heat shock occurs also before hatching, albeit at a much lower rate per embryo and also per nucleus (Sconzo *et al.*, 1995). A nuclear protein which specifically binds the heat shock elements of the hsp70 gene has also been found (Sconzo *et al.*, 1997b). The binding occurs also at noninducible stages and in the absence of heat shock, but at a much lower extent than in inducible stages and following heat shock (Sconzo *et al.*, manuscript in preparation). The homologous heat shock promoter has been successfully used to elicit transcription of other genes from constructs microinjected into the egg (Sconzo *et al.*, 1997b). An antibody against hsp70 has also permitted detection of the constitutive heat shock proteins (hscs) (Sconzo *et al.*, 1997a) and the territorial location in oocytes, eggs, and embryos of both hscs and hsps. That the synthesis of hsp70 RNA is preferentially localized in the ectoderm had already been indicated by *in situ* hybridization (Sconzo *et al.*, 1985) and the ectodermic preferential localization of the hsps by separation of embryonic territories of partially dissociated plutei (Roccheri *et al.*, 1981b). The arisal of hsp synthesis inducibility during development does not depend on cell interactions, since it occurs also in dissociated cells at the same time as in the entire embryos (Sconzo *et al.*, 1983).

Sconzo *et al.* (manuscript in preparation) have found that the hsc70 clearly localizes in the mitotic apparatus during cleavage and that microinjection of antibodies against it blocks cleavage. Also the minor hsps and hscs of *P. lividus* have been studied by two-dimensional electrophoresis, which has distinguished at least 16 of them (Roccheri *et al.*, 1993). The mechanism of acquisition of thermotolerance (Sconzo *et al.,* 1986) has been attributed to the phosphorylation of one of them (38 kDa) by Roccheri *et al.* (1995).

An hsc60 with the function of chaperonine has been found in the *P. lividus* mitochondrial matrix by Roccheri *et al.* (1997). As to other sea urchin species, Maglott (1982) found that in *Arbacia punctulata* the inducibility of the hsp70 synthesis starts already between the 64- and 128-cell stage. This was confirmed for the species *A. lixula* by Roccheri *et al.* (1986), who found that in *Sp. granularis* hsp70 inducibility starts at the same stage as in *P. lividus*. Infante *et al.* (1985) found that the inducible period is also the same in *S. purpuratus* and *L. pictus* at least for the 21-kDa hsp. Bédard and Brandhorst (1986) found by two-dimensional electrophoresis that at least five different hsps are synthesized by *S. purpuratus* embryos following heating at 25°C. The synthesis of the hsp90, studied in greater detail by these authors, is also observed in nonshocked embryos, with a sharp rise at morula because of a shift of preexisting mRNA from a pool of free RNP particles onto polysomes, accompanied by polyadenylation; the heat shock inducibility, however, follows the blastula stage also in these embryos.

B. Ubiquitin Genes

Their transcripts, of 3.2 kb when on polysomes, are in constant amounts from the unfertilized egg to blastula in *S. purpuratus* and then decline (Nemer *et al.*, 1991b). Transcription is transiently stimulated by heat shock and permanently stimulated by Zn treatment. Different rounds of Zn treatment induce transcripts of different sizes in the nuclei. Both ectodermal and mesodermal tissues are active in plutei following induction; it is of interest, however, that isolated tissues cannot be induced.

Gong *et al*. (1991) reported similar results for *L. variegatus* but found no heat shock inducibility in this species. The amino acid sequence, deduced by their cloned cDNAs, is very similar to that of other animals.

XIII. Genes That Are Differentially Expressed along the Animal–Vegetal Axis

We have already seen that several genes, e.g., some of the homeobox-containing ones, are differentially expressed along the oral–aboral axis and that theories have been advanced about their possible role in the establishment of this axis. It is well known that the animal–vegetal axis is already established in the unfertilized egg. A theory was formulated by Runnstrom (1928) on the existence of two opposite gradients of molecules which are responsible for the establishment of this axis. This theory has been challenged by Wilt (1987) and Davison (1989, 1993) in favor of a theory according to which sequential, contact-mediated cell interactions at the 16 to 60-cell stage determine the specification of the main embryonic territories. We shall limit ourselves here to report some facts which describe the difference of gene expression along the animal–vegetal axis, without drawing any final conclusion about the possible role that these facts may have either in the establishment of an animal–vegetal axis or in establishing the conditions which generate the short-range signals which in turn lead to the specification of the different embryonic territories, because there are yet no proofs that this differential territorial expression has a role in either process.

Now to the facts: Di Carlo *et al*. (1996) found a new easy method for establishing which is the animal pole in the unfertilized egg, consisting of looking at the location of the female pronucleus, which is located toward the animal pole in *P. lividus*. Using this method, they were able to show by microsurgery that three previously isolated maternal mRNAs and proteins, the already mentioned **beps** (butanol-extractable proteins) (Di Carlo *et al*., 1990, 1994; Romancino *et al*., 1992), are anchored to the cytoskeleton of the animal side of the egg and that the vegetal part of the developing embryo remains devoid of (or at least much less rich in) them till gastrula, when beps become undetectable. A 54-kDa protein is responsible for the anchorage of the 3′UTRs of these RNAs to the cytoskeleton (Montana *et al*., 1997, 1998; Romancino *et al*., 1998). Splitting the egg by centrifugation into nucleated and nonnucleated halves is a means for enriching the cytoplasm which contains the bep proteins, i.e., the nucleated part (Costa *et al*., 1997). Antibodies against these proteins produce exogastrulation, which is a sign of hyperdevelopment of the vegetal structures. This has represented the first example of differential distribution of maternal gene products along the animal–vegetal axis.

Zygotic gene products are also asymmetrically distributed along this axis: One of them is the **hatching enzyme** (HE), first purified from *P. lividus* by Lepage and Gache (1989, 1990) as a 52-kDa protein of the collagenase family. Similar results were obtained by Roe and Lennarz (1990) on *S. purpuratus* and by Nomura on *H. pulcherrimus* (Nomura *et al.*, 1991, 1997; Nomura and Suzuki, 1993). Lepage *et al.* (1992a) found that the RNA for this enzyme, transcribed in the early blastula, is localized in the animal half of it, in a territory roughly corresponding to the presumptive ectoderm. The border of expression of this gene along the animal–vegetal axis can be shifted toward the animal pole by LiCl treatment (see also Ghiglione *et al.* (1993, 1996) but not micromere implantation onto entire embryos or isolated animal halves, which suggests that at least for this gene the expression pattern along the animal–vegetal axis is not due to a signal cascade starting from micromeres. Lepage *et al.* (1992b) reported the same temporal and territorial distribution for the expression of another gene, whose product shows homologies to the human **BMP1** and to the *Drosophila* tolloid. The same results were independently reported for the territorial distribution in *S. purpuratus* by Reynolds *et al.* 1992).

Stenzel *et al.* (1994) also identified in *S. purpruratus* a gene called **univin** (from the Japanese uni, for sea urchin) encoding a member of the same **TGFβ** superfamily, resembling again BMP1, tolloid, Veg1, and decapentaplegic, with the same territorial distribution at blastula; at gastrulation its RNA is concentrated in the presumptive foregut and in the ciliated band.

There is as yet no proof that this localization of proteolytic enzymes has an influence on the future development of the animal side of the embryo. We shall only recall here that it is an old notion that treatment with trypsin may cause animalization (Hörstadius, 1949; Rieder-Henderson and Rosenbaum, 1979). Emily-Fenouil *et al.* (1998) found that a homologue of **GSK3β/shaggy** gene is present in sea urchins: If this is experimentally overexpressed, there is an apparent animalization, as judged by the shifting toward the vegetal pole of the expression of the HE gene; conversely, inactivation of this enzyme produces vegetalization, with shifting of the HE expression toward the animal pole. This is in agreement with the hypothesis of a role of this enzyme in the regulation of the animal–vegetal potentialities. The role of the **IMPase** in the well-known vegetalizing effect of lithium cannot, however, be dismissed, since Giudice *et al.* (1992) found that this vegetalizing effect can be at least partially counteracted by myoinisitol and that a morphological effect similar to that of this partial vegetalization reversal can be obtained by partial inhibition of IMPase obtained by low doses of L690,330 (Sconzo *et al.*, 1998).

As to the mechanism of regulation of the territorial expression of the *S. purpuratus* hatching enzyme (SpHe), Wei *et al.* (1995, 1997a, 1997b, 1997c) investigated the relevant regulatory sequences of the promoter and found that a 330 bp region upstream of the transcription start site, or even shorter sequences within it, are sufficient to elicit β-galactosidase reporter gene expression in the animal territory. Chimeric constructs containing the SpHe basal promoter in combination with cis elements of the PMC specific SM50 gene elicit reporter gene transcription in the PMCs and constructs containing territorial specific sequences of these two genes elicit ubiquitous transcription of the reporter gene, indicating that there is no reciprocal inhibition of these cis-acting sequences. It is suggested that the spatial regulation of transcription is due to the concentration of maternal trans-acting positive transcription factors in the animal or vegetal territories. *In vivo* footprinting analysis shows that many SpHe cis regulatory sequences are occupied by proteins when the gene is active, i.e., at early blastula, and much less occupied when the gene is inactive, i.e., at late blastula, whereas the corresponding chromatin acquires a nucleosomal conformation; the trans-acting factors are, however, present in the nuclei of both stages, as indicated by *in vitro* footprints. It is therefore suggested that the temporal downregulation is due to impairment of the binding to the gene of some positive trans-acting factors; this impairment may be due to negative factors which bind to regulatory regions that are outside the first -1225 nucleotides, because these are, however, active in *in vivo* assays also after the very early blastula. Kozlowski *et al.* (1996) found that a region within 300 nucleotides upstream of the basal promoter controls the territorial expression of the already mentioned BMP-1; at least five different proteins bind to six sites within it. A general recent description of axis determination in sea urchins can be found in Jeffery (1992), Davidson *et al.* (1995a) and Henry (1998), plus some news about the orientation of the first cleavage, which does not specify the axes of bilateral symmetry in *L. variegatus*, in Summers *et al.* (1996).

XIV. Genes for Arylsulfatase, Myogenic Factors, and Ribosomal Proteins, Genes with an EGF Domain, and Genes for Poly(A)-Binding Proteins

A. Genes for Arylsulfatase

Two such genes were identified, the first in *H. pulcherrimus* and the second in *S. purpuratus*, and called ARSII. The first gene is expressed only in the aboral ectoderm at the gastrula stage, but its expression at earlier stages (blastula) is ubiquitous; at pluteus its protein accumulates along the apical surface of the aboral ectoderm and, to a lesser extent, in part of the oral ectoderm (Yamada *et al.*, 1989, 1992, 1993; Yang *et al.*, 1989a, 1993; Akasaka *et al.*, 1990). Microinjection of constructs with a CAT reporter gene in eggs of *H. pulcherrimus* revealed the importance for the proper temporal expression of a region between -2680 and -1280 bp, which seems to contain at least two positive elements and one negative element. Three restriction fragments within this region, located between -2200 and -2800 bp, specifically bind nuclear proteins, whose concentration changes during development. A G string and the sequence GATCTCCC were identified as sites for this interaction. According to Morokuma *et al.* (1997), the minimum information for temporal expression is contained in the region between −100 and +38 bp from the transcription start site; a strong enhancer is located between −184 and −164 bp. As already mentioned, the gene Otx has been assumed to be involved in the activation of the arylsulfatase gene (Kiyama et al., 1998). As to the function of arylsulfatase, Mitsunaga-Nakatsubo et al. (1998) found that the majority of this protein is distributed on the surface of the aboral ectoderm cells with no apparent enzymatic activity, whereas a minor fraction shows a lysosomal location.

B. Genes for Myogenic Factors

Venuti et al. (1991) identified a factor, called **SUM-1**, with homology to the MyoD family, which starts to accumulate at gastrula. Later, Venuti et al. (1993) demonstrated that the expression of SUM-1 in S. purpuratus embryos is likely to coincide with the earliest commitment of myogenic precursors during gastrulation (see also Wessel et al. (1990) for the immunological identification of myosin heavy chains in the macromere cell lineage).

C. Genes for Ribosomal Proteins

Angerer et al. (1992) were the first to isolate the RNA coding for a ribosomal protein, **SpS24**, in S. purpuratus, which is produced only in the endoderm and oral ectoderm. The authors' conclusion is that the production of this protein is not correlated with that of ribosomal RNA, which, in their opinion, is produced in all embryonic territories. The latter observation is, however, at variance with that of Roccheri et al. (1979), who in P. lividus demonstrated that at least in stages between blastula and pluteus rRNA synthesis occurs almost exclusively in territories which are predominantly represented by archenteron cells. Further data produced by the group at Palermo (Barbieri et al., 1992; Sgroi et al., 1996) strongly suggest that at least in P. lividus the synthesis of the S24 ribosomal protein is coupled with that of ribosomal RNA.

D. Genes with an EGF Domain

Two such genes have been described in S. purpuratus and called SpEGFI and SpEGFII (Yang et al., 1989b; Raff, 1990; Grimwade et al., 1991; Bisgrove et al., 1991). Their synthesis starts when the ectoderm cells start to be determined, more specifically, at the 16- to 32-cell stage for SpEGFI and at the 32- to 64-cell stage for SpEGFII, reaching a peak at mesenchyme blastula and gastrula, respectively. At pluteus their transcripts become localized in the aboral and postoral ectoderm.

SpEGFI is active also during oogenesis. A third gene was identified by Bisgrove and Raff (1993) in *S. purpuratus* and called SpEGFIII: this codes for the third major component of the apical lamina, a member of the fibropellin family. Its expression considerably increases at early blastula and the protein might be involved in gastrulation.

E. Genes for Poly(A)-Binding Proteins

At least two poly(A)-binding proteins have been isolated by Drawbridge *et al.* (1990) from *S. purpuratus*. Each of them has at least five posttranslationally modified forms. It is relevant to their possible role in protein synthesis regulation that they are in a very large excess in the unfertilized or fertilized egg.

XV. Genes for Jelly Proteins and the Egg Receptor for Sperm

A. Genes for Jelly Proteins

Kinoh *et al.* (1994) described in *H. pulcherrimus* a gene encoding the peptides 5SAP-I and 7SAP-I, separated from each other by a single lysine residue. *In situ* hybridization showed its RNA only in the accessorian ovarian cells. Bonnel *et al.* (1994) described the glicoproteic structure of jelly.

B. Genes for the Egg Receptor for Sperm

Foltz and Lennarz (1993, 1994) isolated, cloned, and sequenced a cDNA from *S. purpuratus*, which very probably encodes such a receptor. Its binding domain shows an interesting similarity to the hsp70 and, according to Mauk *et al.* (1997), to the hsp97. Also, Just and Lennarz (1997), after revisiting the sequence of the gene for such a receptor, confirm the similarity to the hsp110 gene family. Ohlendieck *et al.* (1994) found that this receptor in its active form is a disulfide-bonded homomultimer. Stears and Lennarz (1997) mapped the sperm binding domain within it. Giusti *et al.* (1997) after a series of analyses concluded that at least part of this protein is localized on the egg surface as one would expect. Kitazume-Kawaguchi *et al.* (1997) identified the sulfated oligosialic units of the O-linked glycan of the sperm receptor. Correa and Carroll (1997) isolated another of the eight major glycopeptides composing the vitelline envelope of *S. purpuratus* eggs, which inhibits fertilization when preincubated with sperms, therefore showing sperm binding activity. Cameron *et al.* (1996) found an affinity of a recombinant bindin for a putative bindin receptor in *S. purpuratus*.

XVI. Genes Expressed by Coelomocytes

A. Genes for Vitellogenin

Following the earlier work of Shyu *et al.* (1986, 1987), the products of these genes have been studied by Cervello *et al.* (1989, 1994) in *P. lividus* and found in coelomocytes where a 200-kDa glycoprotein precursor of the already mentioned toposome is synthesized; the authors provide evidence that toposome proteins derive from vitellogenin. Cervello *et al.* (1996) found that this protein also has the properties of an agglutinin. The presence of a major yolk protein in phagocytes of both female and male immature gonads has also been reported in *P. depressus* by Unuma *et al.* (1998). Shogomori *et al.* (1997) have shown by immunoelectromicroscopy that the yolk lipoproteins of *H. pulcherrimus* eggs contain gangliosides. Yokota *et al.* (1993) studied yolk degradation in *H. pulcherrimus*.

B. Genes for SpCoel1 (a Sea Urchin Profilin Gene)

SpCoel1 is expressed by most coelomocytes of the adult *S. purpuratus* in response to injuries: its protein shows significant similarities to the profilins from yeast to mammals (Smith *et al.*, 1992). The same group (Smith *et al.*, 1994) found that profilin is present in the egg at a concentration of about 13 µM. Profilin DNA transcription increases at the onset of gastrulation and is localized primarily in the PMCs. Three isoforms of this protein are produced during oogenesis and development.

C. Other Genes

Smith *et al.* (1996), analyzing by expressed sequence tags the genes expressed in activated *S. purpuratus* coelomocytes, found evidence for the synthesis of 55 different proteins, among which are CR/regulatory proteins, clotting factors, many immune effectors (among which is a C protein), cell surface proteins, proteases, and others, all tabulated by the authors. Al-Sharif *et al.* (1998) found that coelomocytes of *S. purpuratus* express a homologue of the complement component C3.

XVII. Miscellaneous Genes

A. Genes for Fascin

Bryan *et al.* (1993) cloned and sequenced a cDNA for fascin, which had long been purified from *T. gratilla* and was known to have a role as an actin bundling protein, for example, as in the formation of microvilli at fertilization. Its sequence shows that it belongs to a separate family of actin binding proteins, which includes the product of the *Drosophila singed* gene.

B. Genes for Calmodulin

A cDNA for calmodulin was first isolated by Floyd *et al.* (1986) from *S. purpuratus*. Transcripts are rare in the unfertilized egg and accumulate about 100-fold during embryogenesis; an enormous amount of calmodulin is, however, accumulated during oogenesis. Two genes for calmodulin were cloned by Hardy *et al.* (1988) from *A. punctulata*. Their transcripts are present in oocytes and eggs. They have a high degree of homology with those of the vertebrate calmodulin genes.

C. Genes Related to Mitochondria

We shall refer only to the most recent work, which has been dedicated to the problem of mitochondrial replication: Roberti *et al.* (1996, 1997) isolated from *P. lividus* a mitochondrial DNA-helicase and a mitochondrial single-stranded DNA-binding protein.

D. Genes for Distrophin-Related Proteins

One such gene has been isolated by Wang *et al.* (1998). Its sequence makes it a good candidate for a phylogenetic precursor of the human genes for distrophin and urotrophin.

E. Genes for Bindin

Experiments of *in situ* hybridization (Cameron *et al.*, 1990; Nishioka *et al.*, 1990) have demonstrated that its synthesis initiates in the second-order spermatocytes and continues in the spermatids. Lopez *et al.* (1993) compared the species specificity of variously mutated bindins and identified two amino acid repeats which are characteristic of *S. purpuratus* and *S. franciscanus*, respectively, and which might be involved in the species recognition mechanism. Also, Minor *et al.* (1993) addressed the same problem by testing the effect of different bindin peptides on fertilization. See also Hofmann and Glabe (1994) and Vacquier *et al.* (1995) for reviews. The evolution of the bindin gene in *Arbacia* has been studied by Metz et al. (1998) by comparison with the mitochondrial phylogeny, the conclusion being that fertilization specificity in *Arbacia* has evolved relatively slowly.

F. Genes for Other Sperm-Associated Proteins

Schulz et al. (1997) demonstrated that *S. purpuratus* sperm during the acrosome reaction shed a **syntaxin** (a cell membrane protein) and a **synaptobrevin** (vesicle-associated membrane protein (**VAMP**). The authors suggest a role for these proteins in the regulation of acrosome reaction, probably at the level of membrane fusion. Fang et al. (1998) added another piece of information to the very long list of studies on sea urchin sperm motors (see Wright and Scholey (1992) for a review) concerning the mechanism of sliding of flagellar axonemes, which can be inhibited by the ATPase inhibitor purealin, and Gingras et al. (1998) cloned a cDNA for the radial spoke head protein of the sperm axoneme. As to the long list of papers concerning the proteins involved in sperm energy metabolism, we shall refer here only to the recent review of Mita and Nakamura (1998).

Shimizu et al. (1994) characterized a putative receptor for a sperm-activating protein in sperms of *H. pulcherrimus*, which adds to the **SapII guanylcyclase receptor** (see Hardy et al. (1994) for a review). The corresponding 71-kDa protein shows a 95% identity to the 77-kDa protein already described in *S. purpuratus* spermatozoa (Dangott and Garbers, 1984, 1987; Dangott et al., 1989). Moy et al. (1996) found that the receptor for jelly proteins which triggers acrosomal reaction in *S. purpuratus* sperm is a modular protein with extensive homology to the human polycystic kidney disease protein.

It would be too long and probably beyond the scope of this review, however, to deal here with all the proteins involved in sea urchin fertilization. I shall therefore limit myself to referring the reader to some of the reviews on the subject, such as those of Spudich (1992), Shen (1995), and Ohlendieck and Lennarz (1996), and to listing here the references of some of the pertinent recent papers:

A. Fertilization-related sperm proteins: Mendoza et al. (1993), Morales et al. (1993), Beltran et al. (1996), Labarca et al. (1996), Gingras et al. (1996), Multigner et al. (1996), Mary et al. (1996), Quest et al. (1997), Osawa et al. (1997), Bracho et al. (1997), Gauss et al. (1998).

B. Fertilization-related egg proteins: Jaffe (1993), Keller and Vacquier (1994), Creton and Jaffe (1995), Mohri et al. (1995), Falugi et al. (1995), Xu and Tashjian (1995), Saito et al. (1995), Wessel (1995), Ciapa et al. (1995), Kinsey (1995, 1996), Lee et al. (1996), Bachman and McClay (1996), Ciapa and Epel (1996), Terasaki et al. (1996), Wilding et al. (1996), Sluder et al. (1996), Alves et al. (1997), de Barry et al. (1997), LaFleur et al. (1998).

G. Genes for Lamin

Holy *et al.* (1995) and Holy (1996) isolated and sequenced such genes from *S. purpuratus* and *L. variegatus* and found that they resemble more the vertebrate lamin B genes than the lamin genes of other invertebrates and that they are constitutively expressed in most embryonic cells. Collas *et al.* (1996) found that a protein from *L. pictus* gametes cross-reacting with an antibody against the human lamin B receptor is probably targeting membranes to sperm chromatin and then anchoring them to the lamina.

H. Genes for snRNAs

Santiago and Marzluff (1989), Yu *et al.* (1991), and Stevenson and Marzluff (1992) described the activation of a gene for a U1 RNA in development. Southgate and Busslinger (1989) studied the activation of U7 RNA synthesis. Stefanovic *et al.* (1991, 1992) isolated and characterized genes for two U2 snRNAs. Morales *et al.* (1997) isolated from *L. variegatus* three variants of a U5 snRNA gene, which are coordinately expressed during development.

I. Genes for Elongation Factors

Peeler *et al.* (1990) found that the e.f. 1-α4 is regulated also at a translational level by a feedback mechanism inhibiting mRNA attachment to ribosomes.

XVIII. Proto-oncogenes and Retroposons

A. Proto-oncogenes

One such gene, corresponding to the ets of the avian erythroblastosis, was described by Chen *et al.* (1988), which is maximally expressed during oogenesis and early development of *S. purpuratus*.

B. Retroposons

Some sequences with retroposon characteristics have been isolated by Nisson *et al.* (1988) in *S. purpuratus* and called SURF1 (for sea urchin retroposon family 1; see also the section dealing with histones). They are interspersed in the genome in about 800 copies and are maximally transcribed at the stage of 128 cells, from

RNA polymerase III. Yamaguchi and Ohba (1997) isolated from *Anthocidaris crassispina* two other cDNAs, called D7 and C2, which are maternally transcribed, perhaps by RNA polymerase II, and whose transcripts persist till late cleavage and gastrula, respectively.

References

Abassi, Y. A., and Foltz, K. R. (1994). Tyrosine phosphorylation of the egg receptor for sperm at fertilization. *Dev. Biol.* **164,** 430–443.
Adelson, D. L., Alliegro, M. C., and McClay, D. R. (1992). On the ultrastructure of the sea urchin embryo extracellular matrix. *J. Cell Biol.* **116,** 1283–1289.
Akasaka, K., Ueda, T., Higashinakagawa, T., Yamada, K., and Shimada, H. (1990). Spatial patterns of arylsulfatase mRNA expression in sea urchin embryo. *Dev. Growth Differ.* **32,** 9–13.
Akhurst, R. J., Calzone, F. J., Lee, J. J., Britten, R. J., and Davidson, E. H. (1987). Structure and organization of the CyIII actin gene subfamily of the sea urchin, *Strongylocentrotus purpuratus*. *J. Mol. Biol.* **194,** 193–203.
Al-Sharif, W. Z., Sunyer, J. O., Lambris, J. D., and Smith, L. C. (1998). Sea urchin coelomocytes specifically express a homologue of the complement component C3. *J. Immunol.* **160,** 2983–2997.
Alexandraki, D., and Ruderman, J. V. (1985). Expression of α- and β-tubulin genes during development of sea urchin embryos. *Dev. Biol.* **109,** 436–451.
Alliegro, M. C., and McClay, D. R. (1988). Storage and mobilization of extracellular matrix proteins during sea urchin development. *Dev. Biol.* **125,** 208–216.
Alliegro, M. C., Ettensohn, C. A., Burdsal, C. A., Erickson, H. O., and McClay, D. R. (1988). Echinonectin: A new embryonic substrate adhesion protein. *J. Cell Biol.* **107,** 2319–2327.
Alliegro, M. C., Burdsal, C. A., and McClay, D. R. (1990). *In vitro* biological activities of echinonectin. *Biochemistry* **29,** 2135–2141.
Alves, A. P., Mulloy, B., Diniz, J. A., and Mourão, P. A. S. (1997). Sulfated polysaccharides from the egg jelly layer are species-specific inducers of acrosomal reaction in sperms of sea urchins. *J. Biol. Chem.* **272,** 6965–6971.
Amemiya, S. (1989). Development of the basal lamina and its role in migration and pattern formation of primary mesenchyme cells in sea urchin embryos. *Dev. Growth Differ.* **31,** 131–145.
Andacht, T., and McClay, D. (1997). Metal binding proteins are involved in fertilization envelope elevation in the sea urchin, *Lytechinus variegatus* (LamarcK). Presented at the XIth Meeting on Developmental Biology of the Sea Urchin, Woods Hole, MA.
Anderson, R., Britten, R. J., and Davidson, E. H. (1994). Repeated sequence target sites for maternal DNA-binding proteins in genes activated in early sea urchin development. *Dev. Biol.* **163,** 11–18.
Angerer, L., Deleon, D., Cox, K., Maxson, R., Kedes, L., Kaumeyer, J., Weinberg, E., and Angerer, R. (1985). Simultaneous expression of early and late histone messenger RNAs in individual cells during development of the sea urchin embryo. *Dev. Biol.* **112,** 157–166.
Angerer, L. M., Kawczynski, G., Wilkinson, D. G., Nemer, M., and Angerer, R. C. (1986). Spatial patterns of metallothionein mRNA expression in the sea urchin embryo. *Dev. Biol.* **116,** 543–547.
Angerer, L. M., Chambers, S. A., Yang, Q., Venkatesan, M., Angerer, R. C., and Simpson, R. T. (1988). Expression of a collagen gene in mesenchyme lineages of the *Strongylocentrotus purpuratus* embryo. *Genes Dev.* **2,** 239–246.
Angerer, L. M., Dolecki, G. J., Gagnon, M. L., Lum, R., Wang, G., Yang, Q., Humphreys, T., and Angerer, R. C. (1989). Progressively restricted expression of a homeobox gene within the aboral ectoderm of developing sea urchin embryos. *Genes Dev.* **3,** 370–383.

Angerer, L. M., Yang, Q., Liesveld, J., Kingsley, P. D., and Angerer, R. C. (1992). Tissue-restricted accumulation of a ribosomal protein mRNA is not coordinated with rRNA transcription and precedes growth of the sea urchin pluteus larva. *Dev. Biol.* **149**, 27–40.

Angerer, R. C., and Davidson, E. H. (1984). Molecular indices of cell lineage specification in sea urchin embryos. *Science* **226**, 1153–1160.

Armstrong, N., and McClay, D. R. (1994). Skeletal pattern is specified autonomously by the primary mesenchyme cells in sea urchin embryos. *Dev. Biol.* **162**, 329–338.

Armstrong, N., Hardin, J., and McClay, D. R. (1993). Cell–cell interactions regulate skeleton formation in the sea urchin embryo. *Development* **119**, 833–840.

Arnone, M. I., and Davidson, E. H. (1997). The hardwiring of development: Organization and function of genomic regulatory systems. *Development* **124**, 1851–1864.

Arnone, M. I., Martin, E. L., and Davidson, E. H. (1998). Cis-regulation downstream of cell type specification: A single compact element controls the complex expression of the CyIIa gene in sea urchin embryos. *Development* **125**, 1381–1395.

Arnone, M., Bogard, L. D., Collazo, A., Kirchhamer, C. V., Cameron, R. A., Rast, J. P., Gregorians, A., and Davidson, E. H. (1997). Green fluorescent protein in the sea urchin: New experimental approaches to transcriptional regulatory analysis in embryos and larvae. *Development* **124**, 4649–4659.

Bachman, E. S., and McClay, D. R. (1996). Molecular cloning of the first metazoan β-1,3 glucanase from eggs of the sea urchin *Strongylocentrotus purpuratus*. *Proc. Natl. Acad. Sci. USA* **93**, 6808–6813.

Bachman, E. S., and McClay, D. R. (1997). Characterization of moesin in the sea urchin. *Lytechinus variegatus*: Redistribution to the plasma membrane following fertilization is inhibited by cytochalasin B. *J. Cell Sci.* **108**, 161–171.

Bai, G., Stuebing, E. W., Parker, H. R., Harlow, P., and Nemer, M. (1993). Combinatorial regulation by promoter and itron 1 regions of the metallothionein SpMTA gene in the sea urchin embryo. *Mol. Cell. Biol* **13**, 993–1001.

Barberis, A., Superti-Furga, G., and Busslinger, M. (1987). Mutually exclusive interaction of the CCAAT-binding factor and of a displacement protein with overlapping sequences of a histone gene promoter. *Cell* **50**, 347–359.

Barberis, A., Superti-Furga, G., Vitelli, L., Kemler, I., and Busslinger, M. (1989). Developmental and tissue-specific regulation of a novel transcription factor of the sea urchin. *Genes Dev.* **3**, 663–675.

Barbieri, R., Izzo, V., Cantone, M., Duro, G., and Giudice, G. (1992). Regulation of ribosomal RNA synthesis in sea urchin embryos. *Rend. Fis. Acc. Lincei* **3**, 369–374.

Becchetti, A., and Whitaker, M. (1997). Lithium blocks cell cycle transitions in the first cell cycles of sea urchin embryos, an effect rescued by myoinositol. *Development* **124**, 1099–1107.

Bèdard, P. A., and Brandhorst, B. P. (1986). Translational activation of maternal mRNA encoding the heat shock protein hsp 90 during sea urchin embryogenesis. *Dev. Biol.* **117**, 286–293.

Bell, J., Char, B. R., and Maxson, R. (1992). Octamer element is required for the expression of the alpha H2B histone gene during the early development of the sea urchin. *Dev. Biol.* **150**, 363–371.

Bellomonte, D., Di Bernardo, M., Russo, R., Caronia, G., and Spinelli, G. (1998). Highly restricted expression at the ectoderm–endoderm boundary of PlHbox 9, a sea urchin homeobox gene related to human HB9 gene. *Mech. Dev.* **74**, 185–188.

Beltrán, C., Zapata, O., and Darszon, A. (1996). Membrane potential regulates sea urchin sperm adenylylcyclase. *Biochemistry* **35**, 7591–7598.

Benink, H., Wray, G., and Hardin, J. (1997). Archenteron precursor cells can organize secondary axial structures in the sea urchin embryo. *Development* **124**, 3461–3470.

Benson, H. C., Sucov, H., Stephens, L., Davidson, E. H., and Wilt, F. (1987). Lineage specific gene encoding a major matrix protein of the sea urchin embryo spicule. I. Authentication of the cloned gene and its developmental expression. *Dev. Biol.* **120**, 499–506.

Benson, S., Smith, L., Wilt, F., and Shaw, R. (1990). The synthesis and secretion of collagen by cultured sea urchin micromeres. *Exp. Cell Res.* **188**, 141–146.

Berg, L. K., and Wessel, G. M. (1997). Cortical granules of the sea urchin translocate early in oocyte maturation. *Development* **124**, 1845–1850.

Berg, L. K., Chen, S. W., and Wessel, G. M. (1996). An extracellular matrix molecule that is selectively expressed during development is important for gastrulation in the sea urchin embryo. *Development* **122**, 703–713.

Berman, A., Addadi, L., and Weiner, S. (1988). Interactions of sea-urchin skeleton macromolecules with growing calcite crystals—A study of intracrystalline proteins. *Nature* **331**, 546–548.

Berman, A., Addadi, L., Kvick, A., Leiserowitz, L., Nelson, M., and Weiner, S. (1990). Intercalation of sea urchin proteins in calcite: Study of a crystalline composite material. *Science* **250**, 664–667.

Birnstiel, M. L., Busslinger, M., and Strub, K. (1985). Transcription termination and 3′ processing. The end is in site! *Cell* **41**, 349–359.

Bisgrove, B. W., and Raff, R. A. (1993). The SpEGF III gene encodes a member of the fibropellins: EGF repeat-containing proteins that form the apical lamina of the sea urchin embryo. *Dev. Biol.* **157**, 526–538.

Bisgrove, B. W., Andrews, M. E., and Raff, R. A. (1991). Fibropellins, products of an EGF repeat-containing gene, form a unique extracellular matrix structure that surrounds the sea urchin embryo. *Dev. Biol.* **146**, 89–99.

Bonder, E. M., and Fishkind, D. J. (1995). Actin-membrane cytoskeletal dynamics in early sea urchin development. *Curr. Top. Dev. Biol.* **31**, 101–137.

Bonnel, B. S., Keller, S. H., Vacquier, V. D., and Chandler, D. E. (1994). The sea urchin egg jelly coat consists of globular glycoproteins bound to a fibrous fucan superstructure. *Dev. Biol.* **162**, 313–324.

Bracho, G. E., Fritch, J. J., and Tash, J. S. (1997). A method for preparation, storage and activation of large populations of immotile sea urchin sperm. *Biochem. Biophys. Res. Commun.* **237**, 59–62.

Brandhorst, B. P., and Klein, W. H. (1992). Territorial specification and control of gene expression in the sea urchin embryo. *Dev. Biol.* **3**, 175–186.

Brandhorst, B. P., Filion, M., Nisson, P. E., and Crain, W. R., Jr. (1991). Restricted expression of the *Lytechinus pictus* Spec1 gene homologue in reciprocal hybrid embryos with *Strongylocentrotus purpuratus*. *Dev. Biol.* **144**, 405–411.

Brandt, W. F., Schwager, S. U. L., Rodriguez, J. A., and Busslinger, M. (1997). Isolation and aminoacid sequence analysis reveal an ancient origin of the cleavage stage (CS) histones of the sea urchin. *Eur. J. Biochem.* **247**, 784–791.

Brennan, C., and Robinson, J. J. (1994). Cloning and characterization of HLC-32, a 32-kDa protein component of the sea urchin extraembryonic matrix, the hyaline layer. *Dev. Biol.* **165**, 556–565.

Britten, R. J., McCormack, T. J., Mears, T. L., and Davidson, E. H. (1995). Gypsy/Ty3-class retrotransposons integrated in the DNA of herring tunicate and echinoderms. *J. Mol. Evol.* **40**, 13–24.

Brooks, J. M., and Wessel, G. M. (1997). Genomic characterization of the SFE-9 oocyte-specific promoter that is active throughout oogenesis. Presented at the XIth Meeting on Developmental Biology of the Sea Urchin, Woods Hole, MA.

Bryan, J., Edwards, R., Matsudaira, P., Otto, J., and Wulfkuhle, J. (1993). Fascin, an echinoid actin-bundling protein, is a homolog of the *Drosophila* singed gene product. *Proc. Natl. Acad. Sci. USA* **90**, 9115–9119.

Burdsal, C. A., Alliegro, M. C., and McClay, D. R. (1991). Tissue-specific, temporal changes in cell adhesion to echinonectin in the sea urchin embryo. *Dev. Biol.* **144**, 327–334.

Busslinger, M., and Barberis, A. (1985). Synthesis of sperm and late histone cDNAs of the sea urchin with a primer complementary to the conserved 3′ terminal palindrome: Evidence for tissue-specific and more general histone gene variants. *Proc. Natl. Acad. Sci. USA* **82**, 5676–5680.

Butler, E., Hardin, J., and Benson, S. (1987). The role of lysyl oxidase and collagen crosslinking during sea urchin development. *Exp. Cell Res.* **173**, 174–182.

Calzone, F. J., Theze, N., Thiebaud, P., Hill, R. L., Britten, R. J., and Davidson, E. H. (1988). Developmental appearance of factors that bind specifically to cis-regulatory sequences of a gene expressed in the sea urchin embryo. *Genes Dev.* **2,** 1074–1088.

Calzone, F. J., Höög, C., Teplow, D. B., Cutting, A. E., Zeller, R. W., Britten, R. J., and Davidson, E. H. (1991). Gene regulatory factors of the sea urchin embryo. I. Purification by affinity chromatography and cloning of P3A2, a novel DNA-binding protein. *Development* **112,** 335–350.

Calzone, F. J., Grainger, J., Coffman, J. A., and Davidson, E. H. (1997). Extensive maternal representation of DNA-binding proteins that interact with regulatory target sites of the *Strongylocentrotus purpuratus* CyIIIa gene. *Mol. Mar. Biol. Biotechnol.* **6,** 79–83.

Cameron, R. A., and Davidson, E. H. (1991). Cell type specification during sea urchin development. *Trends Genet.* **7,** 212–218.

Cameron, R. A., Britten, R. J., and Davidson, E. H. (1989). Expression of two actin genes during larval development in the sea urchin *Strongylocentrotus purpuratus*. *Mol. Reprod. Dev.* **1,** 149–155.

Cameron, R. A., Minor, J. E., Nishioka, D., Britten, R. J., and Davidson, E. H. (1990). Locale and level of bindin mRNA in maturing testis of the sea urchin, *Strongylocentrotus purpuratus*. *Dev. Biol.* **142,** 44–49.

Cameron, R. A., Walkup, T. S., Rood, K., Moore, J. G., and Davidson, E. H. (1996). Specific *in vitro* interaction between recombinant *Strongylocentrotus purpuratus* bindin and a recombinant 45A fragment of the putative bindin receptor. *Dev. Biol.* **180,** 348–352.

Casano, C., Ragusa, M., Cutrera, M., Costa, S., and Gianguzza, F. (1996). Spatial expression of α- and β-tubulin genes in the late embryogenesis of the sea urchin *Paracentrotus lividus*. *Int. J. Dev. Biol.* **40,** 1033–1041.

Cervello, M., and Matranga, V. (1989). Evidence of precursor–product relationship between vitellogenin and toposome, a glycoprotein complex mediating cell adhesion. *Cell Differ. Dev.* **26,** 67–76.

Cervello, M., Arizza, V., Lattuca, G., Parrinello, N., and Matranga, V. (1994). Detection of vitellogenin in a subpopulation of sea urchin coelomocytes. *Eur. J. Cell Biol.* **64,** 314–319.

Cervello, M., Arizza, V., Cammarata, M., Matranga, V., and Parrinello, N. (1996). Properties of sea urchin coelomocytes agglutinins. *Ital. J. Zool.* **63,** 353–356.

Char, B. R., Bell, J. R., Dovala, J., Coffman, J. A., Harrington, M. G., Becerra, J. C., Davidson, E.H., Calzone, F. J., and Maxson, R. (1993). SpOct, a gene encoding the major octamer-binding protein in sea urchin embryos: Expression profile, evolutionary relationships and DNA binding of expressed protein. *Dev. Biol.* **158,** 350–363.

Char, B. R., Tan, H., and Maxson, R. (1994). A POU gene required for early cleavage and protein accumulation in the sea urchin embryo. *Development* **120,** 1929–1935.

Chen, J., Maxson, R., and Jones, P. (1993). Direct induction of DNA hypermethylation in sea urchin embryos by microinjection of 5-methyl dCTP stimulates early histone gene expression and leads to developmental arrest. *Dev. Biol.* **155,** 75–86.

Chen, S. W., and Wessel, G. M. (1996). Endoderm differentiation *in vitro* identifies a transitional period of endoderm ontogeny in the sea urchin embryo. *Dev. Biol.* **175,** 57–75.

Chen, Z. Q., Kan, N. C., Pribyl, L., Lautenberger, J. A., Moudrianakis, E., and Papas, T. S. (1988). Molecular cloning of the ets proto-oncogene of the sea urchin and analysis of its developmental expression. *Dev. Biol.* **125,** 432–440.

Cho, J. W., Partin, J. S., and Lennarz, W. J. (1996). A technique for detecting matrix proteins in the crystalline structure of the sea urchin embryo. *Proc. Natl. Acad. Sci. USA* **93,** 1282–1286.

Ciapa, B., and Epel, D. (1996). An early increase in cGMP follows fertilization of sea urchin eggs. *Biochem. Biophys. Res. Commun.* **223,** 633–636.

Ciapa, B., Allemand, D., and De Renzis, G. (1995). Effect of arachidonic acid on Na^+/H^+ exchange and neutral amino acid transport in sea urchin eggs. *Exp. Cell Res.* **218,** 248–254.

Coffman, J. A., and Davidson, E. H. (1992). Expression of spatially regulated genes in the sea urchin embryo. *Curr. Opin. Genet. Dev.* **2,** 260–268.

Coffman, J. A., and Davidson, E. H. (1994). Regulation of gene expression in the sea urchin embryo. (1994). *J. Mar. Biol. Assoc. U.K.* **74,** 17–26.
Coffman, J. A., and McClay, D. R. (1990). A hyaline layer protein that becomes localized to the oral ectoderm and foregut of sea urchin embryos. *Dev. Biol.* **140,** 93–104.
Coffman, J. A., Moore, J. G., Calzone, F. J., Britten, R. J., Hood, L. E., and Davidson, E. H. (1992). Automated sequential affinity chromatography of sea urchin embryo DNA binding proteins. *Mol. Mar. Biol. Biotechnol.* **1,** 136–146.
Coffman, J. A. Kirchhamer, C. V., Harrington, M. G., and Davidson, E. H. (1996). SpRunt-1, a new member of the Runt domain family of transcription factors, is a positive regulator of the aboral ectoderm-specific CyIIIa gene in sea urchin embryos. *Dev. Biol.* **174,** 43–54.
Coffman, J. A., Kirchhamer, C. V., Harrington, M. G., and Davidson, E. H. (1997). SpMyb functions as an intramodular repressor to regulate spatial expression of CyIIIa in sea urchin embryos. *Development* **123,** 4717–4727.
Cole, D. G., Cande, W. Z., Baskin, R. J., Skoufias, D. A., Hogan, C. J., and Scholey, J. M. (1992). Isolation of a sea urchin egg kinesin-related protein using peptide antibodies. *J. Cell Sci.* **101,** 291–301.
Cole, D. G., Chinn, S. W., Wedaman, K. P., Hall, K., Vuong, T., and Scholey, J. M. (1993). Novel heterotrimeric kinesin-related protein purified from sea urchin eggs. *Nature* **366,** 268–269.
Collas, P., Courvalin, J. C., and Poccia, D. (1996). Targeting of membranes to sea urchin sperm chromatin is mediated by a lamin B receptor-like integral membrane protein. *J. Cell Biol.* **135,** 1715–1725.
Collura, R., and Katula, K. S. (1992). Spatial pattern of expression of CyL actin-β-galactosidase fusion genes injected into sea urchin eggs. *Dev. Growth Differ.* **34,** 635–647.
Conner, S., and Wessel, G. M. (1997). RAB3 functions following cortical granule docking in the sea urchin egg and may function in multiple phases in the developing embryo. Presented at the XIth Meeting on Developmental Biology of the Sea Urchin, Woods Hole, MA.
Conner, S., Leaf, D., and Wessel, G. (1997). Members of the SNARE hypothesis are associated with the cortical granule exocytosis in the sea urchin egg. *Mol. Reprod. Dev.* **48,** 1–13.
Correa, L. M., and Carroll, E. J., Jr. (1997). Identification of a new sea urchin vitelline envelope sperm binding glycoprotein. *Dev. Growth Differ.* **39,** 773–786.
Costa, C., Rinaldi, A. M., Romancino, D. P., Cavalcante, C., Vizzini, A., and Di Carlo, M. (1997). Centrifugation does not alter spatial distribution of "bep4" mRNA in *Paracentrotus lividus* egg. *FEBS Lett.* **410,** 499–501.
Cox, K. H., Angerer, L. M., Lee, J. J., Davidson, E. H., and Angerer, R. C. (1986). Cell lineage-specific programs of expression of multiple actin genes during sea urchin embryogenesis. *J. Mol. Biol.* **188,** 159–172.
Crawford, B. D., and Burke, R. D. (1994). YIGSR domain of laminin binds surface receptors of mesenchyme and stimulates migration during gastrulation in sea urchin. *Development* **120,** 3227–3234.
Créton, R., and Jaffe, L. F. (1995). Role of calcium influx during the latent period in sea urchin fertilization. *Dev. Growth Differ.* **37,** 703–709.
Cserjesi, P., Fang, H., and Brandhorst, B. P. (1997). Metallothionein gene expression in embryos of the sea urchin *Lytechinus pictus*. *Mol. Reprod. Dev.* **47,** 39–46.
Cutting, A. E., Höög, C., Calzone, F. J., Britten, R. J., and Davidson, E. H. (1990). Rare maternal mRNAs code for regulatory proteins that control lineage-specific gene expression in the sea urchin embryo. *Proc. Natl. Acad. Sci. USA* **87,** 7953–7957.
Czerny, T., Bouchard, M., Kozmik, Z., and Busslinger, M. (1997). The characterization of novel *Pax* genes of the sea urchin and *Drosophila* reveal an ancient evolutionary origin of the *Pax2/5/8* subfamily. *Mech. Dev.* **67,** 179–192.
D'Alessio, M., Ramirez, F., Suzuchi, H. R., Solursh, M., and Gambino, R. (1989). Structure and de-

velopment expression of sea urchin fibrillar collagen gene. *Proc. Natl. Acad. Sci. USA* **86**, 9303–9307.

D'Alessio, M., Ramirez, F., Suzuchi, H. R., Solursh, M., and Gambino, R. (1990). Cloning of a fibrillar collagen gene expressed in the mesenchymal cells of the developing sea urchin embryo. *J. Biol. Chem.* **265**, 7050–7054.

Dangott, L. J., and Garbers, D. L. (1984). Identification and partial characterization of the receptor for speract. *J. Biol. Chem.* **259**, 13712–13716.

Dangott, L. J., and Garbers, D. L. (1987). Further characterization of a speract receptor on sea urchin spermatozoa. *Ann. N.Y. Acad. Sci.* **513**, 274–285.

Dangott, L. J., Jordan, J. E., Bellett, R. A., and Garbers, D.L. (1989). Cloning of the mRNA for the protein that crosslinks to the egg the peptide speract. *Proc. Natl. Acad. Sci. USA* **86**, 2128–2132.

Davidson, E. H. (1986). "Gene Activity in Early Development." Academic Press, New York.

Davidson, E. H. (1989). Lineage-specific gene expression and the regulative capacities of the sea urchin embryo: A proposed mechanism. *Development* **105**, 421–445.

Davidson, E. H. (1993). Later embryogenesis: Regulatory circuitry in morphogenetic fields. *Development* **118**, 665–690.

Davidson, E. H., Peterson, K. J., and Cameron, R. A. (1995a). Origin of bilaterian body plans: Evolution of developmental regulatory mechanisms. *Science* **270**, 1319–1325.

Davidson, L. A., Koehl, M. A. R., Keller, R., and Oster, G. F. (1995b). How do sea urchins invaginate? Using biomechanics to distinguish between mechanisms of primary invagination. *Development* **121**, 2005–2018.

De Barry, J., Kawahara, S., Takamura, K., Janoshazi, A., Kirino, Y., Olds, J. L., Lester, D. S., Alkon, D. L., and Yoshioka, T. (1997). Time-resolved imaging of protein kinase C activation during sea urchin egg fertilization. *Exp. Cell Res.* **234**, 115–124.

Decker, G. L., and Lennarz, W. J. (1988). Growth of linear spicules in cultured primary mesenchyme cells of sea urchin embryos is bidirectional. *Dev. Biol.* **126**, 433–436.

Decker, G. L., Valdizan, M. C., Wessel, G. M., and Lennarz, W. J. (1988). Developmental distribution of a cell surface glycoprotein in the sea urchin *Strongylocentrotus purpuratus*. *Dev. Biol.* **129**, 339–349.

De Falco, J., and Childs, G. (1996). The embryonic transcription factor stage specific activator protein contains a potent bipartite activation domain that interacts with several RNA polymerase II basal transcription factors. *Proc. Natl. Acad. Sci. USA* **93**, 5802–5807.

Del Gaudio, R., Fucci, L., and Geraci, G. (1996). Unexpected conservation in the sequence of a H1 histone gene in a sea urchin and in a polychaete worm. Is it a case of horizontal gene transfer? *Ital. J. Biochem.* **45**, 179–180.

DeSimone, D. W., and Spiegel, M. (1985). Micromere-specific cell surface proteins of 16-cell stage sea urchin embryos. *Exp. Cell Res.* **156**, 7–14.

DeSimone, D. W., and Spiegel, M. (1986). Wheat germ agglutinin binding to the micromeres and primary mesenchyme cells of sea urchin embryos. *Dev. Biol.* **114**, 336–346.

DeSimone, B. E., Spiegel, E., and Spiegel, M. (1985). The biochemical identification of fibronectin in sea urchin embryos. *Biochem. Biophys. Res. Commun.* **27**, 183–188.

Di Bernardo, M., Russo, R., Oliveri, P., Melfi, R., and Spinelli, G. (1994). Expression of homeobox-containing genes in the sea urchin (*Paracentrotus lividus*) embryo. *Genetica* **94**, 141–150.

Di Bernardo, M., Russo, R., Oliveri, P., Melfi, R., and Spinelli, G. (1995). Homeobox-containing gene transiently expressed in a spatially restricted pattern in the early sea urchin embryo. *Proc. Natl. Acad. Sci. USA* **92**, 8180–8184.

Di Carlo, M., Montana, G., and Bonura, A. (1990). Analysis of the sequence and expression during sea urchin development of two members of a multigenic family, coding for butanol-extractable proteins. *Mol. Reprod. Dev.* **25**, 28–36.

Di Carlo, M., Romancino, D. P., Montana, G., and Ghersi, G. (1994). Spatial distribution of two maternal messengers in *Paracentrotus lividus* during oogenesis and embryogenesis. *Proc. Natl. Acad. Sci. USA* **91,** 5622–5626.

Di Carlo, M., Romancino, D. P., Ortolani, G., Montana, G., and Giudice, G. (1996). "BEP" RNAs and proteins are situated in the animal side of sea urchin unfertilized egg, which can be recognized by female pronuclear localization. *Biochem. Biophys. Res. Commun.* **229,** 511–517.

Di Liberto, M., Lai, Z. C., Fei, H., and Childs, G. (1989). Developmental control of promoter-specific factors responsible for the embryonic activation and inactivation of the sea urchin early histone H3 gene. *Genes Dev.* **3,** 973–985.

Dobias, S. L., Ma, L., Wu, H., Bell, J. R., and Maxson, R. (1997). The evolution of *Msx* gene function: Expression and regulation of a sea urchin Msx class homeobox gene. *Mech. Dev.* **61,** 37–48.

Dolecki, G. J., and Humphreys, T. (1988). An engrailed class homeobox gene in sea urchins. *Gene* **64,** 21–31.

Dolecki, G. J., Wannakrairoj, S., Lum, R., Wang, G., Riley, H. D,. Carlos, R., Wang, A., and Humphreys, T. (1986). Stage-specific expression of a homeobox-containing gene in the non-segmented sea urchin embryo. *EMBO J.* **5,** 925–930.

Dolecki, G. J., Wang, G., and Humphreys, T. (1988). Stage and tissue-specific expression of two homeobox genes in sea urchin embryos and adults. *Nucleic Acids Res.* **16,** 11543–11558.

Drager, B. J., Harkey, M. A., Iwata, M., and Whiteley, A. H. (1989). The expression of embryonic primary mesenchyme genes of the sea urchin, *Strongylocentrotus purpuratus*, in the adult skeletogenic tissues of this and other species of echinoderms. *Dev. Biol.* **133,** 14–23.

Drawbridge, J., Grainger, J. L., and Winkler, M. M. (1990). Identification and characterization of the poly(A)-binding protein from the sea urchin: A quantitative analysis. *Mol. Cell. Biol.* **10,** 3994–4006.

Egaña, A. L., Godin, R. E., and Ernst, S. G. (1997). Characterization of Wnt and Hedgehog family members during sea urchin development. Presented at the XIth Meeting on Developmental Biology of the Sea Urchin, Woods Hole, MA.

Eldon, E. D., Angerer, L. M., Angerer, R. C., and Klein, W. H. (1987). Spec3: Embryonic expression of a sea urchin gene whose product is involved in ectodermal ciliogenesis. *Genes Dev.* **1,** 1280–1292.

Eldon, E. D., Montpetit, I. C., Nguyen, T., Decker, G., Valdizan, M. C., Klein, W. H., and Brandhorst, B. P. (1990). Localization of the sea urchin Spec3 protein to cilia and Golgi complexes of embryonic ectoderm cells. *Genes Dev.* **4,** 111–122.

Emily-Fenouil, F., Ghiglione, C., Lhomond, G., Lepage, T., and Gache, C. (1998). GSK3beta/shaggy mediates patterning along the animal–vegetal axis of the sea urchin embryo. *Development* **125,** 2489–2498.

Ettensohn, C. A., and Ingersoll, E. P. (1992). Morphogenesis of the sea urchin embryo. *In* Morphogenesis: An Analysis of the Development of Biological Form, (E. Rossomando and S. Alexander, Eds.), pp. 189–262. Dekker, New York.

Ettensohn, C. A., and McClay, D. B. (1996). The regulation of primary mesenchyme cell migration in the sea urchin embryo. Transplantations of cells and later beads. *Dev. Biol.* **117,** 380–391.

Ettensohn, C. A., and McClay, D. B. (1988). Cell lineage conversion in the sea urchin embryo. *Dev. Biol.* **125,** 396–409.

Ettensohn, C. A., Guss, K. A., Hodor, P. G., and Malinda, K. M. (1997). Morphogenesis of the skeletal system of the sea urchin embryo. *In* "Reproductive Biology of Invertebrates" (K. G. Adiyodi and R. G. Adiyodi, Eds.), vol. VII, pp. 225–265. Oxford and IBH Publishing, New Delhi, Calcutta.

Ettensohn, C. A., and Ruffins, S. W. (1993). Mesodermal cell interactions in the sea urchin embryo: Properties of skeletogenic secondary mesenchyme cells. *Development* **117,** 1275–1285.

Ettensohn, C. A., Guss, K. A., Malinda, K. M., Miller, R. N., and Ruffins, S. (1996). Cell interactions in the sea urchin embryo. *Adv. Dev. Biol.* **4,** 47–97.

Evans, T., Rosenthal, E. T., Youngblom, J., Distel, D., and Hunt, T. (1983). Cyclin: A protein specified by maternal mRNA in sea urchin eggs that is destroyed at each cleavage division. *Cell* **33,** 389–396.

Exposito, J. Y., D'Alessio, M., and Ramirez, F. (1992a). Novel amino-terminal propeptide configuration in a fibrillar procollagen undergoing alternative splicing. *J. Biol. Chem.* **267,** 17404–17408.

Exposito, J. Y., D'Alessio, M., Solursh, M., and Ramirez, F. (1992b). Sea urchin collagen evolutionarily homologous to vertebrate pro-α2(I) collagen. *J. Biol. Chem.* **267,** 15559–15562.

Exposito, J. Y., D'Alessio, M., Di Liberto, M., and Ramirez, F. (1993). Complete primary structure of a sea urchin type IV collagen α chain and analysis of the 5' end of its gene. *J. Biol. Chem.* **268,** 5249–5254.

Exposito, J. Y., Suzuki, H., Geourjon, C., Garrone, R., Solursh, M., and Ramirez, F. (1994). Identification of a cell lineage-specific gene coding for a sea urchin α2(IV)-like collagen chain. *J. Biol. Chem.* **269,** 13167–13171.

Exposito, J. Y., Lethias, C., and Garrone, R. (1997). Function of the amino-terminal propeptide of the 2A fibrillar collagen during sea urchin development. Presented at the XIth Meeting on Developmental Biology of the Sea Urchin, Woods Hole, MA.

Falugi, C., Trielli, F., Germano, R., Cappelli, E., and Prestipino, G. (1995). Effects of a neutral Ca^{2+} ionophore on sea urchin block to polyspermy and early development. *Anim. Biol.* **4,** 51–58.

Falugi, C., Iannone, R., Morale, A., Angelini, C., Coniglio, L., and Corte, G. (1997). Inhibition of acetylcholinesterase activity affects primary mesenchyme migration and OTX-2-like protein expression of the sea urchin *Paracentrotus lividus*. Presented at the XIth Meeting on Developmental Biology of the Sea Urchin, Woods Hole, MA.

Fang, H., and Brandhorst, B. P. (1996). Expression of the actin gene family in embryos of the sea urchin *Lythechinus pictus*. *Dev. Biol.* **173,** 306–317.

Fang, Y.-I., Yokota, E., Mabuchi, I., Nakamura, H., and Ohizumi, Y. (1998). Purealin blocks the sliding movement of sea urchin flagellar axonemes by selective inhibition of half the ATPase activity of axonemal dyneins. *Biochemistry* **36,** 15561–15567.

Farach, M. C., Valdizan, M., Park,, H. R., Decker, G. L., and Lennarz, W. J. (1987). Developmental expression of a cell surface protein involved in calcium uptake and skeleton formation in sea urchin embryos. *Dev. Biol.* **122,** 320–331.

Farach-Carson, M. C., Carson, D. D., Collier, J. L., Lennarz, W. J., Park, H. R., and Wright, G. C. (1989). A calcium-binding asparagine-linked oligosaccharide is involved in skeleton formation in the sea urchin embryo. *J. Cell Biol.* **109,** 1289–1299.

Fei, H., and Childs, G. (1993). Temporal embryonic expression of the sea urchin early H1 gene is controlled by sequences immediately upstream and downstream of the TATA element. *Dev. Biol.* **155,** 383–395.

Fink, R. D., and McClay, D. R. (1985). Three cell recognition changes accompany the ingression of sea urchin primary mesenchyme cells. *Dev. Biol.* **107,** 66–74.

Fishkind, D. J., Bonder, E. M., and Begg, D. A. (1990). Sea urchin spectrin in oogenesis and embryogenesis: A multifunctional integrator of membrane cytoskeletal interactions. *Dev. Biol.* **142,** 453–464.

Floyd, E. E., Gong, Z. Y., Brandhorst, B. P., and Klein, W. H. (1986). Calmodulin gene expression during sea urchin development: Persistence of a prevalent maternal protein. *Dev. Biol.* **113,** 501–511.

Flytzanis, C. N., Bogosian, E. A., and Niemeyer, C. C. (1989). Expression and structure of the CyIIIb actin gene of the sea urchin *Strongylocentrotus purpuratus*. *Mol. Reprod. Dev.* **1,** 208–218.

Foltz, K. R. (1994). The sea urchin egg receptor for sperm. *Dev. Biol.* **5,** 243–253.

Foltz, K. R., Partin, J. S., and Lennarz, W. J. (1993). Sea urchin egg receptor for sperm: Sequence similarity of binding domain and hsp 70. *Science* **259,** 1421–1425.

Franks, R. R., Hough-Evans, B. R., Britten, R. J., and Davidson, E. H. (1998a). Spatially deranged through temporally correct expression of *Strongylocentrotus purpuratus* actin gene fusion in transgenic embryos of a different sea urchin family. *Genes Dev.* **2,** 1–12.

Franks, R. R., Hough-Evans, B. R., Britten, R. J., and Davidson, E. H. (1988b). Direct introduction of cloned DNA into the sea urchin zygote nucleus, and fate of injected DNA. *Development* **102,** 287–299.

Franks, R. R., Anderson, R., Moore, J. G., Hough-Evans, B. R., Britten, R. J., and Davidson, E. H. (1990). Competitive titration in living sea urchin embryos of regulatory factors required for expression of the CyIIIa actin gene. *Development* **110,** 31–40.

Frudakis, T. N., and Wilt, F. (1995). Two *cis* elements collaborate to spatially repress transcription from a sea urchin promoter. *Dev. Biol.* **172,** 230–241.

Fucci, L., Forte, A., Mancini, P., Affaitati, A., Branno, M., Aniello, F., and Geraci, G. (1997). The S.//.A.IG amino acid motif is present in a replication dependent late H3 histone variant of *P. lividus* sea urchin. *FEBS Lett.* **407,** 101–104.

Fuhrman, M. H., Suhan, J. P., and Ettensohn, C. A. (1992). Development expression of echinonectin, an endogenous lectin of the sea urchin embryo. *Dev. Growth Differ.* **34,** 137–150.

Fujita, Y., Yamasu, K., Suyemitsu, T., and Ishihara, K. (1994). A protein that binds an exogastrula-inducing peptide, EGIP-D, in the hyaline layer of sea urchin embryos. *Dev. Growth Differ.* **36,** 275–280.

Furuya, S., Kamata, Y., and Yasumasu, I. (1994). ADP-ribosylation of histones in nuclei isolated form embryos of the sea urchin, *Hemicentrotus pulcherrimus*. *Dev. Growth Differ.* **36,** 103–110.

Gagnon, M. L., Angerer, L. M., and Angerer, R. C. (1992). Posttranscriptional regulation of ectoderm-specific gene expression in early sea urchin embryos. *Development* **114,** 457–467.

Gambino, R., Romancino, D. P., Cervello, M., Vizzini, A., Isola, M. G., Virruso, L., and DiCarlo, M. (1997). Spatial distribution of collagen type I mRNA in *Paracentrotus lividus* eggs and embryos. *Biochem. Biophys. Res. Commun.* **238,** 334–337.

Gan, L., and Klein, W. (1993). A positive cis-regulatory element with a bicoid target site lies within the sea urchin Spec2a enhancer. *Dev. Biol.* **157,** 119–132.

Gan, L., Zhang, W., and Klein, W. H. (1990a). Repetitive DNA sequences linked to the sea urchin Spec genes contain transcriptional enhancer-like elements. *Dev. Biol.* **139,** 186–196.

Gan, L., Wessel, G. M., and Klein, W. H. (1990b). Regulatory elements from the related Spec genes of *Strongylocentrotus purpuratus* yield different spatial patterns with a lacZ reporter gene. *Dev. Biol.* **142,** 346–359.

Gan, L., Mao, C.-A., Wikramanayake, A., Angerer, L. M., Angerer, R. C., and Klein, W. H. (1995). An orthodenticle-related protein from *Strongylocentrotus purpuratus*. *Dev. Biol.* **167,** 517–528.

Gauss, R., Seifert, R., and Kaupp, U. B. (1998). Molecular identification of a hyperpolarization-activated channel in sea urchin sperm. *Nature* **393,** 583–584.

George, N. C., Killian, C. E., and Wilt, F. H. (1991). Characterization and expression of a gene encoding a 30.6-kDa *Strongylocentrotus purpuratus* spicule matrix protein. *Dev. Biol.* **147,** 334–342.

Ghersi, G., and Vittorelli, M. L. (1990). Immunological evidence for the presence in sea urchin embryos of an adhesion protein similar to mouse uvomorulin (E-cadherin). *Cell Differ. Dev.* **31,** 67–75.

Ghersi, G., Salamone, M., Dolo, V., Levi, G., and Vittorelli, M. L. (1993). Differential expression and function of cadherin-like proteins in the sea urchin embryo. *Mech. Dev.* **41,** 47–55.

Ghersi, G., Salamone, M., Levi, G., and Vittorelli, M. L. (1996). Cell adhesion-dependent regulation of growth during sea urchin development. *Eur. J. Cell Biol.* **69,** 259–266.

Ghiglione, C., Lhomond, G., Lepage, T., and Gache, C. (1993). Cell-autonomous expression and position-dependent repression by Li^+ of two zygotic genes during sea urchin early development. *EMBO J.* **12,** 87–96.

Ghiglione, C., Emily-Fenouil, F., Chang, P., and Gache, C. (1996). Early gene expression along the animal–vegetal axis in sea urchin embryoids and grafted embryos. *Development* **122,** 3067–3074.

2. Genes and Their Products in Sea Urchin Development

Gianguzza, F., Di Bernardo, M. G., Sollazzo, M., Palla, F., Ciaccio, M., Carra, E., and Spinelli, G. (1989). DNA sequence and pattern of expression of the sea urchin (*Paracentrotus lividus*) α-tubulin genes. *Mol. Reprod. Dev.* **1,** 170–181.

Gianguzza, F., Di Bernardo, M. G., Di Blasi, F., Colombo, P., Fais, M., Ragusa, M., Palla, F., and Spinelli, G. (1992). Pattern of transcription and DNA sequence analysis of β-tubulin cDNA clones of the sea urchin *Paracentrotus lividus*. *Mol. Biol. (Life Sci. Adv.),* **11,** 105–117.

Gianguzza, F., Casano, C., and Ragusa, M. (1995). α-Tubulin marker gene of neural territory of sea urchin embryos detected by whole-mount *in situ* hybridization. *Int. J. Dev. Biol.* **39,** 477–483.

Giga, Y., Okai, A., and Takahashi, K. (1987). The complete amino acid sequence of echinoidin, a lectin from the coelomic fluid of the sea urchin *Anthocidaris crassispina*. Homologies with mammalian and insect lectins. *J. Biol. Chem.* **262,** 6197–6203.

Gingras, D., White, D., Garin, J., Multigner, L., Job, D., Cosson, J., Huitorel, P. Zingg, H., Dumas, F., and Gagnon, C. (1996). Purification, cloning, and sequence analysis of a $M_r = 30,000$ protein from sea urchin axonemes that is important for sperm motility. *J. Biol. Chem.* **271,** 12807–12813.

Gingras, D., White, D., Garin, J., Cosson, J., Huitorel, P., Zingg, H., Cibert, C., and Gagnon, C. (1998). Molecular cloning and characterization of a radial spoke head protein of sea urchin sperm axonemes: Involvement of the protein in the regulation of sperm motility. *Mol. Biol. Cell* **9,** 513–522.

Giudice, G. (1962). Restitution of whole larvae from disaggregated cells of sea urchin embryos. *Dev. Biol.* **5,** 402–411.

Giudice, G. (1973). "Developmental Biology of the Sea Urchin Embryo." Academic Press, New York/London.

Giudice, G. (1986). "The Sea Urchin Embryo. A Developmental Biological System." Springer-Verlag, Berlin/Heidelberg/New York/Tokyo.

Giudice, G. (1989). Heat shock proteins in sea urchin embryos. *Dev. Growth Differ.* **31,** 103–106.

Giudice, G. (1993). Regulation of sea urchin gene expression. *Anim. Biol.* **2,** 61–73.

Guidice, G. (1995). Genes of the sea urchin embryo: An annotated list as of December 1994. *Develop. Growth Differ.* **37,** 221–242.

Guidice, G., Grasso, G., Sconzo, G., Cascino, D., Scardina, G., and Ferraro, M. G. (1992). Myo-inositol contracts the vegetalizing effect of lithium on *Paracentrotus lividus* embryos. *Cell Biol. Intern. Rep.* **16,** 47–52.

Giusti, A. F., Hoang, K. M., and Foltz, K. R. (1997). Surface localization of the sea urchin egg receptor for sperm. *Dev. Biol.* **184,** 10–24.

Godin, R. E., Urry, L. A., and Ernst, S. G. (1996). Alternative splicing of Endo 16 transcript produces differentially expressed mRNAs during sea urchin gastrulation. *Dev. Biol.* **179,** 148–159.

Godin, R. E., Klinzing, D. C., Porcaro, W. A., and Ernst, S. G. (1997). Specification of endoderm in sea urchin embryo. *Mech. Dev.* **67,** 35–47.

Gong, Z., and Brandhorst, B. P. (1987). Stimulation of tubulin gene transcription by deciliation of sea urchin embryos. *Mol. Cell. Biol.* **7,** 4238–4246.

Gong, Z. Y., Cserjesi, P., Wessel, G. M., and Brandhorst, B. P. (1991). Structure and expression of the polyubiquitin gene in sea urchin embryos. *Mol. Reprod. Dev.* **28,** 111–118.

Govindarajan, V., Ramachandran, R. K., George, J. M., Shakes, D. C., and Tomlinson, C. R. (1995). An ECM-bound, PDGF-like growth factor and a TGF-α-like growth factor are required for gastrulation and spiculogenesis in the *Lytechinus* embryo. *Dev. Biol.* **172,** 541–555.

Grant, S. R., Farach, M. C., Decker, G. L., Woodward, H. D., Farach, H. A., Jr., and Lennarz, W. J. (1985). Developmental expression of cell surface (glyco)proteins involved in gastrulation and spicule formation in sea urchin embryos. *Cold Spring Harbor Symp. Quant. Biol.* **50,** 91–98.

Gratwohl, E. K.-M., Kellemberg, E., Lorand, L., and Noll, H. (1991). Storage, ultrastructural targeting and function of toposomes and hyalin in sea urchin. *Mech. Dev.* **33,** 127–138.

Green, G. R., Collas, P., Burrell, A., and Poccia, D. L. (1995). Histone phosphorylation during sea urchin development. *Semin. Cell Biol.* **6,** 219–227.

Grimwade, J. E., Gagnon, M. L., Yang, Q., Angerer, R. C., and Angerer, L. M. (1991). Expression of two mRNAs encoding EGF-related proteins identifies subregions of sea urchin embryonic ectoderm. *Dev. Biol.* **143,** 44–57.

Gross, P. R., and Cousineau, G. H. (1963). Effect of actinomycin D on macromolecule synthesis and early development in sea urchin eggs. *Biochem. Biophys. Res. Commun.* **10,** 321–326.

Grosschedl, R., and Birnstiel, M. L. (1980). Spacer DNA sequences upstream of the T-A-T-A-A-A-T-A sequence are essential for promotion of H2A histone gene transcription *in vivo. Proc. Natl. Acad. Sci.* **77,** 7102–7106.

Grosschedl, R., Mächler, M., Rohrer, U., and Birnstiel, M. L. (1983). A functional component of the sea urchin H2A gene modulator contains an extended sequence homology to a viral enhancer. *Nucleic Acids Res.* **11,** 8123–8136.

Guss, C. A., and Ettensohn, C. A. (1997). Skeletal morphogenesis in the sea urchin embryo: Regulation of the primary mesenchyme gene expression and skeletal rod growth by ectoderm-derived cues. *Development* **124,** 1899–1908.

Haley, S. A., and Wessel, G. M. (1997). A serine protease of the sea urchin cortical granule is important for the block to polyspermy. Presented at the XIth Meeting on Developmental Biology of the Sea Urchin, Woods Hole, MA.

Harada, Y., Akasaka, K., Shimada, H., Peterson, K. J., Davidson, E. H., and Satoh, N. (1996). Spatial expression of a *forkhead* homologue in the sea urchin embryo. *Mech. Dev.* **60,** 163–173.

Hardin, J. (1996). The cellular basis of sea urchin gastrulation. *Curr. Top. Dev. Biol.* **33,** 159–262.

Hardin, J., and Armstrong, N. (1997). Short range cell–cell signals control ectodermal patterning in the oral region of the sea urchin embryo. *Dev. Biol.* **182,** 134–149.

Hardin, J., and McClay, D. R. (1990). Target recognition by the archenteron during sea urchin gastrulation. *Dev. Biol.* **142,** 86–102.

Hardin, P. E., Angerer, L. M., Hardin, S. H., Angerer, R. C., and Klein, W. H. (1988). Spec2 genes of *Strongylocentrotus purpuratus.* Structure and differential expression in embryonic aboral ectoderm cells. *J. Mol. Biol.* **202,** 417–431.

Hardy, D. M., Harumi, T., and Garbers, D. L. (1994). Sea urchin sperm receptors for egg peptides. *Semin. Dev. Biol.* **5,** 217–224.

Hardy, D. O., Bender, P. K., and Kretsinger, R. H. (1988). Two calmodulin genes are expressed in *Arbacia punctulata.* An ancient gene duplication is indicated. *J. Mol. Biol.* **199,** 223–227.

Harkey, M. A., and Whiteley, A. H. (1983). The program of protein synthesis during the development of the micromere-primary mesenchyme cell line in the sea urchin embryo. *Dev. Biol.* **100,** 12–28.

Harkey, M. A., Whiteley, H. R., and Whiteley, A. H. (1988). Coordinate accumulation of five transcripts in the primary mesenchyme during skeletogenesis in sea urchin embryo. *Dev. Biol.* **125,** 381–395.

Harkey, M. A., Klueg, K., Sheppard, P., and Raff, R. A. (1995). Structure, expression, and extracellular targeting of PM27, a skeletal protein associated specifically with growth of the sea urchin larval spicule. *Dev. Biol.* **168,** 549–566.

Harlow, P., Watkins, E., Thornton, R. D., and Nemer, M. (1990). Structure of an ectodermally expressed sea urchin metallothionein gene and characterization of its mental responsive region. *Mol. Cell. Biol.* **9,** 5445–5455.

Harrington, M. G., Coffman, J. A., Calzone, F. J., Hood, L. E., Britten, R. J., and Davidson, E. H. (1992). Complexity of sea urchin embryo nuclear proteins that contain basic domains. *Proc. Natl. Acad. Sci. USA* **89,** 6252–6256.

Harrington, M. G., Coffman, J. A., and Davidson, E. H. (1997). Covalent variation is a general property of transcription factors in the sea urchin embryo. *Mol. Mar. Biol. Biotechnol.* **6,** 153–162.

He, X., and Dembo, M. (1997). On the mechanics of the first cleavage division of the sea urchin egg. *Exp. Cell Res.* **233,** 252–273.

Henry, J. J. (1998). The development of dorsoventral and bilateral axial properties in sea urchin embryos. *Semin. Cell Dev. Biol.* **9,** 43–52.

Henson, J. H., Nesbitt, D., Wright, B. D., and Scholey, J. M. (1992). Immunolocalization of kinesin in sea urchin coelomocytes: Association of kinesin with intracellular organelles. *J. Cell Sci.* **103,** 309–320.

Henson, J. H., Cole, D. G., Terasaki, M., Rashid, D., and Scholey, J. M. (1995). Immunolocalization of the heterotrimeric kinesin-related protein KRP$_{(85/95)}$ in the mitotic apparatus of the sea urchin embryo. *Dev. Biol.* **171,** 182–194.

Hinchcliffe, E. H., and Linck, R. W. (1998). Two proteins isolated from sea urchin flagella: Structural components common to stable microtubules of axonemes and centrioles. *J. Cell Sci.* **111,** 585–595.

Hirada, Y., Yasuo, H., and Satoh, N. (1995). A sea urchin homologue of the chordate Brachiury (T) gene is expressed in the secondary mesenchyme founder cells. *Development* **121,** 2747–2754.

Hofmann, A., and Glabe, C. (1994). Bindin, a multifunctional sperm ligand and the evolution of new species. *Semin. Dev. Biol.* **5,** 233–242.

Holy, J. (1996). Conserved expression of a B-type nuclear lamin in different tissues of the sea urchin. *Biochem. Biophys. Res. Commun.* **222,** 531–536.

Holy, J., Wessel, G., Berg, L., Gregg, R. G., and Schatten, G. (1995). Molecular characterization and expression patterns of a B-type nuclear lamin during sea urchin embryogenesis. *Dev. Biol.* **168,** 464–478.

Höög, C., Calzone, F. J., Cutting, A. E., Britten, R. J., and Davidson, E. H. (1991). Gene regulatory factors of the sea urchin embryo. II. Two dissimilar proteins, P3A1 and P3A2, bind to the same target sites that are required for early territorial gene expression. *Development* **112,** 351–364.

Hörstadius, H. (1973). "Experimental Embryology of Echinoderms." Clarendon, Oxford.

Hörstadius, S. (1935). Uber die Determination im Verlaufe der Eiachse bei Seeigeln. *Pubbl. Stn. Zool. Napoli* **14,** 251–249.

Hörstadius, S. (1939). The mechanics of sea urchin development, studied by operative methods. *Biol. Rev.* **14,** 132–179.

Hörstadius, S. (1949). Experimental research on the development physiology of the sea urchin. *Pubbl. Stn. Zool. Napoli (Suppl.)* **21,** 131–172.

Hoshino, K., Nomura, K., and Suzuki, N. (1997). Cyclic-AMP-dependent activation of an inter-phylum hybrid histone-kinase complex reconstituted from sea urchin sperm-regulatory subunits and bovine heart catalytic subunits. *Eur. J. Biochem.* **243,** 612–623.

Hough-Evans, B. R., Britten, R. J., and Davidson, E. H. (1988). Mosaic incorporation and regulated expression of an exogenous gene in the sea urchin embryo. *Dev. Biol.* **129,** 198–208.

Hough-Evans, B. R., Franks, R. R., Zeller, R. W., Britten, R. J., and Davidson, E. H. (1990). Negative spatial regulation of the lineage specific CyIIIa actin gene in the sea urchin embryo. *Development* **110,** 41–50.

Hung, M. Rosenthal, E., Boblett, B., and Benson, S. (1995). Characterization and localized expression of the laminin binding protein/p40 (LBP/p40) gene during sea urchin development. *Exp. Cell Res.* **221,** 221–230.

Hwang, S. L., Partin, J. S., and Lennarz, W. J. (1994). Characterization of a homolog of human bone morphogenetic protein 1 in the embryo of the sea urchin. *Strongylocentrotus purpuratus. Development* **120,** 559–568.

Hylander, B. L., and Summers, R. G., (1982). An ultrastructural immunocytochemical localization of hyalin in the sea urchin egg. *Dev. Biol.* **93,** 368–380.

Infante, A. A., Akhayat, D., Rimland, J., and Infante, D. (1985). Characterization of the prosome and a cytoplasmic particle containing a 21 kD heat shock protein in sea urchin embryos. *Acta Embryol. Morphol. Exp.* **6,** 151–152.

Ingersoll, E. P., and Wilt, F. H. (1998). Matrix metalloproteinase inhibitors disrupt spicule formation by primary mesenchyme cells in the sea urchin embryo. *Dev. Biol.* **196,** 95–106.

Ingersoll, E. P., and Ettensohn, C. A. (1994). An N-linked carbohydrate-containing extracellular matrix determinant plays a key role in sea urchin gastrulation. *Dev. Biol.* **163,** 351–366.

Ishii, M., Mitsunaga-Nakatsubo, K., Kitajima, T., Kusunoki, S., Akasaka, K., and Shimada, H.

(1997). Establishment of sea urchin aboral ectoderm by HpHox8, a hox type homeobox gene. Presented at the XIth Meeting on Developmental Biology of the Sea Urchin, Woods Hole, MA.
Ito, M., Bell, J., Lyons, G., and Maxson, R. (1988). Synthesis and turnover of late H2B histone mRNA in developing embryos of the sea urchin, *Strongylocentrotus purpuratus. Dev. Biol.* **129**, 147–158.
Iwata, M., and Nakano, E. (1983). Characterization of sea urchin fibronectin. *Biochem. J.* **215**, 205–208.
Iwata, M., and Nakano, E. (1981). Fibronectin from the ovary of the sea urchin *Pseudocentrotus depressus. Roux's Arch. Dev. Biol.* **190**, 83–86.
Iwata, M., and Nakano, E. (1985). Fibronectin-binding acid polysaccharide in the sea urchin embryo. *Roux's Arch. Dev. Biol.* **194**, 377–384.
Jaffe, L. F. (1993). Classes and mechanisms of calcium waves. *Cell Calcium* **14**, 736–745.
Jasinskas, A., Kersulyte, D., Langmore, J., Steponaviciute, D., Jasinskiene, N., and Gineitis, A. (1997). Turnover of histone acetyl groups during sea urchin early development is not required for histone, heat shock and actin gene transcription. *Biochim. Biophys. Acta.* **1351**, 168–180.
Jeffery, W. R. (1992). Axis determination in sea urchin embryos: From confusion to evolution. *Trends Genet.* **8**, 2323–2325.
Just, M. L., and Lennarz, W. J. (1997). Reexamination of the sequence of the sea urchin egg receptor for sperm: Implication with respect to its properties. *Dev. Biol.* **184**, 25–30.
Kabakoff, B., and Lennarz, W. J. (1990). Inhibition of glycoprotein processing blocks assembly of spicules during development of sea urchin embryo. *J. Cell Biol.* **111**, 391–400.
Kabakoff, B., Hwang, S.-P., and Lennarz, W. J. (1992). Characterization of posttranslational modifications common to three primary mesenchyme cell-specific glycoproteins involved in sea urchin embryonic skeleton formation. *Dev. Biol.* **150**, 294–305.
Kamata, Y., Fujiwara, A., Furuya, S., and Yasumasu, I. (1993a). Does ADP-ribosylation of proteins in nuclei contribute to ectoderm cell differentiation in sea urchin embryos? *Dev. Growth Differ.* **35**, 89–98.
Kamata, Y., Furuya, S., and Yasumasu, I. (1993b). Proteins ADP-ribosylated in nuclei and plasma membrane vesicles isolated from sea urchin embryos at various stages of early development. *Dev. Growth Differ.* **35**, 283–291.
Katoh-Fukui, Y., Noce, T., Ueda, T., Fujiwara, Y., Hashimoto, N., Higashinakagawa, T., Killian, C. E., Livingston, B. T., Wilt, F. H., Benson, S. C., Sucov, H., and Davidson, E. H. (1991). The corrected structure of the SM50 spicule matrix protein of *Strongylocentrotus purpuratus. Dev. Biol.* **145**, 201–202.
Katoh-Fukui, Y., Noce, T., Ueda, T., Fujiwara, Y., Hashimoto, N., Tanaka, S., and Higashinakagawa, T. (1992). Isolation and characterization of cDNA encoding a spicule matrix protein in *Hemicentrotus pulcherrimus* micromeres. *Int. J. Dev. Biol.* **36**, 353–361.
Katow, H. (1995). Pamlin, a primary mesenchyme cell adhesion protein, in the basal lamina of the sea urchin embryo. *Exp. Cell Res.* **218**, 469–478.
Katow, H., and Komazaki, S. (1996). Spatio-temporal expression of pamlin during early embryogenesis in sea urchin and importance of N-linked glycosylation for the glycoprotein function. *Roux's Arch. Dev. Biol.* **205**, 371–381.
Katow, H., Yamamoto, Y., and Sofuku, S. (1997). Histological distribution of FR-1, a cyclic RGDS-peptide, binding sites during embryogenesis, and isolation and initial characterization of FR-1 receptor in the sand dollar embryo. *Dev. Growth Differ.* **39**, 207–219.
Kedes, L. H., and Gross, P. R. (1969). Identification in cleaving embryos of three RNA species serving as templates for the synthesis. *J. Mol. Biol.* **42**, 559–576.
Kedes, L. H., Chang, A. C. Y., Houseman, D., and Cohen, S. N. (1975a). Isolation of histone genes from unfractionated sea urchin DNA by subculture cloning in *E. coli. Nature* **225**, 533–538.
Kedes, L. H., Cohn, R. H., Lowry, J. C., Chang, A. C. Y., and Cohen, S. N. (1975b). The organization of sea urchin histone genes. *Cell* **6**, 359–369.

Keller, S. H., and Vacquier, V. D. (1994). N-Linked oligosaccharides of sea urchin egg jelly induce the sperm acrosome reaction. *Dev. Growth Differ.* **36,** 551–556.
Killian, C. E., and Wilt, F. H. (1996). Characterization of the protein comprising the integral matrix of *Strongylocentrotus purpuratus* embryonic spicules. *J. Biol. Chem.* **271,** 9150–9159.
Kinoh, H., Shimizu, T., Fujimoto, H., and Suzuki, N. (1994). Expression of a putative precursor mRNA for sperm-activating peptide I in accessory cells of the ovary in the sea urchin *Hemicentrotus pulcherrimus. Roux's Arch. Dev. Biol.* **203,** 381–388.
Kinsey, W. (1995). Differential phosphorylation of a 57-kDa protein tyrosine kinase during egg activation. *Biochem. Biophys. Res. Commun.* **208,** 204–208.
Kinsey, W. H. (1996). Biphasic activation of fyn kinase upon fertilization of the sea urchin egg. *Dev. Biol.* **174,** 281–287.
Kirchhamer, C. V., and Davidson, E. H. (1996). Spatial and temporal information processing in the sea urchin embryo: Modular and intramodular organization of CyIIIa gene cis-regulatory system. *Development* **122,** 333–348.
Kirchhamer, C. V., Yuh, C.-H., and Davidson, E. H. (1996a). Modular cis-regulatory organization of developmentally expressed genes: Two genes transcribed territorially in the sea urchin embryo, and additional examples. *Proc. Natl. Acad. Sci. USA* **93,** 9322–9328.
Kirchhamer, C. V., Bogarad, L. D., and Davidson, E. H. (1996b). Developmental expression of synthetic cis-regulatory systems composed of spatial control elements from two different genes. *Proc. Natl. Acad. Sci. USA* **93,** 13849–13854.
Kissinger, J. C., Hahn, J. H., and Raff, R. A. (1997). Rapid evolution in a conserved gene family. Evolution of the actin gene family in the sea urchin genus *Heliocidaris* and related genera. *Mol. Biol. Evol.* **14,** 654–665.
Kitazume-Kawaguchi, S., Inoue, S., Inoue, Y., and Lennarz, W. J. (1997). Identification of sulfated oligosialic acid units in the *O*-linked glycan of the sea urchin egg receptor for sperm. *Proc. Natl. Acad. Sci. USA* **94,** 3650–3655.
Kiyama, T., Akasaka, K., Takata, K., Mitsunaga-Nakatsubo, K., Sakamoto, N., and Shimada, H. (1998). Structure and function of a sea urchin orthodenticel-related gene. *Dev. Biol.* **193,** 139–145.
Kiyomoto, M., and Tsukahara, J. (1991). Spicule formation-inducing substance in sea urchin embryo. *Dev. Growth Differ.* **33,** 443–450.
Klueg, M., Harkey, M. A., and Raff, R. A. (1997). Mechanisms of evolutionary changes in timing, spatial expression, and mRNA processing in a direct-developing sea urchin, *Heliocidaris erythrogramma. Dev. Biol.* **182,** 121–133.
Knowles, J. A., and Childs, G. J. (1984). Temporal expression of late histone messenger RNA in the sea urchin *Lytechinus pictus. Proc. Natl. Acad. Sci. USA* **81,** 2411–2415.
Knowles, J. A., Lai, Z. C., and Childs, G. J. (1987). Isolation, characterization, and expression of the gene encoding the late histone subtype H1-gamma of the sea urchin *Strongylocentrotus purpuratus. Mol. Cell. Biol.* **7,** 478–485.
Komatsu, S., Murai, N., Totsukawa, G., Abe, M., Akasaka, K., Shimada, H., and Hosoya, H. (1997). Identification of MAPKAPK homolog (MAPKAPK-4) as a myosin II regulatory light-chain kinase in sea urchin egg extracts. *Arch. Biochem. Biophys.* **343,** 55–62.
Kontrogianni-Konstantopoulos, A., Vlahou, A., Vu, D., and Flytzanis, C. N. (1996). A novel sea urchin nuclear receptor encoded by alternatively spliced maternal RNAs. *Dev. Biol* **177,** 371–382.
Kozlowski, D. J., Gagnon, M. L., Marchant, J. K., Reynolds, S. D., Angerer, L. M., and Angerer, R. C. (1996). Characterization of a SpAN promotor sufficient to mediate correct spatial regulation along the animal–vegetal axis of the sea urchin embryo. *Dev. Biol.* **176,** 95–107.
Kuno, S., Mitsunaga-Nakatsubo, K., Nagura, T., and Yasumasu, I. (1994). Changes in insulin-binding capacity of the plasma membrane fraction during culture *in vitro* of cells derived from micromeres of 16-cell-stage sea urchin embryos. *Dev. Growth Differ.* **36,** 289–298.
Kuriyama, R., and Maekawa, T. (1992). Phosphorylation of a 225-kDa centrosomal component in mitotic CHO cells and sea urchin eggs. *Exp. Cell Res.* **202,** 345–354.

Kurokawa, D., Akasaka, K., Mitsunaga-Nakatsubo, K., and Shimada, H. (1997). Cloning of cyclin E cDNA of the sea urchin, *Hemicentrotus pulcherrimus. Zoolog. Sci.* **14,** 791–794.
Labarca, P., Santi, C., Zapata, O., Morales, E., Beltrán, C., Liévano, A., and Darszon, A. (1996). A cAMP regulated K^+-selective channel from the sea urchin sperm plasma membrane. *Dev. Biol.* **174,** 271–280.
LaFleur, G. J., Jr., Horiuchi, Y., and Wessel, G. M. (1998). Sea urchin ovoperoxidase: Oocyte-specific member of a heme-dependent peroxidase superfamily that functions in the block to polyspermy. *Mech. Dev.* **70,** 77–89.
Lai, Z. C., and Childs, G. (1986). Isolation and characterization of the gene encoding the testis specific histone protein H2B-2 from the sea urchin *Lytechinus pictus. Nucleic Acids Res.* **14,** 6845–6856.
Lai, Z. C., and Childs, G. (1988). Characterization of the structure and transcriptional patterns of the gene encoding the late histone subtype H1-beta of the sea urchin *Strongylocentrotus purpuratus. Mol. Cell. Biol.* **8,** 1842–1844.
Lai, Z. C., Maxson, R., and Childs, G. (1988). Both basal and ontogenic promoter elements affect the timing and level of expression of a sea urchin H1 gene during early embryogenesis. *Genes Dev.* **2,** 173–183.
Lai, Z. C., Deangelo, D. J., Diliberto, M., and Childs, G. (1989). An embryonic enhancer determines the temporal activation of a sea urchin late H1 gene. *Mol. Cell. Biol.* **9,** 2315–2321.
Laidlaw, M., and Wessel, G. M. (1994). Cortical granule biogenesis is active throughout oogenesis in sea urchins. *Development* **120,** 1325–1333.
Lane, M. C., Koehl, M. A. R., Wilt, F., and Keller, R. (1993). A role for regulated secretion of apical extracellular matrix during epithelial invagination in the sea urchin. *Development* **117,** 1049–1060.
La Rosa, M., Sconzo, G., Giudice, G., Roccheri, M. C., and Di Carlo, M. (1990). Sequence of a sea urchin hsp70 gene and its 5′ flanking region. *Gene* **96,** 295–300.
Leaf, D. S., Anstrom, J. A., Chin, J. E., Harkey, M. A., Showman, R. M., and Raff, R. A. (1987). Antibodies to a fusion protein identify a cDNA clone encoding msp130, a primary mesenchyme-specific cell surface protein of the sea urchin embryo. *Dev. Biol.* **121,** 29–40.
Lee, I. J., Tung, L., Bumcrot, D. A., and Weinberg, E. S. (1991). UHF-1, a factor required for maximal transcription of early and late sea urchin histone H4 genes: Analysis of promoter-binding sites. *Mol. Cell. Biol.* **11,** 1048–1061.
Lee, J. J., Calzone, F. J., and Davidson, E. H. (1992). Modulation of sea urchin actin mRNA prevalence during embryogenesis: Nuclear synthesis and decay rate measurements of transcripts from five different genes. *Dev. Biol.* **149,** 415–431.
Lee, S. J., Christenson, L., Martin, T., and Shen, S. S. (1996). The cyclic GMP-mediated calcium release pathway in sea urchin eggs is not required for the rise in calcium during fertilization. *Dev. Biol.* **180,** 324–335.
Lepage, T., and Gache, C. (1989). Purification and characterization of the sea urchin embryo hatching enzyme. *J. Biol. Chem.* **264,** 4787–4793.
Lepage, T., and Gache, C. (1990). Early expression of a collagenase-like hatching enzyme gene in the sea urchin embryo. *EMBO J.* **9,** 3003–3012.
Lepage, T., Sardet, C., and Gache, C. (1992a). Spatial expression of the hatching enzyme gene in the sea urchin embryo. *Dev. Biol.* **150,** 23–32.
Lepage, T., Ghiglione, C., and Gache, C. (1992b). Spatial and temporal expression pattern during sea urchin embryogenesis of a gene coding for a protease homologous to the human protein BMP-1 and to the product of the *Drosophila* dorsal–ventral patterning gene tolloid. *Development* **114,** 147–163.
Li, X., Chuang, C.-K., Mao, C.-A., Angerer, L. M., and Klein, W. H. (1997). Two Otx proteins generated from multiple transcripts of a single gene in *Strongylocentrotus purpuratus. Dev. Biol.* **187,** 253–266.
Lieber, T., Weisser, K., and Childs, G. (1986). Analysis of histone gene expression in adult tissues of

2. Genes and Their Products in Sea Urchin Development

the sea urchins *Strongylocentrotus purpuratus* and *Lytechinus pictus*: Tissue-specific expression of sperm histone genes. *Mol. Cell. Biol.* **6**, 2602–2612.

Livant, D. L., Cutting, A. E., Britten, R. J., and Davidson, E. H. (1988). An *in vitro* titration of regulatory factors required for expression of a fusion gene in transgenic sea urchin embryos. *Proc. Natl. Acad. Sci. USA* **85**, 7607–7611.

Livant, D. L., Hough-Evans, B. R., Moore, J. G., Britten, R. J., and Davidson, E. H. (1991). Differential stability of expression of similarly specified endogenous and exogenous genes in the sea urchin embryo. *Development* **113**, 385–398.

Livingston, B. T., and Wilt, F. H. (1992). Phorbol esters alter cell fate during development of sea urchin embryos. *J. Cell Biol.* **119**, 1641–1648.

Livingston, B. T., Shaw, R., Bailey, A., and Wilt, F. (1991). Characterization of a cDNA encoding a protein involved in formation of the skeleton during development of the sea urchin *Lytechinus pictus*. *Dev. Biol.* **148**, 473–480.

Livingston, B. T., Van Winkle, C. E., and Kinsey, W. H. (1998). Protein tyrosine kinase activity following fertilization is required to complete gastrulation, but not for initial differentiation of endoderm and mesoderm in the sea urchin embryo. *Dev. Biol.* **193**, 90–99.

Logan, C. Y., and McClay, D. R. (1997a). The allocation of the early blastomeres to the ectoderm and endoderm is variable in sea urchin embryo. *Development* **124**, 2213–2223.

Logan, C. Y., and McClay, D. R. (1997b). Cell fate specification of the endoderm in the sea urchin embryo. *Dev. Biol.* **186**, 344 (abstract).

Logan, C. Y., Miller, J. R., Ferkowicz, M. J., and McClay, D. R. (1999). Nuclear β-catenin is required to specify vegetal cell fates in the sea urchin embryo. *Development* **126**, in press.

Lopez, A., Miraglia, S. J., and Glabe, C. G. (1993). Structure/function analysis of the sea urchin sperm adhesive protein bindin. *Dev. Biol.* **156**, 24–33.

Lowe, C. J., and Wray, G. A. (1997). Radical alterations in the roles of homeobox genes during echinoderm evolution. *Nature* **389**, 718–721.

Lozano, J.-C., Schatt, P., Marquez, F., Peaucellier, G., Fort, P., Féral, J.-P., Genevière, A. M., and Picard, A. (1998). A presumptive developmental role for a sea urchin cyclin B splice variant. *J. Cell Biol.* **140**, 283–293.

Luke, N. H., Killian, C. E., and Livingston, B. T. (1997). Spfkhl 1 encodes a transcription factor implicated in gut formation during sea urchin development. *Dev. Growth Differ.* **39**, 285–294.

MacIsaac, R., Ng, E. Y. W., Nocente-McGrath, C., and Ernst, S. G. (1992). Histone H2A.F/Z mRNA is stored in the egg cytoplasm and basally regulated in the sea urchin embryo. *Dev. Biol.* **153**, 402–406.

MacNally, F. J., and Vale, R. D. (1993). Identification of katanin, an ATPase that severs and disassembles stable microtubules. *Cell* **75**, 419–429.

Maglott, D. R. (1982). Heat shock response in the Atlantic sea urchin *Arbacia punctulata*. *Experientia* **39**, 268–270.

Makabe, K. W., Kirchhamer, C. V., Britten, R. J., and Davidson, E. H. (1995). Cis regulatory control of the SM50 gene, an early marker of skeletogenic lineage specification in the sea urchin embryo. *Development* **121**, 1957–1970.

Malinda, K. M., Fisher, G. W., and Ettensohn, C. A. (1995). Four-dimensional microscopic analysis of the filopodial behavior of primary mesenchyme cells during gastrulation in the sea urchin embryo. *Dev. Biol.* **172**, 552–566.

Mandl, B., Brandt, W. F., Superti-Furga, G., Graninger, P. G., Birnstiel, M. L., and Busslinger, M. (1997). The five cleavage-stage (CS) histones of sea urchin are encoded by a maternally expressed family of replacement histone genes: Functional equivalence of the CS H1 and frog H1M (B4) proteins. *Mol. Cell. Biol.* **17**, 1189–1200.

Mao, C.-A., Wikramanayake, A., Gan, L., Chuang, C.-K., Summers, R. G., and Klein, W. H. (1996). Altering cell fates in sea urchin embryos by overexpressing SpOtx, an orthodenticle-related protein. *Development* **122**, 1489–1498.

Marsden, M., and Burke, R. D. (1997). Cloning and characterization of novel β-integrin subunits from a sea urchin. *Dev. Biol.* **181,** 234–245.

Martinez, P., and Davidson, E. H. (1997). SpHmx, a sea urchin homeobox gene expressed in embryonic pigment cells. *Dev. Biol.* **181,** 213–222.

Martinez, P., Lee, J. C., and Davidson, E. H. (1997). Complete sequence of SpHox8 and its linkage in the single Hox gene cluster of *Strongylocentrotus purpuratus*. *J. Mol. Evol.* **44,** 371–377.

Mary, J., Redeker, V., Le Caer, J.-P., Rossier, J., and Schmitter, J.-M. (1996). Posttranslational modifications in the C-terminal tail of axonemal tubulin from sea urchin sperm. *J. Biol. Chem.* **271,** 9928–9933.

Matese, J. C., Balck, S., and McClay, D. R. (1997). Regulated exocytosis and sequential construction of the extracellular matrix surrounding the sea urchin zygote. *Dev. Biol.* **186,** 16–26.

Matranga, V., Kuwasaki, B., and Noll, H. (1986). Functional characterization of toposomes from sea urchin blastula embryos by a morphogenetic cell aggregation assay. *EMBO J.* **5,** 3125–3132.

Matranga, V., Di Ferro, D., Cervello, M., Zito, F., and Nakano, E. (1991). Adhesion of sea urchin embryonic cells to substrate coated with cell adhesion molecules. *Biol. Cell* **71,** 289–291.

Matranga, V., Di Ferro, D., Zito, F., Cervello, M., and Nakano, E. (1992). A new extracellular matrix protein of the sea urchin embryo with properties of a substrate adhesion molecule. *Roux's Arch. Dev. Biol.* **201,** 173–178.

Matranga, V., Yokota, Y., Zito, F., Tesoro, V., and Nakano, E. (1995). Biochemical and immunological relationships among fibronectin-like proteins from different sea urchin species. *Roux's Arch. Dev. Biol.* **204,** 423–427.

Matsuda, R., Kitajima, T., Oshinata, H., Katoh, Y., and Higashinakagawa, T. (1988). Micromere differentiation in sea urchin embryo: Two-dimensional gel electrophoretic analysis of newly synthesized proteins. *Dev. Growth Differ.* **30,** 25–33.

Mauk, R., Jaworski, D., Kamei, N., and Glabe, C. G. (1997). Identification of a 97-kDa heat shock protein from *S. franciscanus* ovaries with 94% amino acid identity to the *S. purpuratus* egg surface receptor for sperm. *Dev. Biol.* **184,** 31–37.

Mauro, V. P., and Edelman, G. M. (1997). rRNA-like sequences occur in diverse primary transcripts: Implications for the control of gene expression. *Proc. Natl. Acad. Sci. USA* **94,** 422–427.

Maxson, R., Mohern, T., Gormezano, G., Childs, G., and Kedes, Z. (1983). Distinct organizations and patterns of expression of early and late histone gene sets in the sea urchin. *Nature* **301,** 120–125.

Maxson, R., Ito, M., Balcells, S., Thayer, M., French, M., Lee, F., and Etkin, L. (1988). Differential stimulation of sea urchin early and late H2B histone gene expression by a gastrula nuclear extract after injection into *Xenopus laevis* oocytes. *Mol. Cell. Biol.* **8,** 1236–1246.

McClay, D. R., and Logan, C. Y. (1996). Regulative capacity of the archenteron during gastrulation in the sea urchin. *Development* **122,** 607–616.

McClay, D. R., Cannon, G. W., Wessel, G. M., Fink, R. D., and Marchase, R. B. (1983). Patterns of antigenic expression in early sea urchin development. *In* "Time, Space, and Pattern in Embryonic Development" (W. Jefferies and R. Raff, Eds.), pp. 157–169. A. R. Liss, New York.

McCoon, P. E., Angerer, R. C., and Angerer, L. M. (1996). SpFGFR, a new member of the fibroblast growth factor receptor family, is developmentally regulated during early sea urchin development. *J. Biol. Chem.* **271,** 20119–20125.

Meijer, L., Borgne, A., Mulner, O., Chong, J. P., Blow, J. J., Inagaki, N., Inagaki, M., Delcros, J. G., and Moulinoux, J. P. (1997). Biochemical and cellular effects of roscovitine, a potent and selective inhibitor of the cyclin-dependent kinases cdc2, cdk2 and cdk5. *Eur. J. Biochem.* **243,** 527–536.

Mendoza, L. M., Nishioka, D., and Vacquier, V. D. (1993). A GPI-anchored sea urchin sperm membrane protein containing EGF domains is related to human uromodulin. *J. Cell Biol.* **121,** 1291–1297.

Metz, E. C., Gomez-Gutierrez, G., and Vacquier, V. D. (1998). Mitochondrial DNA and bindin gene

sequence evolution among allopatric species of the sea urchin genus *Arbacia. Mol. Biol. Evol.* **15,** 185–195.
Miller, J. R., and McClay, D. R. (1997a). Changes in the pattern of adherens junction-associated β-catenin accompany morphogenesis in the sea urchin embryo. *Dev. Biol.* **192,** 310–322.
Miller, J. R., and McClay, D. R. (1997b). Characterization of the role of cadherin in regulating cell adhesion during sea urchin development. *Dev. Biol.* **192,** 323–339.
Miller, J. R., and Moon, R. T. (1996). Signal transduction through β-catenin and specification of cell fate during embryogenesis. *Genes Dev.* **10,** 2527–2539.
Miller, J., Fraser, S. E., and McClay, D. (1995). Dynamics of thin filopodia during sea urchin gastrulation. *Development* **121,** 2507–2511.
Miller, R. N., Dalamags, D. G., Kingsley, P. D., and Ettensohn, C. A. (1996). Expression of S9 and actin CyIIa mRNAs reveals dorsoventral polarity and mesodermal sublineages in the vegetal plate of sea urchin embryo. *Mech. Dev.* **60,** 3–12.
Minor, J. E., Britten, R. J., and Davidson, E. H. (1993). Species-specific inhibition of fertilization by a peptide derived from the sperm protein bindin. *Mol. Biol. Cell* **4,** 375–387.
Mita, M., and Nakamura, M. (1998). Energy metabolism of sea urchin spermatozoa: An approach based on Echinoid phylogeny. *Zool. Sci.* **15,** 1–10.
Mitsunaga, K., Fujino, Y., and Yasumasu, I. (1987). Distributions of H^+,K^+-ATPase and Cl^-,HCO_3^--ATPase in micromere-derived cells of sea urchin embryos. *Differentiation* **35,** 190–196.
Mitsunaga, K., Fujiwara, A., Fujino, Y., and Yasumasu, I. (1989). Changes in the activities of H^+,K^+-ATPase and Na^+,K^+-ATPase in cultured cells derived from micromeres of sea urchin embryos with special reference to their roles in spicule rod formation. *Dev. Growth Differ.* **31,** 171–178.
Mitsunaga-Nakatsubo, K., Kanda, M., Yamazaki, K., Kawashita, H., Fujiwara, A., Yamada, K., Akasaka, K., Shimada, H, and Yasumasu, I. (1992). Expression of Na^+,K^+-ATPase α-subunit in animalized and vegetalized embryos of the sea urchin, *Hemicentrotus pulcherrimus. Dev. Growth Differ.* **34,** 677–684.
Mitsunaga-Nakatsubo, K., Akasaka, K., Kiyama, T., and Shimada, H. (1997). Functional analysis of sea urchin orthodenticle related proteins (HpOtxE and HpOtxL) during early development. Presented at the XIth Meeting on Developmental Biology of the Sea Urchin, Woods Hole, MA.
Mitsunaga-Nakatsubo, K., Akasaka, K., Akimoto, Y., Akiba, E., Kitajima, T., Tomita, M., Hirano, H., and Shimada, H. (1998). Arylsulfatase exists as non-enzymatic cell surface protein in sea urchin embryos. *J. Exp. Zool.* **180,** 220–230.
Mohri, T., Ivonnet, P. I., and Chambers, E. L. (1995). Effect on sperm-induced activation current and increase of cytosolic Ca^{2+} by agents that modify the mobilization of $[Ca^{2+}]_i$. *Dev. Biol.* **172,** 139–157.
Montana, G., Bonura, A., Romancino, D. P., Sbisà, E., and Di Carlo, M. (1997). A 54-kDa protein specifically associates the 3′ untranslated region of three maternal mRNAs with the cytoskeleton of the animal part of the *Paracentrotus lividus* egg. *Eur. J. Biochem.* **247,** 183–187.
Montana, G., Sbisà, E., Romancino, D. P., Bonura, A., and Di Carlo, M. (1998). Folding and binding activity of the 3′UTRs of *Paracentrotus lividus* bep messengers. *FEBS Lett.* **425,** 157–160.
Moore, K. L., and Kinsey, W. H. (1994). Identification of an Abl-related protein tyrosine kinase in the cortex of the sea urchin egg: Possible role at fertilization. *Dev. Biol.* **164,** 444–455.
Morales, E., De La Torre, L., Moy, G. W., Vacquier, V. D., and Darszon, A. (1993). Anion channels in the sea urchin sperm plasma membrane. *Mol. Reprod. Dev.* **36,** 174–182.
Morales, J., Borrero, M., Sumerel, J., and Santiago, C. (1997). Identification of developmentally regulated sea urchin U5 snRNA genes. *DNA Seq.* **7,** 243–259.
Morokuma, J., Akasaka, K., Mitsunaga-Nakatsubo, K., and Shimada, H. (1997). A cis-regulatory element within the 5′ flanking of arylsulfatase gene of sea urchin, *Hemicentrotus pulcherrimus. Dev. Growth Differ.* **39,** 469–476.
Morris, R. L., and Scholey, J. M. (1997). Heterotrimeric kinesin-II is required for the assembly of motile 9+2 ciliary axonemes on sea urchin embryos. *J. Cell Biol.* **138,** 1009–1022.

Morris, V. B., Brammall, J., Byrne, M., and Frommer, M. (1997). Hox-type and non-Hox homeobox gene sequences in genomic DNA of the sea urchin *Holopneustes purpurescens. Gene* **201**, 107–110.

Mous, J., Stunnenberg, H., Georgiev, O., and Birnstiel, M. L. (1985). Stimulation of sea urchin H2B histone gene transcription by a chromatin-associated protein fraction depends on gene sequences downstream of the transcription start site. *Mol. Cell. Biol.* **5**, 2764–2769.

Moy, G. W., Mendoza, L. M., Schultz, J. R., Swanson, W. J., Glabe, C. G., and Vacquier, V. D. (1996). The sea urchin sperm receptor for egg jelly is a modulator protein with extensive homology to the human polycystic kidney disease protein, PDK1. *J. Cell Biol.* **133**, 809–817.

Multigner, L., Pignot-Paintrand, I., Saoudi, Y., Job, D., Plessmann, U., Rüdiger, M., and Weber, K. (1996). The A and B tubules of the outer doublets of sea urchin sperm axonemes are composed of different tubulin variants. *Biochemistry* **35**, 10862–10871.

Nakajima, Y., and Burke, R. D. (1996). The initial phase of gastrulation in sea urchins is accompanied by the formation of bottle cells. *Dev. Biol* **179**, 436–446.

Nakano, E., and Iwata, M. (1982). Collagen binding proteins in sea urchin eggs and embryos. *Cell Differ.* **11**, 339–340.

Nemer, M., and Harlow, P. (1988). Sea-urchin RNAs displaying differences in developmental regulation and in complementarity to a collagen exon probe. *Biochim. Biophys. Acta* **950**, 445–449.

Nemer, M., and Lindsay, D. T. (1969). Evidence that the s polysomes of early sea urchin embryos may be responsible for the synthesis of chromosomal histones. *Biochem. Biophys. Res. Commun.* **35**, 156–160.

Nemer, M., and Stuebing, E. W. (1996). WEE1-like CDK tyrosine kinase mRNA level is regulated temporally and spatially in sea urchin embryos. *Mech. Dev.* **58**, 75–88.

Nemer, M., Travaglini, E. C., Rondinelli, E., and D'Alonzo, J. (1984). Developmental regulation, induction and embryonic tissue specificity of sea urchin metallothionein gene expression. *Dev. Biol.* **102**, 471–482.

Nemer, M., Wilkinson, D. G., Travaglini, E. C., Sternberg, E. J., and Butt, T. R. (1985). Sea urchin metallothionein sequence: Key to an evolutionary diversity. *Proc. Natl. Acad. Sci. USA* **82**, 4992–4994.

Nemer, M., Rondinelli, E., Infante, D., and Infante, A. A. (1991b). Polyubiquitin RNA characteristics and conditional induction in sea urchin embryos. *Dev. Biol.* **145**, 255–265.

Nemer, M. Stuebing, E. W., Bai, G., and Parker, H. R., (1995). Spatial regulation SpMTA metallothionein gene expression in sea urchin embryos by a regulatory cassettte in intron 1. *Mech. Dev.* **50**, 131–137.

Nemer, M., Thornton, R. D., Stuebing, E. W., and Harlow, P. (1991a). Structure, spatial,and temporal expression of two sea urchin metallothionein genes, SpMTB1 and SpMTA. *J. Biol. Chem.*, **266**, 6586–6593.

Niemeyer, C. C., and Flytzanis, C. N. (1993). Upstream elements involved in the embryonic regulation of the sea urchin CyIIIb actin gene: Temporal and spatial specific interactions at a single *cis*-acting element. *Dev. Biol.* **156**, 293–302.

Nishioka, D., Ward, R. D., Poccia, D., Kostacos, C., and Minor, J. E. (1990). Localization of bindin expression during sea urchin spermatogenesis. *Mol. Reprod. Dev.* **27**, 181–190.

Nisson, P. E., Hickey, R. J., Boshar, M. F., and Crain, W. R., Jr. (1988). Identification of a repeated sequence in the genome of the sea urchin which is transcribed by RNA polymerase III and contains the features of a retroposon. *Nucleic Acids Res.* **16**, 1431–1452.

Nisson, P. E., Gaudette, M. F., Brandhorst, B. P., and Crain, W. R. (1992). Mutually exclusive expression of the *Strongylocentrotus purpuratus* Spec1 gene and its *Lytechinus pictus* homologue in cells of hybrid embryos. *Development* **114**, 193–201.

Nocente-McGrath, C., Brenner, C. A., and Ernst, S. G. (1989). Endo16, a lineage-specific protein of the sea urchin embryo, is first expressed just prior to gastrulation. *Dev. Biol.* **136**, 264–272.

Noll, H., Matranga, V., Cascino, D., and Vittorelli, L. (1979). Reconstitution of membranes and embryonic development in dissociated blastula cells of the sea urchin by reinsertion of aggregation promoting membrane proteins extracted with butanol. *Proc. Natl. Acad. Sci. USA* **76,** 288–292.

Noll, H., Matranga, V., Palma, P., Cutrono, F., and Vittorelli, L. (1981). Species-specific dissociation into single cells of live sea urchin embryos by Fab against membrane components of *Paracentrotus lividus* and *Arbacia lixula*. *Dev. Biol.* **87,** 229–241.

Noll, H., Matranga, V., Cervello,, M., Humphreys, T., Kuwasaki, B., and Adelson, D. (1985). Characterization of toposome from sea urchin blastula cells: A cell organelle mediating cell adhesion and expressing positional information. *Proc. Natl. Acad. Sci. USA* **82,** 8062–8066.

Nomura, K., and Suzuki, N. (1993). Stereo-specific inhibition of sea urchin envelysin (hatching enzyme) by a synthetic autoinhibitor peptide with a cysteine-switch consensus sequence. *FEBS Lett.* **321,** 84–88.

Nomura, K., Tanaka, H., Kikkawa, Y., Yamaguchi, M., and Suzuki, N. (1991). The specificity of sea urchin hatching enzyme (envelysin) places it in the mammalian matrix metalloproteinase family. *Biochemistry* **30,** 6115–6123.

Nomura, K., Shimizu, T., Kinoh, H., Sendai, Y., Inomata, M., and Suzuki, N. (1997). Sea urchin hatching enzyme (envelysin): cDNA cloning and deprivation of protein substrate specificity by autolytic degradation. *Biochemistry* **36,** 7225–7238.

Norrander, J. M., Linck, R. W., and Stephens, R. E. (1995). Transcriptional control of tektin A mRNA correlates with cilia development and length determination during sea urchin embryogenesis. *Development* **121,** 1615-1623.

Ohlendieck, K., and Lennarz, W. J. (1996). Molecular mechanisms of gamete recognition in sea urchin fertilization. *Curr. Top. Dev. Biol.* **32,** 39–58.

Ohlendieck, K., Partin, J. S., and Lennarz, W. J. (1994). The biologically active form of the sea urchin egg receptor for sperm is a disulfide-bonded homo-multimer. *J. Cell. Biol.* **125,** 817–824.

Okazaki, K. (1975). Spicule formation by isolated micromeres of the sea urchin embryo. *Am. Zool.* **15,** 567–581.

Onodera, M. S., Sato, M., Suyemitsu, T., and Yamasu, K. (1997). Expression of a SRC-type tyrosine kinase, AcSrc1, in the sea urchin embryo. Presented at the XIth Meeting on Developmental Biology of the Sea Urchin, Woods Hole, MA.

Osawa, M., Kaneko, N., Terakawa, A., Kitani, T., Kuroda, R., and Kuroda, H. (1997). Studies on the mechanism for Ca_i-transients in sea urchin zygotes caused by refertilization and external application of sperm extract. *Exp. Cell Res.* **231,** 104–111.

Ozeki, Y., Matsui, T., Suzuki, M., and Titani, K. (1991). Amino acid sequence and molecular characterization of a D-galactoside-specific lectin purified from sea urchin (*Anthocidaris crassispina*) eggs. *Biochemistry* **30,** 2391–2394.

Ozeki, Y., Yokota, Y., Kato, K. H., and Matsui, T. (1995). Developmental expression of D-galactoside binding lectin in sea urchin (*Anthocidaris crassispina*) eggs. *Exp. Cell Res.* **216,** 318–324.

Page, L., and Benson, S. (1992). Analysis of competence in cultured sea urchin micromeres. *Exp. Cell Res.* **203,** 305–311.

Palla, F., Casano, C., Albanese, I., Anello, L., Gianguzza, F., Di Bernardo, M. G., Bonura, C., and Spinelli, G. (1989). Cis-acting elements of the sea urchin histone H2A modulator bind transcriptional factors. *Proc. Natl. Acad. Sci. USA* **86,** 6033–6037.

Palla, F., Bonura, C., Anello, L., Casano, C., and Ciaccio, M. (1993). Sea urchin early histone H2A modulator binding factor 1 is a positive transcription factor also for the early histone H3 gene. *Proc. Natl. Acad. Sci. USA* **90,** 6854–6858.

Palla, F., Bonura, C., Anello, L., Di Gaetano, L., and Spinelli, G. (1994). Modulator factor binding sequence of the sea urchin early histone H2A promoter acts as an enhancer element. *Proc. Natl. Acad. Sci. USA* **91,** 12322–12326.

Palla, F., Melfi, R., Anello, L., Di Bernardo, M., and Spinelli, G. (1997). Enhancer blocking activity located near the 3′ end of the sea urchin early H2A histone gene. *Proc. Natl. Acad. Sci. USA* **94,** 2272–2277.

Papaioannu, V. E., and Silver, L. M. (1998). The T-box gene family. *BioEssays* **20,** 9–19.
Patel, R., Wright, E. M., and Whitaker, M. (1997). Caffeine overrides the S-phase cell cycle block in sea urchin embryos. *Zygote* **5,** 127–138.
Peaucellier, G., Shartzer, K., Jiang, W., Maggio, K., and Kinsey, W. (1993). Anti-peptide antibody identifies a 57kDa protein tyrosine kinase in the sea urchin egg cortex. *Dev. Growth Differ.* **35,** 199–208.
Peeler, M. T., Kelso-Winemiller, L., Wu, M. F., Skipper, J. K., and Winkler, M. M. (1990). Counterproductive transcriptional and translational regulation of elongation factor 1-α synthesis during early development in sea urchins. *Dev. Biol.* **142,** 486–488.
Pesando, D., Dominice, C., Dufour, M. N., Guillon, G., Jouin, P., and Ciapa, B. (1995). Effect of nordidemnin on the cell cycle of sea urchin embryos. Role in synthesis and phosphorylation of proteins and in polyphosphoinositide turnover in mitosis progression. *Exp. Cell Res.* **220,** 18–28.
Pfeffer, P. L., and von Holt, C. (1991). Stage- and adult tissue-specific expression of a homeobox gene in embryo and adult *Parechinus angulosus* sea urchins. *Gene* **108,** 219–226.
Pfohl, R. J., and Giudice, G. (1967). The role of cell interaction in the control of enzyme activity during embryogenesis. *Biochim. Biophys. Acta* **142,** 263–266.
Pucci-Minafra, I., Casano, C., and La Rosa, C. (1972). Collagen synthesis and spicule formation in sea urchin embryos. *Cell Differ.* **1,** 157–165.
Pucci-Minafra, I., Minafra, S., Gianguzza, F., and Casano, C. (1975). Amino acid composition of collagen extracted from the spicules of sea urchin embryo (*Paracentrotus lividus*). *Boll. Zool.* **42,** 201–204.
Pucci-Minafra, I., Galante, R., and Minafra, S. (1978). Identification of collagen in the Aristotle's lanternae of *Paracentrotus lividus*. *J. Submicrosc. Cytol.* **10,** 53–63.
Pudles, J., Moudjou, M., Hisanaga, S. I., Maruyama, K., and Sakai, H. (1990). Isolation, characterization and immunochemical properties of a giant protein from sea urchin egg cytomatrix. *Exp. Cell Res.* **189,** 253–260.
Quest, A. F. G., Harvey, D. J., and McIlhinney, R. A. J. (1997). Myristoylated and nonmyristoylated pools of sea urchin sperm flagellar creatine kinase exist side-by-side: Myristoylation is necessary for efficient lipid association. *Biochemistry* **36,** 6993–7002.
Quigley, J. P., Braithwaite, R. S., and Armstrong, P. B. (1993). Matrix metalloproteases of the developing sea urchin embryo. *Differentiation* **54,** 19–23.
Raff, R. A. (1990). A sea urchin gene containing multiple EGF repeats is similar to vertebrate EGF-containing genes. *Mol. Reprod. Dev.* **27,** 66–72.
Rakow, T. L., and Shen, S. S. (1994). Molecular cloning and characterization of protein kinase C from the sea urchin *Lytechinus pictus*. *Dev. Growth Differ.* **36,** 489–497.
Ramachandran, R. K., Seid, C. A., Lee, H., and Tomlinson, C. R. (1993). PDGF-BB and TGF-α rescue gastrulation, spiculogenesis, and LpS1 expression in collagen-disrupted embryos of the sea urchin genus *Lytechinus*. *Mech. Dev.* **44,** 33–40.
Ramachandran, R. K., Wikramanayake, A. H., Uzman, J. A., Govindarajan, V., and Tomlinson, C. R. (1997). Disruption of gastrulation and oral–aboral ectoderm differentiation in the *Lytechinus pictus* embryo by a dominant/negative PDGF receptor. *Development* **124,** 2355–2364.
Raman, V., Andrews, M. E., Harkey, M. A., and Raff, R. A. (1993). Protein–DNA interactions at putative regulatory regions of two coordinately expressed genes, msp 130 and PM27, during skeletogenesis in sea urchin embryos. *Int. J. Dev. Biol.* **37,** 499–507.
Ransick, A., and Davidson, E. H. (1993). A complete second gut induced by transplanted micromeres in the sea urchin embryo. *Science* **259,** 1134–1138.
Ransick, A., and Davidson, E. H. (1995). Micromeres are required for normal vegetal plate specification in sea urchin embryos. *Development* **121,** 3215–3222.
Ransick, A., and Davidson, E. H. (1998). Late specification of Veg$_1$ lineages to endodermal fate in the sea urchin embryo. *Dev. Biol.* **195,** 38–48.
Ransick, A., Ernst, S., Britten, R. J., and Davidson, E. H. (1993). Whole mount *in situ* hybridization shows *Endo 16* to be a marker for the vegetal plate territory in sea urchin embryos. *Mech. Dev.* **42,** 117–124.

Reynolds, S. D., Angerer, L. M., Palis, J., Nasir, A., and Angerer, R. C. (1992). Early mRNAs, spatially restricted along the animal–vegetal axis of sea urchin embryos, include one encoding a protein related to tolloid and BMP-1. *Development* **114,** 769–786.
Rieder-Henderson, M. A., and Rosenbaum, J. L. (1979). Ciliary elongation in blastulae of *Arbacia punctulata* induced by trypsin. *Dev. Biol.* **70,** 500–508.
Roberti, M., Musicco, C., Loguercio-Polosa, P., Gadaleta, M. N., and Cantatore, P. (1996). DNA-helicase activity from sea urchin mitochondria. *Biochem. Biophys. Res. Commun.* **219,** 134–139.
Roberti, M., Musicco, C., Loguercio-Polosa, P., Gadaleta, M. N., Quagliariello, E., and Cantatore, P. (1997). Purification and characterization of a mitochondrial, single-stranded-DNA-binding protein from *Paracentrotus lividus* eggs. *Eur. J. Biochem.* **247,** 52–58.
Robinson, J. J., Taylor, L., and Ananthanarayanan, V. S. (1988). Role of calcium in stabilizing the structure of hyalin, a major protein component of the sea urchin extraembryonic hyaline layer. *Biochem. Biophys. Res. Commun.* **152,** 830–836.
Roccheri, M. C., Di Bernardo, M. G., and Giudice, G. (1979). Archenteron cells are responsible for the increase in ribosomal RNA synthesis in sea urchin gastrulae. *Cell Biol. Int. Rep.* **3,** 733–737.
Roccheri, M. C., Di Bernardo, M. G., and Giudice, G. (1981a). Synthesis of heat-shock proteins in developing sea urchin. *Dev. Biol.* **83,** 173–177.
Roccheri, M. C., Sconzo, G., Di Bernardo, M. G., Albanese, I., Di Carlo, M., and Giudice, G. (1981b). Heat shock proteins in sea urchin embryos. Territorial and intracellular location. *Acta Embryol. Morphol. Exp.* **2,** 91–99.
Roccheri, M. C., Sconzo, G., La Rosa, M., Oliva, D., Abrignani, A., and Giudice, G. (1986). Response to heat shock of different sea urchin species. *Differentiation* **22,** 175–178.
Roccheri, M. C., Cascino, D., and Giudice, G. (1993). Two-dimensional electrophoretic analysis of stress proteins in *Paracentrotus lividus*. *J. Submicrosc. Cytol. Pathol.* **25,** 173–179.
Roccheri, M. C., Isola, M. G., Bosco, L., Cascino, D., and Giudice, G. (1995). Achievement of thermotolerance through hsps phosphorylation in sea urchin embryos. *Cell Biol. Int.* **19,** 137–141.
Roccheri, M. C., Bosco, L., Ristuccia, M. E., Cascino, D., Giudice, G., Oliva, A. O., and Rinaldi, A. M. (1997). Sea urchin mitochondrial matrix contains a 56-kDa chaperonine-like protein. *Biochem. Biophys. Res. Commun.* **234,** 646–650.
Roe, J. L., and Lennarz, W. J. (1990). Biosynthesis and secretion of the hatching enzyme during sea urchin embryogenesis. *J. Biol. Chem.* **265,** 8704–8711.
Roe, J. L., Park, H. R., Strittmatter, W. J., and Lennarz, W. J. (1989). Inhibitors of metalloproteases block spiculogenesis in sea urchin primary mesenchyme cells. *Exp. Cell Res.* **181,** 545–550.
Romancino, D. P., Ghersi, G., Montana, G., Bonura, A., Perriera, S., and Di Carlo, M. (1992). Characterization of bep1 and bep4 antigens involved in cell interactions during *Paracentrotus lividus* development. *Differentiation* **50,** 67–74.
Romancino, D., Montana, G., and Di Carlo, M. (1998). Involvement of the cytoskeleton in the localization of *Paracentrotus lividus* maternal BEP mRNAs and proteins. *Exp. Cell Res.* **238,** 101–109.
Romano, G. (1992). Histone variants during sea urchin development. *Cell Biol. Int. Rep.* **16,** 197–206.
Ruffins, S. W., and Ettensohn, C. A. (1996). A fate map of the vegetal plate of the sea urchin (*Lytechinus variegatus*) mesenchyme blastula. *Development* **122,** 253–263.
Runnström, J. (1928). Plasmbau und determination bei dem Ei von *Paracentrotus lividus* LK. *Wilhem Roux' Arch. Entwickllungsmech. Org.* **133,** 554–581.
Saito, K., Imoto, M., Nagano, N., Toriyama, M., and Nakajima, T. (1995). Such hydrophobic peptides as dansylated mastoparan can elevate the fertilization membrane of sea urchin eggs. *Biochem. Biophys. Res. Commun.* **215,** 828–834.
Saitta, B., Buttice, G., and Gambino, R. (1989). Isolation of a putative collagen-like gene from the sea urchin *Paracentrotus lividus*. *Biochem. Biophys. Res. Commun.* **158,** 633–639.
Sakamoto, N., Akasaka, K., Mitsunaga-Nakatsubo, K., Takata, K., Nishitani, T., and Shimada, H. (1997). Two isoforms of orthodenticle-related proteins (HpOtx) bind to the enhancer element of sea urchin arylsulfatase gene. *Dev. Biol.* **181,** 284–295.

Sakuma, M., Onodera, H., Suyemitsu, T., and Yamasu, K. (1997). The protein tyrosine kinases of the sea urchin *Anthocidaris crassispina*. *Zool. Sci.* **14,** 941–946.
Santiago, C., and Marzluff, W. F. (1989). Expression of the U1 RNA gene repeat during early sea urchin development: Evidence for a switch in U1 RNA genes during development. *Proc. Natl. Acad. Sci. USA* **86,** 2572–2576.
Sasaki, H., and Aketa, K. (1981). Purification and distribution of a lectin in sea urchin (*Anthocidaris crassispina*) egg before and after fertilization. *Exp. Cell Res.* **135,** 15–19.
Scholey, J. M., Porter, M. E., Grissom, P. M., and McIntosh, J. R. (1985). Identification of kinesin in sea urchin eggs, and evidence for its localization in the mitotic spindle. *Nature* **318,** 483–486.
Schulz, J. R., Wessel, G. M., and Vacquier, V. D. (1997). The exocytosis regulatory proteins syntaxin and VAMP are shed from sea urchin sperm during the acrosome reaction. *Dev. Biol.* **191,** 80–87.
Sciarrino, S., and Matranga, V. (1995). Effects of retinoic acid and dimethylsulfoxide on the morphogenesis of the sea urchin embryo. *Cell Biol. Int.* **19,** 675–679.
Sconzo, G., Roccheri, M. C., Di Carlo, M., Di Bernardo, M. G., and Giudice, G. (1983). Synthesis of heat shock proteins in dissociated sea urchin embryonic cells. *Cell Differ.* **12,** 317–320.
Sconzo, G., Roccheri, M. C., Oliva, D., La Rosa, M., and Giudice, G. (1985). Territorial localization of heat shock mRNA production in sea urchin gastrulae. *Cell Biol. Int. Rep.* **9,** 877–881.
Sconzo, G., Roccheri, M. C., La Rosa, M., Oliva, D., Abrignani, A., and Giudice, G. (1986). Acquisition of thermotolerance in sea urchin embryos correlates with the synthesis and age of the heat shock proteins. *Cell. Differ.* **19,** 173–177.
Sconzo, G., Scardina, G., and Ferraro, M. G. (1992). Characterization of a new member of the sea urchin *Paracentrotus lividus hsp70* gene family and its expression. *Gene* **121,** 353–358.
Sconzo, G., Ferraro, M. G., Amore, G., Giudice, G., Cascino, D., and Scardina, G. (1995). Activation by heat shock of hsp70 gene transcription in sea urchin embryos. *Biochem. Biophys. Res. Commun.* **217,** 1032–1038.
Sconzo, G., Fasulo, G., Romancino, D., Cascino, D., and Giudice, G. (1996). Effect of retinoic acid and valproate on sea urchin development. *Pharmazie* **3,** 175–180.
Sconzo, G., Amore, G., Capra, G., Giudice, G., Cascino, D., and Ghersi, G. (1997a). Identification and characterization of a constitutive hsp75 in sea urchin embryos. *Biochem. Biophys. Res. Commun.* **234,** 24–29.
Sconzo, G., Geraci, F., Melfi, R., Cascino, D., Spinelli, G., Giudice, G., and Sirchia, R. (1997b). Sea urchin HSF activity *in vitro* and in transgenic embryos. *Biochem. Biophys. Res. Commun.* **240,** 436–441.
Sconzo, G., Cascino, D., Amore, G., Geraci, F., and Giudice, G. (1998). Effect of the IMPase inhibitor L 690,330 on sea urchin development *Cell Biol. Int.* **22,** 91–94.
Scudiero, R., Capasso, C., De Prisco, P. P. Capasso, A., Filosa, S., and Parisi, E. (1994). Metal-binding proteins in eggs of various sea urchin species. *Cell Biol. Int.* **18,** 47–48.
Scudiero, R., Capasso, C., Carginale, V., Riggio, M., Capasso, A., Ciaramella, M., Filosa, S., and Parisi, E. (1997). PCR amplification and cloning of metallothionein complementary DNAs in temperate and Antarctic sea urchin characterized by a large difference in egg metallothionein content. *Cell. Mol. Life Sci.* **53,** 472–477.
Seid, C. A., George, J. M., Sater, A. K., Kozlowski, M. T., Lee, H., Govindarajan, V., Ramachandran, R. K., and Tomlinson, R. C. (1996). USF in the *Lytechinus* sea urchin embryo may act as a transcriptional repressor in non-aboral ectoderm cells for the cell lineage-specific expression of the LpS1 genes. *J. Mol. Biol.* **264,** 7–19.
Seid, C. A., Ramachandran, R. K., George, J. M., Govindarajan, V., Gonzalez-Rimbau, M. F., Flytzanis, C. N., Tomlinson, C. R. (1997). An extracellular matrix response element in the promoter of the LpS1 genes of the sea urchin *Lytechinus pictus*. *Nucleic Acids Res.* **25,** 3175–3182.
Sgroi, A., Colombo, P., Duro, G., Fried, M., Izzo, V., and Giudice, G. (1996). cDNA sequence analysis and expression of the ribosomal protein S24 during oogenesis and embryonic development of the sea urchin *Paracentrotus lividus*. *Biochem. Biophys. Res. Commun.* **221,** 361–367.

Shen, S. S. (1995). Mechanisms of calcium regulation in sea urchin eggs and their activities during fertilization. *Curr. Top. Dev. Biol.* **30,** 65–101.
Sherwood, D. R., and McClay, D. R. (1997). Identification and localization of a sea urchin Notch homologue: insight into vegetal plate regionalization and Notch receptor regulation. *Development* **124,** 3363–3374.
Shimizu, K., Noro, N., and Matsuda, R. (1988). Micromere differentiation in the sea urchin embryo: Expression of primary mesenchyme cell specific antigen during development. *Dev. Growth Differ.* **30,** 35–47.
Shimizu, T., Yoshino, K., and Suzuki, N. (1994). Identification and characterization of putative receptors for sperm-activating peptide I (SAP-I) in spermatozoa of sea urchin *Hemicentrotous pulcherrimus. Dev. Growth Differ.* **36,** 209–221.
Shogomori, H., Chiba, K., and Hoshi, M. (1997). Association of the major ganglioside in sea urchin eggs with yolk lipoproteins. *Glycobiology* **7,** 391–398.
Showman, R. M., Wells, D. E., Anstrom, J., Hursh, D. A., and Raff, R. A. (1982). Message-specific sequestration of maternal histone mRNA in the sea urchin egg. *Proc. Natl. Acad. Sci. USA* **79,** 5944–5947.
Shuster, C. B., Sloboda, R. D., and Burgess, D. R. (1997). A tyrosine-kinase substrate phosphorylated during cytokinesis is a component of the mitotic apparatus. Presented at the XIth Meeting on Developmental Biology of the Sea Urchin. Woods Hole, MA.
Shyu, A. B., Raff, R. A., and Blumenthal, T. (1986). Expression of the vitellogenin gene in female and male sea urchin. *Proc. Natl. Acad. Sci. USA* **83,** 3865–3869.
Shyu, A. B., Blumenthal, T., and Raff, R. A. (1987). A single gene encoding vitellogenin in the sea urchin *Strongylocentrotus purpuratus*: Sequence at the 5′ end. *Nucleic Acids Res.* **15,** 10405–10417.
Skoufias, D. A., Cole, D. G., Wedaman, K. P., and Scholey, J. M. (1994). The carboxyl-terminal domain of kinesin heavy chain is important for membrane binding. *J. Biol. Chem.* **269,** 1477–1485.
Sluder, G., and Rieder, C. L. (1996). Controls for centrosome reproduction in animal cells: Issues and recent observations. *Cell Motil. Cytoskeleton* **33,** 1–5.
Sluder, G., Miller, F. J., Thompson, E. A., and Wolf, D. E. (1994). Feedback control of the metaphase–anaphase transition in sea urchin zygotes: Role of maloriented chromosomes. *J. Cell Biol.* **126,** 189–198.
Sluder, G., Thompson, E. A., Rieder, C. L., and Miller, F. J. (1995). Nuclear envelope breakdown is under nuclear not cytoplasmic control in sea urchin zygotes. *J. Cell Biol.* **129,** 1447–1458.
Smith, L. C., Britten, R. J., and Davidson, E. H. (1992). SpCoel1: A sea urchin profilin gene expressed specifically in coelomocytes in response to injury. *Mol. Biol. Cell* **3,** 403–414.
Smith, L. C., Harrington, M. G., Britten, R. J., and Davidson, E. H. (1994). The sea urchin profilin gene is specifically expressed in mesenchyme cells during gastrulation. *Dev. Biol.* **164,** 463–474.
Smith, L. C., Chang, L., Britten, R. J., and Davidson, E. H. (1996). Sea urchin genes expressed in activated coelomocytes are identified by expressed sequence tags. *J. Immunol.* **156,** 593–602.
Soltysik-Espanola, M., Klinzing, D. C., Pfarr, K., Burke, R. D., and Ernst, S. G. (1994). Endo16, a large multidomain protein found on the surface and ECM of endodermal cells during sea urchin gastrulation, binds calcium. *Dev. Biol.* **165,** 73–85.
Southgate, C., and Busslinger, M. (1989). *In vivo* and *in vitro* expression of U7 snRNA genes: cis- and trans-acting elements required for RNA polymerase II-directed transcription. *EMBO J.* **8,** 539–549.
Spiegel, E., and Spiegel, M. (1979). The hyaline layer is a collagen-containing extracellular matrix in sea urchin embryos and reaggregating cells. *Exp. Cell Res.* **123,** 434–441.
Spinelli, G., and Albanese, I. (1990). *In* "Reproductive Biology of Invertebrates" (K. G. Adiyodi and R. G. Adiyodi, Eds.), Vol. 4, pp. 283–390. Wiley, New York.
Spinelli, G., Casano, C., Gianguzza, F., Di Bernardo, M. G., Palla, F., Anello, L., and Ciaccio, M. (1991). Developmental studies of promoter-binding proteins of early H3 and H2A histone genes of sea urchin *Paracentrotus lividus*. *In* "Proceedings of the Third European Conference on Echinoderms, Lecce, Italy, 9–12 Sept. 1991," pp. 147–152.

Springer, M. S., Tusneem, N. A., Davidson, E. H., and Britten, R. J. (1995). Phylogeny, rates of evolution, and patterns of codon usage among sea urchin retroviral-like elements, with implications for the recognition of horizontal transfer. *Mol. Biol. Evol.* **12,** 219–230.
Spudich, A. (1992). Actin organization in the sea urchin egg cortex. *Curr. Top. Dev. Biol.* **26,** 9–21.
Stears, R. L., and Lennarz, W. J. (1997). Mapping sperm binding domains on sea urchin egg receptor for sperm. *Dev. Biol.* **187,** 200–208.
Stefanovic, B., and Marzluff, W. F. (1992). Characterization of two developmentally regulated sea urchin U2 small nuclear RNA promoters: A common required TATA sequence and independent proximal and distal elements. *Mol. Cell. Biol.* **12,** 650–660.
Stefanovic, B., Li, J. M., Sakallah, S., and Marzluff, W. F. (1991). Isolation and characterization of developmentally regulated sea urchin U2 snRNA genes. *Dev. Biol.* **148,** 284–294.
Stenzel, P., Angerer, L. M., Smith, B. J., Angerer, R. C., and Vale, W. W. (1994). The univin gene encodes a member of the transforming growth factor-β superfamily with restricted expression in the sea urchin embryo. *Dev. Biol.* **166,** 149–158.
Stephens, R. E. (1995). Ciliogenesis in sea urchin embryos—A subroutine in the program of development. *BioEssays* **17,** 331–340.
Stephens, R. E. (1997). Synthesis and turnover of embryonic sea urchin ciliary proteins during selective inhibition of tubulin synthesis and assembly. *Mol. Biol. Cell.* **8,** 2187–2198.
Stevenson, K. A., Yu, J., and Marzluff, W. F. (1992). A conserved region in the sea urchin U1 snRNA promoter interacts with a developmentally regulated factor. *Nucleic Acids Res.* **20,** 351–357.
Sucov, H. M., Benson, S., Robinson, J. J., Britten, R. J., Wilt, F., and Davidson, E. H. (1987). A lineage specific gene encoding a major matrix protein of the sea urchin embryo spicule. *Dev. Biol.* **120,** 507–519.
Sucov, H. M., Hough-Evans, B. R., Franks, R. R., Britten, R. J., and Davidson, E. H. (1988). A regulatory domain that directs lineage-specific expression of a skeletal matrix protein gene in the sea urchin embryo. *Genes Dev.* **2,** 1238–1250.
Summers, R. G., Piston, D. W., Harris, K. M., and Morrill, J. B. (1996). The orientation of first cleavage in the sea urchin embryo, *Lytechinus variegatus,* does not specify the axes of bilateral symmetry. *Dev. Biol.* **175,** 177–183.
Suprenant, K. A., and Foltz Daggett, M. A. (1995). Sea urchin microtubules. *Curr. Top. Dev. Biol.* **31,** 65–99.
Suzuki, H. R., Reiter, R. S., D'Alessio, M., Di Liberto, M., Ramirez, F., Exposito, J. Y., Gambino, R., and Solursh, M. (1997). Comparative analysis of fibrillar and basement membrane collagen expression in embryos of the sea urchin, *Strongylocentrotus purpuratus. Zool. Sci.* **14,** 449–454.
Tamboline, C. R., and Burke, R. D. (1989). Ontogeny and characterization of mesenchyme antigens of the sea urchin embryo. *Dev. Biol.* **136,** 75–86.
Terasaki, M., Jaffe, L. A., Hunnicutt, G. R., and Hammer, J. A., III. (1996). Structural change of the endoplasmic reticulum during fertilization: Evidence for loss of membrane continuity using the green fluorescent protein. *Dev. Biol.* **179,** 320–328.
Thatcher, T. H., and Gorovsky, M. A. (1994). Phylogenetic analysis of the core histones H2A, H2B, H3 and H4. *Nucleic Acids Res.* **22,** 174–179.
Theze, N., Calzone, F. J., Thiebaud, P., Hill, R. L., Britten, R. J., and Davidson, E. H. (1990). Sequences of the CyIIIa actin gene regulatory domain bound specifically by sea urchin embryo nuclear proteins. *Mol. Reprod. Dev.* **25,** 110–122.
Thurmond, F. A., and Trotter, J. A. (1994). Native collagen fibrils from echinoderms are molecularly bipolar. *J. Mol. Biol.* **235,** 73–79.
Tomlinson, C. R., and Klein, W. H. (1990). Temporal and spatial transcriptional regulation of the aboral ectoderm-specific Spec genes during sea urchin embryogenesis. *Mol. Reprod. Dev.* **25,** 328–338.
Tomlinson, C. R., Kozlowsy, M. T., and Klein, W. H. (1990). Ectoderm nuclei from sea urchin embryos contain a Spec DNA binding protein similar to the vertebrate transcription factor USF. *Development* **119,** 259–272.

Tung, L., Morris, G. F., Yager, L. N., and Weinberg, E. S. (1989). Sea urchin early and late H4 histone genes bind a specific transcription factor in a stable preinitiation complex. *Mol. Cell. Biol.* **9,** 1476–1487.

Tung, L., Lee, I. J., Rice, H. L., and Weinberg, E. S. (1990). Positive and negative transcriptional regulatory elements in the early H4 histone gene of the sea urchin, *Strongylocentrotus purpuratus*. *Nucleic Acids Res.* **18,** 7339–7348.

Unuma, T., Suzuki, T., Kurokawa, T., Yamamoto, T., and Akiyama, T. (1998). A protein identical to the yolk protein is stored in the testis in male red sea urchin, *Pseudocentrotus depressus*. *Biol. Bull.* **194,** 92–97.

Vacquier, V. D., Swanson, W. J., and Hellberg, M. (1995). What have we learned about sea urchin sperm bindin? *Dev. Growth Differ.* **37,** 1–10.

Vafa, O., Goetzl, L., Poccia, D., and Nishioka, D. (1996). Localization and characterization of blastocoelic extracellular matrix antigens in early sea urchin embryos and evidence for their proteolytic modification during gastrulation. *Differentiation* **60,** 129–138.

Vater, C. A., and Jackson, R. C. (1990). Immunolocalization of hyalin in sea urchin eggs and embryos using an antihyalin-specific monoclonal antibody. *Mol. Reprod. Dev.* **25,** 215–226.

Venezky, D. L., Angerer, L. M., and Angerer, R. C. (1981). Accumulation of histone repeat transcripts in the sea urchin egg pronucleus. *Cell* **24,** 385–391.

Venkatesan, M., De Pablo, F., Vogeli, G., and Simpson, R. T. (1986). Structure and developmentally regulated expression of a *Strongylocentrotus purpuratus* collagen gene. *Proc. Natl. Acad. Sci. USA* **83,** 3351–3355.

Venuti, J. M., Goldberg, L., Chakraborty, T., Olson, E. N., and Klein, W. H. (1991). A myogenic factor from sea urchin embryos capable of programming muscle differentiation in mammalian cells. *Proc. Natl. Acad. Sci. USA* **88,** 6219–6223.

Venuti, J. M., Gan, L., Kozlowski, M. T., and Klein, W. H. (1993). Developmental potential of muscle cell progenitors and the myogenic factor SUM-1 in the sea urchin embryo. *Mech. Dev.* **41,** 3–14.

Vitelli, L., Kemler, I., Lauber, B., Birnstiel, M. L., and Busslinger, M. (1988). Developmental regulation of micro-injected histone genes in sea urchin embryos. *Dev. Biol.* **127,** 54–63.

Vlahou, A., Gonzalez-Rimbau, M., and Flytzanis, C. N. (1996). Maternal mRNA encoding the orphan steroid receptor SpCOUP-TF is localized in sea urchin eggs. *Development* **122,** 521–526.

Wang, A. V. T., Angerer, L. M., Dolecki, G. J., Lum, R., Wang, G. V. L., Carlos, R., Angerer, R. C., and Humphreys, T. (1994). Distinct pattern of embryonic expression of the sea urchin CyI actin gene in *Tripneustes gratilla*. *Dev. Biol.* **165,** 117–125.

Wang, D. G. W., Kirchhamer, C. V., Britten, R. J., and Davidson, E. H. (1995a). SpZ12-1, a negative regulator required for spatial control of the territory-specific CyIIIa gene in the sea urchin embryo. *Development* **121,** 1111–1122.

Wang, D. G. W., Britten, R. J., and Davidson, E. H. (1995b). Maternal and embryonic provenance of a sea urchin embryo transcription factor, SpZ12-1. *Mol. Mar. Biol. Biotechnol.* **4,** 148–153.

Wang, W., Wikramanayake, A. H., Gonzalez-Rimbau, M., Vlahou, M., Flytzanis, C. N., and Klein, W. H. (1996). Very early and transient vegetal-plate expression of SpKrox1, a Kruppel/Krox gene from *Strongylocentrotus purpuratus*. *Mech. Dev.* **60,** 185–195.

Wang, Y., Mackay, E. A., Zerbe, O., Hess, D., Hunziker, P. E., Vasak, M., and Kagi, J. H. (1995). Characterization and sequential localization of the metal clusters in sea urchin metallothionein. *Biochemistry* **34,** 7460–7467.

Wang, Y., Pansky, A., Venuti, J. M., Yaffe, D., and Nudel, U. (1998). A sea urchin gene encoding distrophin-related proteins. *Hum. Mol. Gen.* **7,** 581–588.

Wei, Z., Angerer, L. M., Gagnon, M. L., and Angerer, R. C. (1995). Characterization of the SpHE promoter that is spatially regulated along the animal–vegetal axis of the sea urchin embryo. *Dev. Biol.* **171,** 195–211.

Wei, Z., Angerer, L. M., and Angerer, R. C. (1997a). Multiple positive *cis* elements regulate the asymmetric expression of the SpHE gene along the sea urchin embryo animal–vegetal axis. *Dev. Biol.* **187,** 71–78.

Wei, Z., Kenny, A. P., Angerer, L. M., and Angerer, R. C. (1997b). Spatial and temporal regulation of SpHE transcription presented at the XIth Meeting on Developmental Biology of the Sea Urchin, Woods Hole, MA.

Wei, Z., Kenny, A. P., Angerer, L. M., and Angerer, R. C. (1997c). The SpHE gene is down regulated in sea urchin late blastulae despite persistence of multiple positive factors sufficient to activate its promoter. *Mech. Dev.* **67,** 171–178.

Wells, D. E., Showman, R. M., Klein, W. H., and Raff, R. A. (1981). Delayed recruitment of maternal histone H3 mRNA in sea urchin embryos. *Nature* **292,** 477–478.

Wessel, G. M. (1995). A protein of the sea urchin cortical granules is targeted to the fertilization envelope and contains an LDL-receptor-like motif. *Dev. Biol.* **167,** 388–397.

Wessel, G. M., and Berg, L. (1995). A spatially restricted molecule of the extracellular matrix is contributed both maternally and zygotically in the sea urchin embryo. *Dev. Growth Differ.* **37,** 517–527.

Wessel, G. M., and Chen, S. W. (1993). Transient, localized accumulation of α-spectrin during sea urchin morphogenesis. *Dev. Biol.* **155,** 161–171.

Wessel, G. M., and McClay, D. R. (1985). Sequential expression of germ-layer specific molecules in the sea urchin embryo. *Dev. Biol.* **111,** 451–463.

Wessel, G. M., Goldberg, L., Lennarz, W. J., and Klein, W. H. (1989). Gastrulation in the sea urchin is accompanied by the accumulation of an endoderm-specific mRNA. *Dev. Biol.* **136,** 526–536.

Wessel, G., Zhang, W., and Klein, W. H. (1990). Myosin heavy chain accumulates in dissimilar cell types of the macromere lineage in the sea urchin embryo. *Dev. Biol.* **140,** 447–454.

Wessel, G. M., Etkin, M., and Benson, S. (1991). Primary mesenchyme cells of sea urchin embryo require an autonomously produced, non fibrillar collagen for spiculogenesis. *Dev. Biol.* **148,** 261–272.

Wessel, G. M., Berg, L., Adelson, D. L., Cannon, G., and McClay, D. R. (1998). A molecular analysis of hyalin-A substrate for cell adhesion in the hyalin layer of the sea urchin embryo. *Dev. Biol.* **193,** 115–126.

Wikramanayake, A. H., and Klein, W. H. (1997a). Is a Wnt/β-catenin pathway responsible for patterning the animal–vegetal axis in sea urchins? *Dev. Biol.* **186,** 260 (abstract).

Wikramanayake, A. H., and Klein, W. H. (1997b). Multiple signal events specify ectoderm and pattern the oral–aboral axis in the sea urchin embryo. *Development* **124,** 13–20.

Wikramanayake, A. H., Brandhorst, B. P., and Klein, W. H. (1995). Autonomous and non-autonomous differentiation of ectoderm in different sea urchin species. *Development* **121,** 1497–1505.

Wikramanayake, A. H., Huang, L., and Klein, W. H. (1998). β-catenin is essential for patterning the maternally specified animal–vegetal axis in the sea urchin embryo. *Proc. Natl. Acad. Sci, USA* **95,** 9343–9348.

Wilding, M., Wright, E. M., Patel, R., Ellis-Davies, G., and Whitaker, M. (1996). Local perinuclear calcium signals associated with mitosis-entry in early sea urchin embryos. *J. Cell Biol.* **135,** 191–199.

Wilkinson, D. G., and Nemer, M. (1987). Metallothionein genes MTa and MTb expressed under distinct quantitative and tissue-specific regulation in sea urchin embryos. *Mol. Cell. Biol.* **7,** 48–58.

Wilt, F. H. (1987). Determination and morphogenesis in the sea urchin embryo. *Development* **100,** 559–575.

Wilt, F. H. (1997). Looking into the sea urchin embryo you can see local cell interactions regulate morphogenesis. *BioEssays* **19,** 665–668.

Wright, B. D., and Scholey, J. M. (1992). Microtubule motors in the early sea urchin embryo. *Curr. Top. Dev. Biol.* **26,** 71–91.

Wright, B. D., Henson, J. H., Wedman, K. P., Willy, P. J., Morand, J. N., and Scholey, J. M. (1991). Subcellular localization and sequence of sea urchin kinesin heavy chain: Evidence for its association with membranes in mitotic apparatus and interphase cytoplasm. *J. Cell Biol.* **113,** 817–833.

Wright, B. D., Terasaki, M., and Scholey, J. M. (1993). Roles of kinesin and kinesin-like proteins in sea urchin embryonic cell division: Evaluation using antibody microinjection. *J. Cell Biol.* **123**, 681–689.

Xian, J., Harrington, M. G., and Davidson, E. H. (1996). DNA-protein binding assays from a single sea urchin egg: A high-sensitivity capillary electrophoresis method. *Proc. Natl. Acad. Sci. USA* **93**, 86–90.

Xiang, M., Lu, S. Y., Musso, M., Karsenty, G., and Klein, W. H. (1991). A G-string positive cis-regulatory element in the LpS1 promotor binds two distinct nuclear factors distributed non-uniformly in *Lytechinus pictus* embryos. *Development* **113**, 1345–1355.

Xu, Y., and Tashjian, A. H., Jr. (1995). Cyclic ADP-ribose-induced calcium release in sea urchin egg homogenates is a cooperative process. *Biochemistry* **34**, 2815–2818.

Xu, N., Niemeyer, C. C., Gonzalez-Rimbau, M. G., Bogosian, E. A., and Flytzanis, C. N. (1996). Distal cis-acting elements restrict expression of the CyIIIb actin gene in the aboral ectoderm of the sea urchin embryo. *Mech. Dev.* **60**, 151–162.

Yamada, K., Akasaka, K., and Shimada, H. (1989). Structure of the sea urchin arylsulfatase gene. *Eur. J. Biochem.* **186**, 405–410.

Yamada, K., Eguchi, S., Yamamoto, T., Akasaka, K., and Shimada, H. (1992). Cis-acting elements for proper ontogenic expression of arylsulfatase gene of sea urchin embryo. *Dev. Growth Differ.* **34**, 719–729.

Yamada, K., Yamamoto, T., Akasaka, K., and Shimada, H. (1993). Cis-elements and protein factors related to regulation of transcription of arylsulfatase gene during sea urchin development. *Dev. Growth Differ.* **35**, 703–710.

Yamaguchi, M., and Ohba, Y. (1997). Transcripts containing the sea urchin retroposon family 1 (SURF1) in embryos of the sea urchin *Anthocidaris crassispina*. *Zool. Sci.* **14**, 947–952.

Yamazaki, K., Maruyama, T., Ihara, T., and Yasumasu, I. (1997a). Behavior of Na^+/K^+-ATPase alpha-subunit alleles in sea urchin eggs upon fertilization. *Dev. Growth Differ.* **39**, 105–109.

Yamazaki, K., Okamura, C., Ihara, T., and Yasumasu, I. (1997b). Two types of Na^+/K^+-ATPase alpha subunit gene transcript in embryos of the sea urchin, *Hemicentrotus pulcherrimus*. *Zool. Sci.* **14**, 469–473.

Yang, Q., Angerer, L. M., and Angerer, R. C. (1989a). Structure and tissue-specific developmental expression of a sea urchin arylsulfatase gene. *Dev. Biol.* **135**, 53–65.

Yang, Q., Angerer, L. M., and Angerer, R. C. (1989b). Unusual pattern of accumulation of mRNA encoding EGF-related protein in sea urchin embryos. *Science* **246**, 806–808.

Yang, Q., Kingsley, P. D. Kozlowski, D. J., Angerer, R. C., and Angerer, L. M. (1993). Immunochemical analysis of arylsulfatase accumulation in sea urchin embryos. *Dev. Growth Differ.* **35**, 139–151.

Yazaki, I. (1993). A novel substance localizing on the apical surface of the ectodermal and the esophageal epithelia of the sea urchin embryos. *Dev. Growth Differ.* **35**, 671–682.

Yazaki, I., Tosti, E., and Dale, B. (1995). Cytoskeletal elements link calcium channel activity and the cell cycle in early sea urchin embryos. *Development* **121**, 1827–1831.

Ye, X., and Sloboda, R. D. (1997). Molecular characterization of p62, a mitotic apparatus protein required for mitotic progression. *J. Biol. Chem.* **272**, 3606–3614.

Yokota, Y., Kato, K. H., and Mita, M. (1993). Morphological and biochemical studies on yolk degradation in the sea urchin, *Hemicentrotus pulcherrimus*. *Zool. Sci.* **10**, 661–670.

Yoshigaki, T. (1997). Accumulation of WGA receptors in the cleavage furrow during cytokinesis of sea urchin eggs. *Exp. Cell Res.* **236**, 463–471.

Yu, J. C., Wendelburg, B., Sakallah, S., and Marzluff, W. F. (1991). The U1 snRNA gene repeat from the sea urchin (*Strongylocentrotus purpuratus*): The 70 kilobase tandem repeat ends directly 3′ to a U1 gene. *Nucleic Acids Res.* **19**, 1093–1098.

Yuh, C.-H., and Davidson, E. H. (1996). Modular cis-regulatory organization of Endo 16, a gut specific gene of the sea urchin embryo. *Development* **122**, 1069–1082.

Yuh, C. H., Ransick, A., Martinez, P., Britten, R. J., and Davidson, E. H. (1994). Complexity and organization of DNA–protein interactions in the 5′-regulatory region of an endoderm-specific marker gene in the sea urchin embryo. *Mech. Dev.* **47,** 165–186.

Yuh, C.-H., Moore, J. G., and Davidson, E. H. (1996). Quantitative functional interrelations within the cis-regulatory system of the *S. purpuratus* Endo 16 gene. *Development* **122,** 4045–4056.

Yuh, C.-H., Bolouri, H., and Davidson, E. H. (1998). Genomic cis-regulatory logic: Experimental and computational analysis of a sea urchin gene. *Science* **279,** 1896–1902.

Zeller, R. W., Britten, R. J., and Davidson, E. H. (1995a). Developmental utilization of SpP3A1 and SpP3A2: Two proteins which recognize the same DNA target site in several sea urchin gene regulatory regions. *Dev. Biol.* **170,** 75–82.

Zeller, R. W., Coffman, J. A., Harrington, M. G., Britten, R. J., and Davidson, E. H. (1995b). SpGCF1, a sea urchin embryo DNA-binding protein, exists as five nested variants encoded by a single mRNA. *Dev. Biol.* **169,** 713–727.

Zeller, R. W., Griffith, J. D., Moore, J. G., Kirchhamer, C. V., Britten, R. J., and Davidson, E. H. (1995c). A multimerizing transcription factor of sea urchin embryos capable of looping DNA. *Proc. Natl. Acad. Sci. USA* **92,** 2989–2993.

Zhao, A. Z., Colin, A. M., Bell, J., Baker, M., Char, B. R., and Maxson, R. (1990). Activation of a late H2B histone gene in blastula-stage sea urchin embryos by an unusual enhancer element located 3′ of the gene. *Mol. Cell. Biol.* **10,** 6730–6741.

Zhao, A. Z., Vansant, G., Bell, J., Humphreys, T., and Maxson, R. (1991). Activation of the L1 late H2B histone gene in blastula-stage sea urchin embryos by *Antennapedia*-class homeoprotein. *Mech. Dev.* **34,** 21–28.

Zito, F., Tesoro, V., McClay, D. R., Nakano, E., and Matranga, V. (1998). Ectoderm cell–ECM interaction is essential for sea urchin embryo skeletogenesis *Dev. Biol.* **196,** 184–192.

3

The Organizer of the Gastrulating Mouse Embryo

Anne Camus and Patrick P. L. Tam
Embryology Unit
Children's Medical Research Institute
Wentworthville
New South Wales 2145, Australia

I. Introduction
  A. Body Pattern and Lineage Diversity
  B. The Organizer
II. The Mouse Gastrula Organizers
  A. Patterning Activity
  B. Cell Fate
  C. Morphogenetic Movement
  D. Molecular Properties
  E. Interplay of Positive and Negative Regulatory Signals
III. Separate Head and Trunk Organizing Activity
  A. Regionalization within the Organizer
  B. Genetic Evidence from Mutant Embryos
  C. Alternative Sources of Organizing Activity
IV. Tissue Patterning and the Specification of the Gastrula Organizer
  A. Searching for the Nieuwkoop Signal
  B. Insight from Mutational and Transgenic Studies
  C. The Role of Extraembryonic Tissues
  D. Emergence of Pattern Asymmetry and Formation of the Organizer
V. An Organizer for All Seasons?
  References

I. Introduction

A. Body Pattern and Lineage Diversity

Early embryogenesis is accomplished by coordinating three developmental processes: the specification of diverse tissue lineages, the delineation of the body plan, and the allocation of the tissues to different parts of the body. In the mouse, lineage differentiation first takes place in the preimplantation embryo leading to the segregation of the trophectoderm and the inner cell mass. The trophectoderm gives rise to the trophoblasts and the extraembryonic ectoderm of the peri-implantation embryo. The inner cell mass differentiates to form the epiblast from which the three definitive germ layers are derived. The inner cell mass also contributes to some extraembryonic tissues such as the primitive endoderm and the extraembryonic mesoderm. The appearance of histologically distinct cell types is accompanied by the expression of lineage-specific molecules and a restriction in developmental potency of the embryonic cells (reviewed by Davidson et al., 1999). At the onset of gastrulation, the mouse embryo displays distinct morphological and molecular characteristics that herald the establishment of the three orthogonal embryonic axes (reviewed by Tam and Behringer, 1997). However, recent experimental evidence suggests that the anterior–posterior axis may be established well before gastrulation (reviewed by Beddington and Robertson, 1998).

Analyses of lineage differentiation have revealed that cells localized to different regions of the germ layer in the pregastrulation and gastrulating embryos display a predictable developmental fate and morphogenetic movement during gastrulation. However, at this stage, cells are not irreversibly committed or restricted to any tissue lineage. In particular, epiblast and ectodermal cells can acquire different fates when transplanted to heterotopic sites in another embryo. This apparent paradox of a predictable fate versus the developmental plasticity of germ layer cells implies that lineage determination is a late event that takes place after the completion of gastrulation.

B. The Organizer

The seminal studies by Spemann and Mangold that demonstrated the ability of a specific population of cells in the dorsal blastoporal lip of the *Triturus* embryos to induce the formation of a complete secondary axis introduce the concept of an organizer responsible for the patterning of tissues in the embryonic body (reviewed by Gilbert and Saxen, 1993). More recent transplantation experiments performed on *Xenopus laevis* embryos show that, similar to the urodele, the dorsal blastoporal tissues of the gastrula also act as an organizer. With the use of markers that could distinguish between the host and graft-derived tissues, the *Xenopus* experiments confirm that in the induced axis, the blastoporal tissues contribute predominantly to the notochord whereas the neural tube and somites are derived almost entirely from the host (Gimlich and Cooke, 1983; Smith and Slack, 1983). The blastoporal tissues therefore exert an instructive influence on the differentiation of the surrounding host tissues and provide global morphogenetic cues that regulate the dimension, orientation, and positioning of the induced tissues along the anterior–posterior axis. This unique combination of both an inductive and a patterning activity distinguishes the organizer from other cell populations in the embryo which appear to only induce the differentiation but not the morphogenesis of tissues (Nieuwkoop, 1985).

The same experimental paradigm of axis induction has been used to identify organizer activity in embryos of other vertebrates such as zebrafish, bird, and mouse

3. The Organizer of the Gastrulating Mouse Embryo

(e.g., Storey *et al.*, 1992, 1995; Beddington, 1994; Shih and Fraser, 1996; Tam *et al.*, 1997a, 1997b). Apart from their ability in organizing the anterior–posterior axis, the vertebrate organizers also display some common developmental and functional properties:

(i) The organizer is present in the embryo at the onset of gastrulation. It constitutes a minor proportion of the cells in the gastrula. It is localized to the dorsal region of the epiblast of the zebrafish gastrula and the dorsal blastoporal lip of the amphibian gastrula. In the chick, cells with organizing activity are found first in the posterior region of the blastoderm in the vicinity of the Koller's sickle and are later incorporated into the Hensen's node at the anterior end of the elongating primitive streak. In the mouse, the organizer is found in the posterior epiblast of the early-streak stage embryo (early gastrula organizer, EGO), and later as the node of the late-streak embryo.

(ii) Analysis of the developmental fate reveals that descendants of the organizer are found in the axial mesoderm and floor plate, as well as in other tissues such as gut endoderm and paraxial mesoderm (Table I). Contributions to the notochord and paraxial mesoderm are a feature common to all organizers. However, the *Xenopus* blastoporal cells do not contribute to the neural tissues. When the cell fates of the avian gastrula are examined collectively for HH stages 3–9, descendants of the avian organizer cover the full range of tissue type, with the exception of the head mesenchyme. After transplantation to an ectopic site, the organizer cells will give rise to the same types of tissues, suggesting that differentiation of the organizer cells is regulated autonomously.

(iii) The organizer cells, and often the axial mesoderm derived from the organizer, express genes encoding transcription factors or molecules associated with cell–cell signaling and growth regulation. In the zebrafish, *Xenopus*, and chick embryo, ectopic expression of some of these genes mimics the patterning activity of a transplanted organizer and leads to the duplication of an embryonic axis. In the mouse, the effect of the ectopic expression of organizer-specific genes has not been tested. However, the loss of organizer gene activity produced by targeted mutation can cause aberrant body patterning. Interestingly, some of these so-called "organizer genes" are also expressed outside the organizer population in the gastrula. This raises an important issue of whether the organizer population is the only source of patterning information in the gastrulating embryo.

(iv) The patterning activity is generally believed to be mediated by the inductive signals emanating from the organizer. Recent molecular and experimental evidence suggests that the vertebrate organizer is also a source of antagonistic signals. The current view is that some aspects of organizer activity, such as neural induction and dorsalization of mesoderm, are the result of the combined action of both instructive and antagonistic factors.

(v) The vertebrate organizer is a dynamic entity, constantly changing its cellular composition, structural organization, and molecular characteristics during embryonic development. In the chick and *Xenopus* embryo, changes in the cellular

Table I. Comparison of Cell Fates of Organizers of Vertebrate Gastrulae[a]

Organizer	Tissue contribution					Reference
	Neurectoderm	Axial mesoderm	Paraxial mesoderm	Endoderm		
Mouse						
EGO (early-streak stage)	Floor plate	Head process, notochord, and node	Head mesenchyme, heart, and somites	Foregut and trunk endoderm		Tam et al., 1997b
Node (late-streak stage)	Floor plate	Notochord	Somites	Trunk endoderm		Beddington, 1994; Sulik, et al., 1994
Zebrafish (50% epiboly)						
Embryonic shield	Ventral part of the neural tube	Chordamesoderm	Head mesenchyme and somites	Pharyngeal endoderm		Melby et al., 1996; Shih and Fraser, 1995, 1996
Xenopus (stage 10)						
Blastopore lip	No contribution	Prechordal plate, notochord	Head mesenchyme and somites	Pharyngeal endoderm		Lane and Keller, 1997
Chick						
Hensen's node HH stage 3–3+	Floor plate	Prechordal plate, notochord	Somites	Foregut and midgut endoderm		Selleck and Stern, 1991
HH stage 5–9	Floor plate	Notochord	Somites	No contribution		

[a] Abbreviations: EGO, early gastrula organizer; HH, Hamburger and Hamilton staging system.

composition and the gene activity of the organizer can be correlated with its ability to induce the differentiation of tissues with different anterior–posterior characteristics. This finding has led to the concept of separate organizers for head and trunk (body and tail) of the embryo.

In this review, we provide a synopsis of the current knowledge on the lineage and functional properties of the mouse gastrula organizer and highlight its unique features. We argue that most of the information required for patterning the embryonic body is endowed in the organizer, but the realization of this patterning activity requires an interaction with tissues that are predisposed (or competent) to respond to the patterning signals. This acquisition of competence may be mediated by signals that are found outside the organizer. Finally, we propose a model for the specification of the mouse gastrula organizer.

II. The Mouse Gastrula Organizers

A. Patterning Activity

Although vertebrate embryos reach the stage of gastrulation in different sizes and shapes, fate-mapping studies reveal that there is a remarkable conservation of the body plan among the gastrulae (Lawson *et al.*, 1991; Tam and Quinlan, 1996). The topographical relationship of the progenitors of major germ layer derivatives and, in particular, the location of the organizer relative to these progenitors are essentially comparable between species. In the zebrafish and *Xenopus* early gastrula, the organizer is located on the dorsal side at the border of the animal and vegetal hemispheres. In the chick, which gastrulates by cellular ingression through the primitive streak, the organizer of the early gastrula is found in the posterior region of the epiblast at the junction of the prospective extraembryonic and embryonic progenitor tissues (Izpisua-Belmonte *et al.*, 1993).

In the mouse, the earliest stage when cells displaying organizer activity can be identified is shortly after the appearance of the primitive streak (early-streak stage of Downs and Davies, 1993). These cells (EGO) are localized in the posterior–proximal epiblast within a zone 50–100 μm distal from the boundary of the epiblast and extraembryonic ectoderm (Figs. 1A and 1B; see color plate). This population contains about 40 cells and constitutes approximately 6–7% of the total epiblast population. There is no distinctive morphological feature distinguishing these cells from others elsewhere in the epiblast. During gastrulation, the primitive streak extends anteriorly and reaches the distal region of the cylindrical mouse gastrula (Figs. 1C and 2). At the anterior extremity of the primitive streak, about 100 cells, which constitutes about 0.5% of the total cell population in the late-streak embryo, congregate into a compact structure called the node (Fig. 1D). The endodermal side of the node is marked by a crescent-shaped depression known as the archenteron whereas the epiblast side is elevated into a hillock (Fig. 1D; Poelmann,

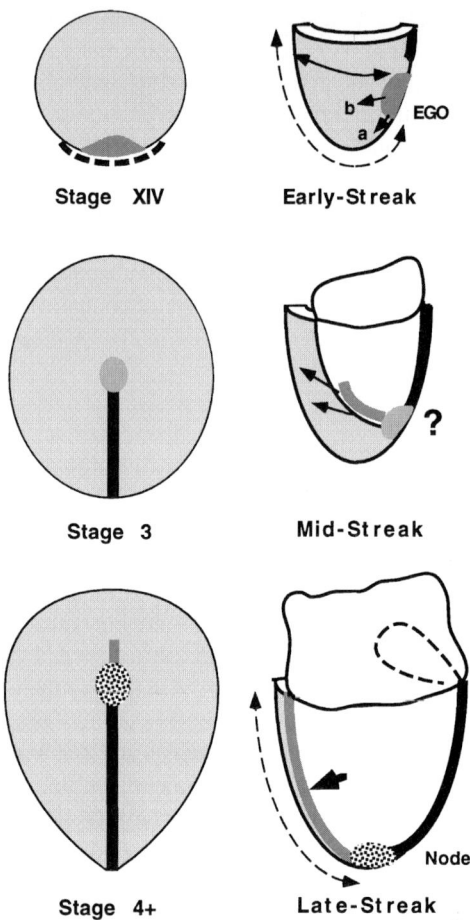

Fig. 2 A comparison of the mouse and avian gastrulae shows that the organizer is located at similar sites during gastrulation. The mouse embryos are shown in their lateral view with anterior to the left and distal to the bottom. The avian embryos are viewed from the dorsal side. The arrows in the early-streak mouse embryo show the movement of cells emerged from the early gastrula organizer (EGO) (a) distally along the midline and (b) laterally to the newly formed mesoderm. The double-headed arrows show the distance of the EGO to the anterior region of the epiblast along the anterior–posterior axis (220 μm, dashed line) and along the girth of the embryo (150 μm, solid line). In the mid-streak embryo, cells derived from the EGO are localized to the distal part of the newly formed mesoderm and the distal region of the primitive streak. Cells in the mesoderm are translocated anteriorly as the mesoderm expands toward the anterior side of the embryo. The mesodermal layers on the left and the right sides of the embryo converge at the anterior midline where the anterior axial mesoderm forms (the thick shaded line indicated by the short arrow). Whether the cells in the anterior end of the primitive streak of the mid-streak embryo [?] are acting as the organizer is not known yet. In the late-streak embryo, the node is about 250 μm (double-headed arrow) from the most rostral end of the embryo.

3. The Organizer of the Gastrulating Mouse Embryo

1981; Sulik *et al.*, 1994). In the avian embryo, a group of cells localized to the posterior blastoderm in the vicinity of the Koller's sickle and the posterior marginal zone are found to express the organizer-specific *goosecoid* (*Gsc*) gene. These cells are subsequently incorporated into the anterior region of the newly formed primitive streak and are subsequently displaced anteriorly with the extension of the primitive streak and finally contribute to the Hensen's node of the late gastrula (Fig. 2). When tested by heterotopic transplantation, tissue fragments containing these *Gsc*-expressing cells can induce an ectopic embryonic axis (Izpisua-Belmonte *et al.*, 1993). In the mouse embryo, the EGO, which also expresses *Gsc* (Fig. 1B), is found at a site equivalent to where the organizer cells are first found in the avian blastoderm. Whether cells in the anterior region of the elongating primitive streak of the mid-streak mouse embryo will display any organizer activity has not yet been thoroughly tested. In the late gastrula, the node of the mouse may be considered homologous to the Hensen's node of the bird (Fig. 2).

Both the early gastrula organizer and the node of the mouse embryo can induce the formation of a secondary axis after heterotopic transplantation to the posterior lateral region of the late-streak embryo (Beddington, 1994; Tam *et al.*, 1997b; Table II). About 25% of the embryos receiving an EGO graft and 25–38% of those receiving node graft display evidence of axis duplication. Based on the histological and molecular characteristics, the ectopic axis contains neural tissues (expressing *NCAM*) derived mostly but not exclusively from the host tissues. Tissues in the neural groove or the presumptive floor plate of the neural tube are sometimes also derived from the graft. Duplication of the axis involves not just the neural tissues but also the formation of extra somites (*Mox1*-expressing cells) and the proliferation of the gut endoderm, both of which are derived mostly from the host tissues. Consistent with the prospective fate of the EGO and the node (Table I), the organizer graft differentiates to give rise to the axial mesoderm (presumed to be notochord based on *T* activity). Thus, an organizer appears to be present in embryos at both the early and late stages of gastrulation (Figs. 3A and 3B; see color plate).

Common to both transplantations, the duplicated embryonic axis lacks anterior (head) structures. Histological characteristics of the neural tissues in the duplicated axis imply that they are reminiscent of trunk neural tube (Beddington, 1994; Tam *et al.*, 1997b). This raises the concern that even the organizer of the early gastrula may be lacking the head organizing activity. However, transplantation of the EGO to the prospective cranial mesoderm of the late-streak embryo can induce the ectopic expression of *Otx2* and *En2* in the hindbrain neuroectoderm and the adjacent surface ectoderm. The EGO therefore appears to be capable of inducing anterior neural differentiation but not morphogenesis of these structures (Tam *et al.*, 1997b). It is possible that the lack of head morphogenesis may be due to the disparity between the developmental stage of the organizer and the host. However, EGO transplanted to the anterior region of the early-streak embryo self-differentiates into ectodermal and mesodermal tissues with posterior characteristics. No

Table II. Functional Assays of the Mouse Gastrula Organizer

Functional assay	Experiment		Outcome	Reference
Tissue patterning	Heterotopic transplantation	EGO or node → lateral region of late-streak (7.5-day) embryo	Partial secondary axis: neural plate and somites, lacked anterior (head) structure	Beddington, 1994
		EGO to anterior epiblast of early-streak (6.5-day) embryo	Suppressed anterior development of host embryo, graft differentiated to extraembryonic mesenchyme and colonized the anterior epiblast and amnion	Tam et al., 1977b
Neural induction	Heterotopic transplantation	EGO → prospective hindbrain mesoderm of late-streak embryo	Induced expression of Otx2 and En2 in the hindbrain and adjacent surface ectoderm	Tam et al., 1977b
Heterospecific assay	Einsteck experiment in Xenopus	Node-containing fragments → blastocoel (einsteck assay)	Secondary anterior organs: eye, cement gland, neural and auditory vesicles, and gut	Blum et al., 1992
	Transplantation to chick limb bud	Node-containing fragments → posterior region of the limb bud	Digit duplication, mimicking the patterning activity of tissues of the zone of polarizing activity	Hogan et al., 1992

3. The Organizer of the Gastrulating Mouse Embryo 125

axis duplication is observed and the host embryo lacks anterior structures (Tam et al., 1997b). Whether the EGO possesses any head organizing activity remains elusive.

The patterning activity of the mouse organizer has also been tested by heterospecific assays (Table II). An embryonic fragment containing the node can induce digit duplication in the avian limb bud, mimicking the zone of polarizing activity (Hogan et al., 1992). The distal region of the 6.75-day embryo, which contains cells expressing *Gsc*, can induce the formation of structures resembling eye vesicles and cement glands when transplanted to the blastocoel of the *Xenopus* gastrula for the einsteck assay (Blum et al., 1992). A proper interpretation of these results requires further information on (i) whether the inductive effect is a consequence of *goosecoid* activity and (ii) the nature of the putative organizer tissues that are tested. The intriguing outcome of these heterospecific tests suggests that the mouse organizer shares some common mechanism of tissue patterning with the *Xenopus* organizer and the signaling center of the limb bud.

B. Cell Fate

Lineage analysis studies have shown that the posterior epiblast of the early-streak embryo containing the EGO contributes to the axial mesoderm of the head folds (head process mesoderm), the notochord, the floor plate, the gut endoderm, and the cranial mesoderm. In addition, some contribution to heart, lateral plate, and extraembryonic mesoderm is also found (Table I; Lawson et al., 1991; Tam et al., 1997b). Such a diversity of cell fate is similar to that displayed by the early organizer of other vertebrate gastrulae (chicken: Selleck and Stern, 1991; zebrafish: Shih and Fraser, 1995; Melby et al., 1996; *Xenopus*: Lane and Keller, 1997). It is not known if there are cells in the EGO that will contribute only to the axial mesoderm and the node. Lineage analysis of single-cell clone has shown that it is possible that such lineage may exist (Camus and Tam, unpublished).

The node of the late mouse gastrula displays a cell fate similar to that of the EGO (Table I). Node-derived cells are present in the floor plate, the notochord, and the gut endoderm. However, a differing contribution of node cells to the cranial mesoderm and the somites is found in fate-mapping experiments that employ either a dye-labeling of cells *in situ* or cell transplantation (Beddington, 1981; Tam and Beddington, 1987; Tam, 1989; Beddington, 1994; Sulik et al., 1994). Despite this discrepancy, the data are generally consistent with the notion that cells in the node are predominantly fated to become notochordal, neural, and endodermal cells. In contrast to the EGO cells, the node cells appear to have a more restricted fate and do not appear to contribute significantly to nonaxial mesoderm. It has been shown that some descendants of the EGO are found in the node of the late-streak embryo (Tam et al., 1997b), which in turn contribute to the node of the early-somite embryo (Beddington, 1994; Wilson and Beddington, 1996; Tam et al.,

1997b). There may be, therefore, a direct genealogical relationship between these organizer cell populations at successive stages of development.

Studies in avian embryos have shown that Hensen's node and its predecessor contribute to different parts of the axial mesoderm. The prechordal mesoderm is derived from the progenitor cells of the node in the early gastrula whereas the notochord is derived from the Hensen's node of the late gastrula (Selleck and Stern, 1991, 1992; Lemaire and Kessel, 1997; Lemaire et al., 1997). In the mouse, cells derived from the EGO are found along the entire length of the axial mesoderm of the early-somite embryo (Tam et al., 1997b). In contrast, cells of the node colonize the notochord of the trunk extensively, but apparently not the axial mesoderm in the head region (Beddington, 1994). These differences in cell fate may reflect a progressive change in the composition of progenitor cells in the organizer during gastrulation. The precise axial level and embryonic stage at which a transition of the source of the axial mesoderm from the EGO to the node source is not known (Fig. 3C).

C. Morphogenetic Movement

The finding that EGO descendants contribute to the axial mesoderm and neuroectoderm in the most rostral part of the mouse embryo raises the question of what morphogenetic mechanism may bring these cells to their anterior destination. The demonstration of a genealogical relationship between the early gastrula organizer and the node poses another morphogenetic enigma regarding how early organizer cells are translocated with the cells of the elongating primitive streak during gastrulation.

The morphogenetic movement of cells in the gastrulating mouse embryo has been studied by tracking the distribution of clonal descendants of the epiblast cells (Lawson et al., 1991). During gastrulation, epiblast cells are recruited to the primitive streak where they ingress to form the mesoderm and the definitive endoderm. Epiblast cells recruited to the primitive streak are added not at the poles but to the body of the primitive streak such that cells which are originally close together are displaced further apart along the anterior–posterior axis. This leads to the apparent anterior (distal for the cylindrical embryo) displacement of the organizer population during gastrulation. Thus EGO cells destined for the node are apparently translocated about 200 μm to reach the distal point of the cylindrical embryo (Figs. 1A, 1C, and 2: early-streak). At the early-streak stage, the EGO population is about 220 μm away along the midline from the most anterior point of the embryo assuming a midline trajectory. The node in the late gastrula is approximately 250 μm away from the most anterior point of the embryonic axis (Fig. 2: late-streak). As the node and the EGO are at a similar distance from the anteriormost point of the axis, this suggests that the organizer has stayed almost at the same location in the axis during the course of gastrulation. There is therefore little net anterior extension of the primitive streak or displacement of the EGO and the node. Rather the continuous addition of cells to the primitive streak posterior to the organizer seems sufficient to keep pace with the apparent anterior extension of the primitive streak.

In the avian gastrula, the current view on the formation of the axial mesendoderm underlying the cephalic neural tube (the prechordal plate and the head process) arises from an anterior protrusion of cells from the Hensen's node (Lemaire *et al.*, 1997). As previously estimated, the EGO population is positioned about 220 μm along the midline and about 150 μm along the girth of the cylinder from the anteriormost point of the embryo (Fig. 2: early-streak). During gastrulation, EGO-derived cells destined for the prechordal and cranial mesoderm would have to travel both of these two distances (Figs. 2 and 3). Assuming that the organizer remains stationary relative to the anterior pole of the embryonic axis, the formation of the axial mesoderm may require the migratory mesodermal cells to travel at least 200 μm anteriorly along the midline. Besides the axial mesoderm, EGO also contributes cells to the floor plate of the midbrain and hindbrain and these cells would therefore have to travel a similar distance. Circumstantial evidence derived from mapping the distribution of clonal descendants of epiblast cells supports a displacement of ectodermal cells toward the anterior regions of the prospective neural plate (Lawson *et al.*, 1991; Quinlan *et al.*, 1995; Tam and Behringer, 1997). It would not be difficult to envisage a similar anterior displacement of EGO-derived axial mesoderm occurring if the mesoderm moves in concert with the overlying ectoderm.

An alternative pattern of morphogenetic movement of cells destined for the axial mesoderm has also been observed by tracking the movement of EGO cells. EGO-derived cells are displaced along the midline at the anterior end of the extending primitive streak but may also move laterally to be part of the mesoderm layer (Fig. 2: early-streak). Mapping the position of these EGO-derived cells in the mid-streak stage embryo (12 h after labeling at the early-streak stage) has revealed that they are localized to the distal region of the newly formed mesodermal layer and in the anterior end of the primitive streak (Fig. 2: mid-streak). When these mesodermal cells are tracked for a further 12–36 h of development to the early-somite stage, they are found positioned in the most rostral axial and paraxial mesoderm associated with the forebrain and midbrain. This segment of the axial mesoderm is likely to be established by the meeting of the expanding mesoderm layers from the two sides of the embryo along the midline. Cells that were previously in the anterior primitive streak also contribute to the axial mesoderm but are localized more caudally in the hindbrain (Tam *et al.*, 1997b; Quinlan, Camus, and Tam, unpublished). Taken together, these observations suggest that the axial mesoderm may be formed by a combination of medial convergence of the mesoderm and the anterior extension of cells derived from the organizer. However, it must be noted that these patterns of cell movement are reconstructed from observations of cell distribution at sequential stages and they have yet to be verified by direct visualization of cell movement *in situ*.

Concomitant with the development of the head folds, the embryonic axis rostral to the node increases rapidly in length (Jacobson and Tam, 1982; Tuckett and Morriss-Kay, 1985), leading to an apparent posterior displacement of the node. It is not clear how much of this posterior displacement can be attributed to the regression of the node similar to that displayed by the Hensen's node in the avian embryo (Catala et al., 1996). As the axis lengthens, the regressing node deposits cells along its track, which then undergo convergence and extension movement to form the notochord (Catala et al., 1996). In the mouse, the length of the primitive streak does appear to shorten progressively during early organogenesis (Lawson et al., 1991), consistent with the regression of the node. At later stages, in the organogenesis-stage avian and mouse embryos, tissues associated with the remnant of the primitive streak and the tail bud continue to contribute to the notochord and the floor plate (Wilson and Beddington, 1996; Knezevic et al., 1998). The chick tail bud is further shown to retain the organizing activity and can specify the appropriate posterior embryonic structures when it is transplanted heterotopically (Knezevic et al., 1998).

D. Molecular Properties

Several approaches have been employed successfully to identify the molecular determinants of the vertebrate organizer. These involve the search for organizer-specific transcripts by differential screening of cDNA libraries prepared from the organizer cells as well as by expression-cloning of molecules that can induce a secondary axis or neural and/or mesodermal differentiation in *Xenopus* embryos (Cho et al., 1991; Taira et al., 1992; Harland and Gerhart, 1997; He et al., 1997). For the mouse embryo, organizer-specific genes are similarly identified by screening for transcripts expressed specifically by the mouse gastrula and its germ layers (Harrison et al., 1995). Alternatively, genes can be isolated by screening for transcripts that are expressed in differentiating embryonic stem cells (Thomas and Rathjen, 1992; Thomas et al., 1995) or by fishing for the orthologue of organizer genes found in other vertebrate organizers.

Intuitively, the inductive and patterning activities of the organizer are expected to be mediated by molecules that are involved in intercellular signaling. Genes that are expressed by the mouse organizer generally encode three major types of molecules: transcription factors, secreted growth factors, and signaling molecules (Table III). Many of these genes are expressed successively in the region of posterior epiblast that contains the EGO, in cells at the anterior region of the elongating primitive streak prior to the formation of the node, in the node, and then in the axial mesendoderm derived from the organizer. The expression of similar genetic activity by these tissues provides corroborative evidence that they are genealogically related and may therefore possess similar inductive and patterning activity. It is possible that organizer activity is not exclusively confined to the organizer

3. The Organizer of the Gastrulating Mouse Embryo

Table III. Expression of Organizer Genes in the Mouse Gastrula[a,b]

	Early gastrula		Late gastrula	
	EGO (epiblast)	Visceral endoderm	Node	Organizer derivatives
Secreted proteins				
Shh	N	N	Yes	Notochord and head process
Nodal	Yes	Yes	Perinodal endoderm, asymmetrical	N
Noggin	N	N	Yes	Notochord
Chordin	N	N	Yes	Notochord
Transcription factors				
Otx2	Yes	Yes	N	Notochord and anterior mesendoderm
Lim1	Yes	AVE	Ventral node	Head process
Hnf3β	Yes	Yes	Both ventral and dorsal layers	Notochord and head process
T	Yes	N	Ventral layer	Notochord and head process
Gsc	Yes	AVE	N	Head process

[a] Abbreviations: N, not known to be expressed; AVE, anterior visceral endoderm.
[b] References: *Brachyury* (Herrmann, 1991), *Chordin* (Shawlot, unpublished), *Gsc* (Blum, *et al.*, 1992; Filosa *et al.*, 1997), *Hnf3β* (Sasaki and Hogan, 1993; Ang and Rossant, 1994; Weinstein *et al.*, 1994; Filosa *et al.*, 1997), *Lim1* (Barnes *et al.*, 1994; Shawlot and Berhinger, 1995); *nodal* (Zhou *et al.*, 1993; Collignon *et al.*, 1996), *Noggin* (McMahon *et al.*, 1998), *Otx2* (Simeone *et al.*, 1993; Acampora *et al.*, 1995; Matsuo *et al.*, 1995; Ang *et al.*, 1996; Filosa *et al.*, 1997); *Shh* (Chiang *et al.*, 1996).

cells but may be perpetuated by their descendants, and as such, the assessment of organizer activity in the embryo should take into account the activity displayed by the derivatives of the organizer.

E. Interplay of Positive and Negative Regulatory Signals

The classical view of the role of the organizer is to provide a source of instructive signals for mesoderm patterning and neural induction. Four secreted proteins (noggin, chordin, follistatin, and cerberus) are expressed in the *Xenopus* dorsal blastopore cells and when ectopically expressed they can mimic the ability of the organizer to induce the differentiation of neural tissues and dorsal mesoderm (reviewed in Harland and Gerhart, 1997; Hsu *et al.*, 1998). It is now known that these molecules may act as antagonists for BMP signaling by binding the BMPs with an affinity comparable to the cellular BMP receptor. BMPs bound by the antagonist are thus prevented from interacting with their receptors, and this modifies their activity to induce ventral mesoderm or endoderm (Sasai *et al.*, 1994, 1996; Piccolo *et al.*, 1996; Zimmerman *et al.*, 1996; Fainsod *et al.*, 1997; Jones and Smith, 1998).

Other secreted molecules encoded by the *Xnr3*, *Frzb*, and *dkk-1* genes also display antagonistic activity against BMP or WNT factors (Hansen *et al.*, 1997; Leyns *et al.*, 1997; Wang *et al.*, 1997; Glinka *et al.*, 1998). An inhibition of BMP and WNT signaling by these antagonists is a plausible mechanism whereby the organizer influences mesoderm patterning, neural differentiation, and endoderm specification (Harland and Gerhart, 1997; Heasman, 1997).

In the mouse, the function of several members of the TGF-β family (e.g., *nodal* and *BMP4*) and their antagonists (*follistatin* and *noggin*) have been analyzed by loss-of-function experiments. The phenotypes of the mutant embryos that are deficient for these protein factors are complex and are not immediately amenable to an understanding of organizer function. Mutation of the *follistatin* and *noggin* genes has provided little insight into their role in body plan organization. Neither the *follistatin*-deficient embryos (Matzuk *et al.*, 1995) nor the noggin-deficient embryos (McMahon *et al.*, 1998) show any abnormality of gastrulation or anterior–posterior patterning. Shortly after implantation (5.5 dpc), a low level of *nodal* transcripts is detected in the epiblast and the associated overlying primitive endoderm. By early-primitive-streak stage, *nodal* becomes restricted posteriorly in the epiblast at the site of primitive streak formation. As the primitive streak elongates, *nodal* expression is rapidly downregulated and is confined to the endoderm around the node (Varlet *et al.*, 1997). The null mutation of the *nodal* gene leads to defective epiblast differentiation, early arrest of gastrulation, and the failure to maintain the primitive streak (Conlon *et al.*, 1994; Varlet *et al.*, 1997). The functional status of the organizer in the *nodal* mutant has not yet been tested. *Bmp4* mRNAs are localized first in the posterior epiblast of the early gastrula and later in the posterior segment of the primitive streak, the allantois, and the amnion (Winnier *et al.*, 1995). Most BMP4-deficient embryos do not develop beyond the egg cylinder stage and have little or no mesoderm. The loss of function of the *Bmpr* gene (encoding a type I TGF-β family receptor for *Bmp2* and *Bmp4*) seems to result in similar developmental arrest at gastrulation (Mishina *et al.*, 1995). Some homozygous *Bmp4*−/− mutant embryos develop beyond organogenesis stage, displaying disorganized or truncated posterior structures, suggesting that, similar to its role in *Xenopus* development, *Bmp4* may play a role in the differentiation of ventral posterior mesoderm derivatives (Winnier *et al.*, 1995).

The patterning of embryonic structures at gastrulation seems to be accomplished by balancing the inductive signals with their specific antagonists that emanate from the organizer. In the chick embryo, ectopic expression of *Bmp4* near the prospective site of the primitive streak or in the blastoderm tissue that expresses chordin inhibits the formation of the primitive streak. In addition, when *Bmp4* is misexpressed next to the Hensen's node, axis development is arrested. The embryo also loses the node and the notochord (Streit *et al.*, 1998). In the *Xenopus* embryo, in contrast to the expression of *Bmp4* in the ventral marginal zone away from the organizer, *Bmp2* and *Bmp7* are expressed in the organizer. This necessitates the presence of an additional mechanism that may antagonize these BMP activities within the organizer. *Xnr3* that is expressed in the organizer seems to be able to suppress the potential ventralizing effect of organizer-derived BMPs by interfering with their synthesis or secretion (reviewed by Thomsen, 1997). Another TGF-β superfamily molecule, the antidorsalizing morphogenetic protein (ADMP), is also expressed in the dorsal blastopore lip. Overexpression of this molecule downregulates the activity of dorsalizing factors and suppresses axis development. It is postulated that the function of the ADMP is to counteract the widespread dorsalizing activity of the organizer in *Xenopus* (Moos *et al.*, 1995).

In the chick embryo, a suppressive activity of the Hensen's node on the surrounding blastoderm tissues has been implicated by the response of the embryonic tissues to surgical ablation of the organizer. In the absence of the Hensen's node, blastoderm tissues can reconstitute the organizer and regenerate the progenitors of the axial mesoderm and floor plate of the neural tube (Yuan *et al.*, 1995a, 1995b; Psychoyos and Stern, 1996). This reconstitutive ability can be suppressed by the presence of the Hensen's node in blastoderm isolates (Yuan and Schoenwolf, 1998). Surprisingly, zebrafish embryos whose dorsal embryonic shield has been ablated develop a remarkably well-organized body axis. This is achieved in the absence of the notochord, which implies that a complete organizer may not be present (Shih and Fraser, 1996). In the mouse, ablation of the node of the late gastrula does not disturb the development of the body axis and no consistent evidence of reconstitution of the organizer or the notochordal progenitors is found. The embryonic tissues after gastrulation seem to have acquired sufficient patterning information to sustain normal axis development without further input from the node (Davidson and Tam, unpublished observations). The loss of *Lim1* and Axin (a potential intercellular Wnt antagonist) activity in the mouse leads to duplication of axis (Perry *et al.*, 1995; Shawlot and Behringer, 1995; Zeng *et al.*, 1997). This duplication may be due to the loss of the suppressive activity of the node that normally will curtail the formation or the activity of supernumerary organizer. It is possible that the emanation of suppressive activity from the host node may account for the low efficiency of axis duplication following organizer transplantation in the mouse embryo. It is predicted that the induction of an extra embryonic axis by the transplanted organizer may be more effective in host embryos that lack a functional organizer. An examination of the molecular events utilizing this experimental paradigm may reveal more of the intricate positive–negative interplay of organizer signals for body patterning.

III. Separate Head and Trunk Organizing Activity

A. Regionalization within the Organizer

Much attention has been given to the existence of separate organizers for the head and trunk of the vertebrate embryo. In the *Xenopus* and the bird, the organizer taken from embryos of progressively more advanced stages of development has been shown to induce neural tissues of more posterior characteristics (reviewed by Dias and Schoenwolf, 1990; Storey *et al.,* 1992; Doniach, 1993; Harland and Gerhart, 1997). Recently, compelling molecular evidence for separate organizers of the body parts has been obtained for the *Xenopus* embryo (Vodicka and Gerhart, 1995). These studies reveal changes not only in the cell fate and the morphology of the organizer during development but also in the gene expression domains and inductive activity of the organizer.

Fate-mapping analysis has shown that the composition of mesodermal progenitor cells in the organizer changes during gastrulation (Lane and Keller, 1997). Cells destined for the prechordal mesoderm and head mesenchyme are found only in the early gastrula organizer and are replaced by those of the notochord and somites in the late gastrula. The ability to form axial mesoderm of different segmental levels may underlie the head versus body patterning activity of these cells when tested by heterotopic transplantation. In addition to these developmental changes in the type of organizer-derived tissues that may confer different inductive activity, a compartmentalization of separate head and body patterning activity may also exist within the organizer. Evidence in support of this notion comes from the revelation of regionalized gene expression and neural inductive activity in the dorsal blastopore lip of the *Xenopus* late blastula and early gastrula (Zoltewicz and Gerhart, 1997). The organizer can be subdivided into anterior (vegetal) and posterior (animal) domains defined respectively by the expression of *Gsc* (expressed later by the head process) and *Xnot* (with specific expression in the notochord), with some overlapping between *gsc* and *Xbra* (later found in the notochord) in the medial region (Vodicka and Gerhart, 1995). During development, the *gsc*-expressing cells disappear from the dorsal blastopore lip, leaving behind the *Xnot*- and *Xbra*-expressing cells. Such changes in expression are compatible with the transition from an early dorsal blastopore lip exhibiting anterior (head) organizer activity to a late dorsal blastopore lip showing posterior (trunk) organizer activity (Gerhart *et al.*, 1991; reviewed by Harland and Gerhart, 1997). To test if gene activity indeed marks the separate functional compartments of the organizer, further studies were performed on isolated halves (fragments of 20 by 25 cells) of the blastopore lip containing either the anterior (*gsc*) or the posterior (*Xnot*) domains. After orthotopic transplantation to the organizer region of the host embryo, the two fragments display distinctively different fates and differentiate respectively into prechordal mesoderm and notochord plus somites. When cultured as isolates, only the posterior fragment differentiates into somites and notochord. Results of culturing anterior fragment are not informative because differentiation *in vitro* is limited and prechordal mesoderm differentiation has not been assayed. When the isolates were tested by coculturing with animal cap ectoderm, the anterior fragment induced the expression of anterior-specific genes (*XAG-1* and *otxA*), whereas the posterior fragment induced the expression of posterior neural gene (*HoxB9*) and only weak activity of *otxA* gene (Zoltewicz and Gerhart, 1997). These findings provide the most compelling evidence for the segregation of region-specific neuralizing activity within the organizer. The detection of regionalized gene expression in the progenitors of the blastopore tissue suggests that the segregation of patterning activity may exist prior to the morphogenesis of the blastopore lip.

A subdivision of the organizer of the mouse gastrula into functional domains has not yet been examined in the same detail as in *Xenopus*. Nothing is known of any regionalization of gene activity in the early organizer (Fig. 4A; see color plate). In the node of the late gastrula, *Brachyury* (*T*) and *Lim1* transcripts are localized to the cells in the ventral part (endodermal aspect) of the node, *nodal* is found in the endoderm on the periphery, and *Hnf3β* is expressed uniformly in the node (Fig. 4B; see color plate). A comparison of the profile of genetic activity reveals that *Gsc* is expressed only in the EGO and in the prechordal and head process mesoderm of the late gastrula but not in the node. Genes such as *chordin*, *noggin*, *cordon bleu*, and *Shh* are expressed in the node and the notochord but not in the EGO (Table III; Gasca *et al.*, 1995; Tam and Behringer, 1997; McMahon *et al.*, 1998). It is not known if the molecular heterogeneity may correlate with distinct fate or the patterning activity of the organizer cells.

B. Genetic Evidence from Mutant Embryos

In the mouse, mutational analysis has delineated different categories of embryonic phenotypes showing specific deficiency of head and trunk structures (Table IV). *Otx2* −/− and *Lim1* −/− mutant embryos lack anterior structures and show a truncation of the embryonic axis at the level of the upper hindbrain (Acampora *et al.*, 1995; Shawlot and Behringer, 1995; Matsuo *et al.*, 1995; Ang *et al.*, 1996). None of these embryos has a morphologically definable node or anterior axial mesoderm, and the absence of these tissues with inductive activity may underlie the severe anterior defects (reviewed by Bally-Cuif and Boncinelli, 1997; Pera and Kessel, 1997). There is, however, mounting evidence that the lack of anterior patterning may also be due to an absence of signals from the visceral endoderm (Bouwmeester and Leyns, 1997; Beddington and Robertson, 1998; see later section). *Hnf3β* −/− mutant embryo has a rudimentary embryonic axis, which by molecular criteria still displays some anterior–posterior neural tube organization. No node or notochord is formed and morphogenesis of the neural tube is defective. Brain structures are not formed even though appropriate segment-specific genes are transcribed (Ang and Rossant, 1994; Weinstein *et al.*, 1994). The craniocaudal gradient of *Col2a1* activity in the paraxial mesoderm is reversed, and *Col2a1* and *Sox9* genes are misexpressed in the lateral mesoderm (Cheah *et al.*, unpublished). In the three aforementioned mutant embryos, the expression of genes that are normally associated with the organizer and the axial mesoderm is disrupted (Table IV).

Mutation of the *eed* gene leads to poor development of anterior structures but

Table IV. Mutations Affecting the Organizer and the Body Plan during Gastrulation and Early Organogenesis[a]

Gene	Mutant phenotype — Embryo	Mutant phenotype — Chimera/compound mutant/explant	Organizer-gene expression	Functional role of gene activity
Hnf3β	Gastrulation defect, notochord absent, lacking dorsoventral patterning in rudimentary neural tube, no somite patterning	4N↔−/−: rescues primitive streak, general patterning defects; *Gsc*−/−; *Hnf3β*+/−: anterior defects	Node absent; no expression of *Lim1*, *Otx2*, *Shh*, and *T* in axial mesoderm, no perinodal expression of *nodal*, diffused *Gsc* expression in proximal epiblast adjacent to the primitive streak and in the extraembryonic mesoderm	Epiblast: formation of the organizer Primitive endoderm: formation of the primitive streak and normal gastrulation Interaction with *Gsc* for dorsoventral patterning of forebrain
Lim1	Loss of anterior (head) structures	4N↔−/−: rescues gastrulation, anterior defects Germ layer recombination: defective mesendoderm in neural induction assay	No structural node; weak expression of *Hnf3β* and *nodal* in anterior part of primitive streak, *T* restricted to primitive streak, diffuse expression of *Gsc* in proximal epiblast	Morphogenesis of the organizer and specification/differentiation of the anterior mesendoderm
Otx2	Loss of head structures in null mutants; malformed craniofacial structures in +/− mutant, affected by genetic background	2N↔−/−: rescues gastrulation, anterior defects, selection against mutant cells in anterior mesendoderm Germ layer recombination: defective ectoderm and mesendoderm in neural induction assay	Node present; expression of *Hnf3β*, *Lim1*, *chordin*, and *noggin*, no expression of *Shh*, and diminished *Gsc* activity; none of these genes are expressed anteriorly	Differentiation of anterior axial mesendoderm and competence of anterior neural tissue Partial impact on organizer gene activity Required by primitive endoderm to sustain gastrulation and for rostral neural induction
nodal	Arrested at early gastrulation, failure to maintain the primitive streak, degeneration of the epiblast	2N↔−/−: rescues gastrulation, anterior defects	Effects on organizer unknown, lacks *Gsc* activity, restricted and transient expression of *T*	Formation of the primitive streak (morphogenetic function of the primitive endoderm), anterior patterning and body asymmetry

eed	Gastrulation defect, lacks embryonic mesoderm, excess extraembryonic mesoderm	Not reported	Specification and migration of embryonic mesoderm, anterior–posterior body patterning
Smad2	Arrested gastrulation (null mutant), excess formation of posterior and extraembryonic mesoderm (hypomorphic mutant?), craniofacial deficiency (+/− mutant)	2N↔−/−: rescues gastrulation, general developmental retardation and anterior tissue deficiency Smad2+/−; nodal+/−: defective early organogenesis and laterality defects	Initiation of gastrulation (function of primitive endoderm) Establishment of anterior–posterior polarity and left–right asymmetry
T	Poor development of the body after the early-somite stage, lacks notochord	2N↔T/T: cell-autonomous defective migration of the mutant cells 2N↔T/T with restored T activity: rescues mesodermal migration defect	No structural node, effects on early gastrula organizer not known Specification or differentiation of the notochordal lineage Cell migration

[a]References: *Hnf3β* (Ang and Rossant, 1994; Weinstein et al., 1994; Filosa et al., 1997; Dufort et al., 1998), *Lim1* (Shawlot and Behringer, 1995, unpublished), *Otx2* (Simeone et al., 1993; Acampora et al., 1995; Matsuo et al., 1995; Ang et al., 1996; Rhinn et al., 1998; Biben et al., 1998), *nodal* (Conlon et al., 1994; Varlet et al., 1997), *eed* (Faust et al., 1995; Schumacher et al., 1996), *Smad2* (Nomura and Li, 1998; Waldrip et al., 1998), *T* (Wilson et al., 199; Wilson and Beddington, 1997; Rashbass et al., 1991, 1994; Beddington et al., 1992).

overproduction of extraembryonic mesoderm (Table IV). No axial and paraxial mesoderm is formed in the mutant embryo which shows ectopic *Hnf3β* and *nodal* gene expression (Table IV; Faust *et al.*, 1995). The aberrant or absence of specific gene expression is generally consistent with a defective or absent head process mesoderm and a depleted population of potential organizer cells.

Another category of mutant embryos show apparently normal anterior development but defective trunk morphogenesis. *Bmp4*−/−, *Wnt3a*−/−, and *T/T* mutants show either dysplasia or loss of trunk structures posterior to the prospective cervical region (6–10th somite level) (Table IV; Rashbass *et al.*, 1994; Takada *et al.*, 1994; Wilson *et al.*, 1995; Winnier *et al.*, 1995; Greco *et al.*, 1996). Whether the phenotypes of these mutant embryos are caused by defects in the trunk organizer is not known. The formation of anterior structures in these embryos as contrasted with those with anterior truncation nevertheless suggests patterning of head and trunk structures could be separately regulated by different genetic activities, some of which may be traced to the organizer function. The patterning activity of the organizer of these mutant embryos has yet to be tested directly to ascertain if there are indeed functional entities that display differential head and trunk organizer activity.

C. Alternative Sources of Organizing Activity

1. The Primitive Endoderm for Anterior Patterning

In the mouse, transplantation of the node to the late-streak embryo can only induce a partial axis. It may be argued, on the basis of gene activity and cell fate, that the node may be the mouse equivalent of the late-gastrula organizer of the amphibian and avian embryos, which functions as the trunk organizer. The finding that even the early gastrula organizer is unable to induce a secondary axis complete with anterior neural characteristics raises the question of whether there are signals besides those of the organizer that are also required for anterior (head) patterning.

Studies of the postimplantation development of chimeric embryos containing different combinations of normal and mutant cells in the extraembryonic and embryonic compartments have shown that the primitive endoderm might be critical for normal anterior patterning of the mouse. In these experiments, chimeras are made either by introducing mutant embryonic stem (ES) cells into wild type blastocysts or conversely by introducing wild type ES cells into mutant blastocysts. Because of the restricted potency of the ES cells, chimeras so produced contain in the trophectodermal tissues cells that are derived from the blastocyst whereas the epiblast and the primitive endoderm are of mixed contributions by the blastocyst and the ES cells (Beddington and Robertson, 1989). Another approach to produce chimeras is to aggregate mutant ES cells with tetraploid wild type blastocyst. The resultant chimeras will be constituted of wild type trophectoderm and primitive

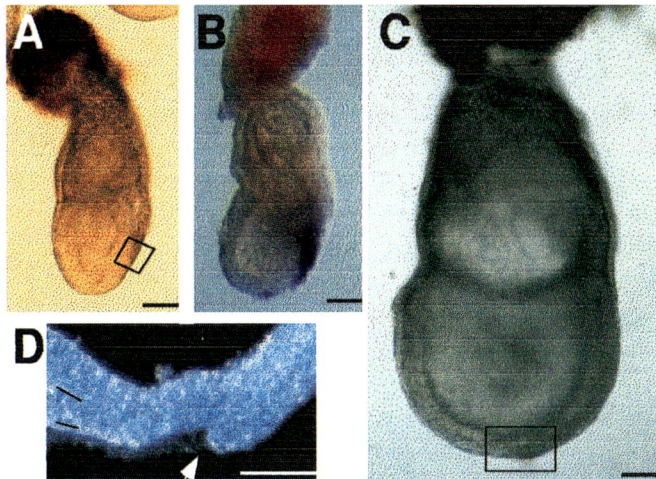

Ch. 3, Fig. 1 The mouse gastrula organizers. (A) Lateral view of an early-streak stage embryo showing the domain (outlined) in the posterior epiblast occupied by the early gastrula organizer (EGO). (B) *Gsc* transcripts are localized predominantly to the posterior epiblast containing the EGO. (C) Lateral view of an early neural plate stage embryo showing the localization of the node (outlined) in the anterior region of the primitive streak and (D) a sagittal section of the anterior region of the primitive streak [equivalent to the outlined region in (C)] showing the separate germ layers (the interfaces are marked by lines) anterior to the node (arrow points to the archenteron on its ventral side) and the contiguity of tissues of the node and the primitive streak. Bars = 50 μm.

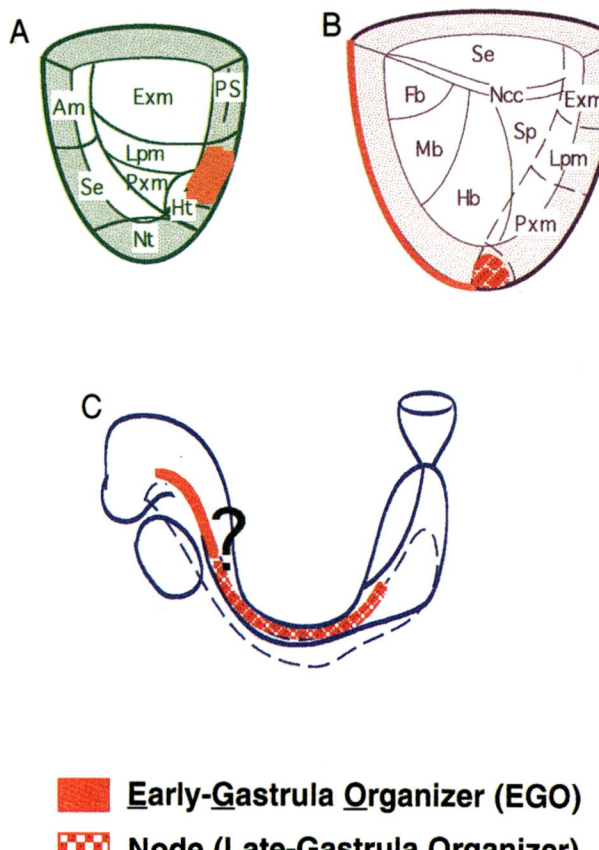

Ch. 3, Fig. 3 The location of the gastrula organizer in the fate map of (A) early-streak and (B) late-streak stage embryo. (C) the early-somite stage embryo showing the presumptive level [?] in the axial mesoderm where the node first contributes to the notochord. Am, Amnion ectoderm; Exm, extraembryonic mesoderm; Fb, forebrain; Hb, hindbrain; Ht, heart; Lpm, lateral plate mesoderm; Mb, midbrain; Ncc, neural crest cells; Nt, neural tube; PS, primitive streak; Pxm, paraxial mesoderm; Se, surface ectoderm; Sp, spinal cord.

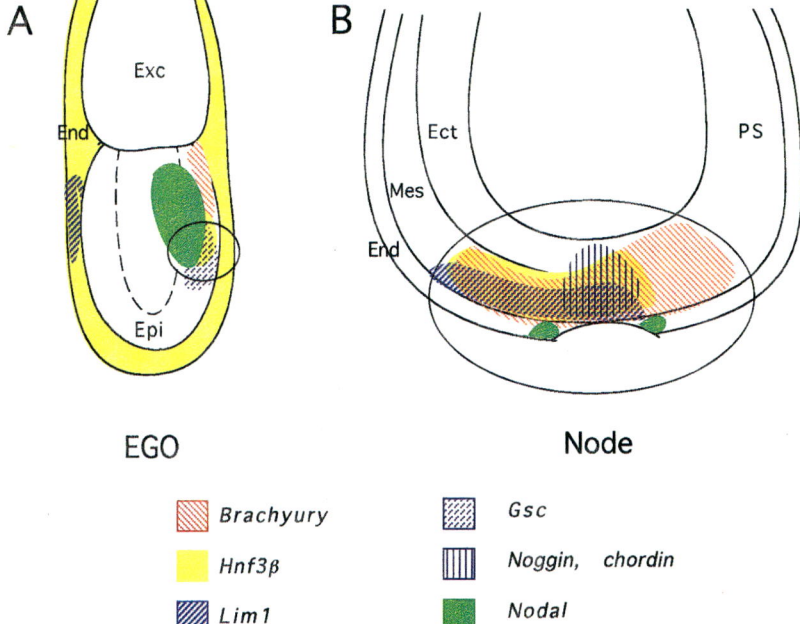

Ch. 3, Fig. 4 Schematic diagrams showing the expression domains of organizer gene in (A) the germ layers of the early-streak gastrula and (B) the node region of the late-streak gastrula. The circles mark the tissues containing the early gastrula organizer (EGO) and the node. Epi, Epiblast; Exc, extraembryonic ectoderm; End, primitive endoderm; Ect, ectoderm; Mes, mesoderm, PS, primitive streak.

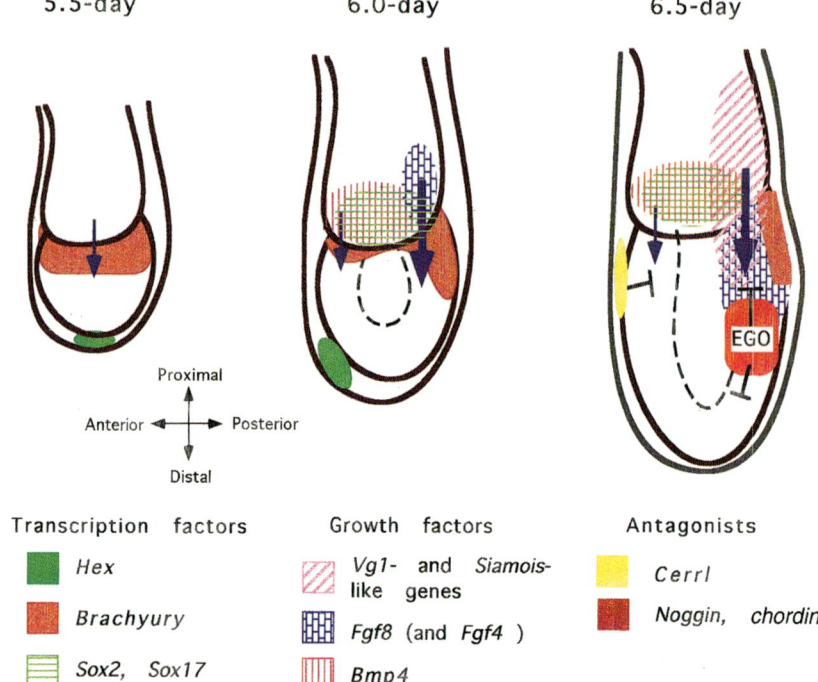

Ch. 3, Fig. 5 A model of the anterior–posterior patterning and the specification of the gastrula organizer in the 5.5- to 6.5-day mouse embryo (see text in Section IV, D). The anterior–posterior and proximal–distal axes of the embryo are shown by the arrows. At 5.5 days, the inductive interaction with the extraembryonic ectoderm activates the *Brachyury (T)* gene in the proximal cell of the pre-cavitated epiblast. The anisotropic growth of the cylindrical embryo leads to greater tissue expansion on the prospective posterior side of the embryo. This is associated with the anterior displacement of the *Hex*-expressing domain in the visceral endoderm. An anterior–posterior polarity is established in the epiblast by 6.0 days. This is brought about by the inductive activity of *BMP4* and *FGF8*, which differs in strength between the anterior and posterior sides of the embryo (indicated by the different boldness of the arrows). This leads to the asymmetrical activity of the *T* gene in the epiblast, which predisposes the mesodermal fate of the posterior epiblast. At the early-streak stage (6.5 days), the extraembryonic ectoderm and the posterior visceral endoderm act as the mouse equivalent of the Nieuwkoop center. It is predicted these tissues will express genes that are homologous to the chick *Vg1* and the *Xenopus Siamois* genes, and they induce the formation of the primitive streak (marked by T expression) and the early gastrula organizer (*Gsc*-expressing EGO, Fig. 1B) in the posterior epiblast. The EGO upon its incorporation into the primitive streak expresses factors that are antagonistic to BMP, WNT, and FGF activity (Piccolo *et al.*, 1996; Zimmerman *et al.*, 1996; Leyns *et al.*, 1997; Wang *et al.*, 1997; Hsu *et al.*, 1998). Together with the anti-BMP activity emanated from the anterior visceral endoderm (Belo *et al.*, 1997; Biben *et al.*, 1998), the gastrula organizer will delimit an area of the epiblast with a diminished BMP influence, resulting in the specification of the ectodermal fate (neural and surface ectoderm). Posteriorly, an interaction of the organizer and the FGFs (FGF4 and FGF8) may specify the mesodermal fates of the posterior epiblast.

endoderm but the embryo will be entirely formed from the mutant ES cells (Nagy et al., 1990, 1993).

The nodal and the SMAD2 proteins are involved in the TGF-β signal transduction pathway. Mutations that lead to a loss of the activity of the *nodal* and *Smad2* genes in the mouse result in an arrest of embryonic development at early gastrulation (Conlon et al., 1991, 1994; Nomura and Li, 1998). Defective craniofacial development is found in the *Smad2*+/− embryos and in embryos heterozygous for *nodal* and *Smad2* genes (Nomura and Li, 1998). *Smad2* gene has also been disrupted by another mutation replacing the first coding exon with a neomycin expression cassette. Although the consequence on the *Smad2* activity is not known, the mutant phenotype suggests a misappropriation of epiblast cells to extraembryonic tissues (Waldrip et al., 1998). A reduction of *Smad2* activity in the primitive endoderm causes the excessive differentiation of the wild type epiblast cells to extraembryonic mesoderm (Waldrip et al., 1998). This finding highlights the critical requirement of *Smad2* activity in the primitive endoderm for proper germ layer differentiation. Development of the *nodal*−/− ES-derived embryo is normal at least up to early organogenesis when wild type extraembryonic tissues and primitive endoderm are present in the chimera. In the chimera that lacks *nodal* activity in the primitive endoderm, severe retardation and loss of rostral craniofacial structures are found in the embryo derived from wild type ES cells (Varlet et al., 1997). *Nodal* activity that is expressed in the primitive endoderm is therefore critical for the patterning of embryonic structures, especially of the anterior region.

An endodermal role in the patterning of forebrain and midbrain structures has also been demonstrated for the *Otx2* mutant. In the presence of wild type tetraploid primitive endoderm that presumably still expresses *Otx2* activity (Acampora et al., 1995), the Otx2−/− chimeric embryo forms neural tissues that display anterior molecular characteristics, but further morphogenesis to forebrain and midbrain is still defective. The mutant Otx2−/− ectoderm also fails to respond to the neural induction when cultured in the presence of wild type mesendoderm (Rhinn et al., 1998). *Otx2* function is therefore required in the primitive endoderm and the ectoderm for proper neural differentiation in the anterior region of the embryo. It is not clear if the mutant phenotype is also caused by the impact of the mutation on the differentiation or the inductive activity of the cranial axial mesoderm.

2. Molecular Regionalization of the Primitive Endoderm

Besides the *nodal*, *Smad2*, and *Otx2* genes, an increasing number of genes have now been found to be expressed in the primitive endoderm of the gastrulating mouse embryo (Beddington and Robertson, 1998; Table III). Many of these genes such as *nodal*, *Gsc*, *Hnf3β*, *Otx2*, and *Lim1* are also expressed in the epiblast population that contains the early gastrula organizer (Table III, Fig. 4). Others such as *Cerrl*, *Hesx1*, *VE1*, *Hex*, *Smad2*, *Hnf4*, *Gata4*, and *Mrg1* are either specific to the primitive endoderm or in germ layer tissues other than the organizer. Of particular significance is the localized expression of some of these genes to the cells in the anterior visceral endoderm that are destined to become the endoderm of the extraembryonic yolk sac (Thomas and Beddington, 1996). The anterior expression of *Hesx1*, *Cerrl*, and *Lim1* are lost in the *Smad2* mutant (Waldrip et al., 1998) whereas *Cerrl* is localized ectopically to the distal part of the most severe *Lim1*−/− and *Otx2*−/− mutant gastrula (Biben et al., 1998; Shawlot et al., 1998). It is postulated that the activity of these genes reveals the existence of an anterior–posterior pattern in the endoderm which may impart upon the epiblast and its germ layer derivatives during gastrulation and early organogenesis. A hint of such interaction is revealed by the dependence of the forebrain ectoderm on the primitive endoderm to maintain appropriate expression of transcription factors encoded by either *Otx2* or *Hesx1/Rpx* (Acampora et al., 1995; Hermesz et al., 1996; Thomas and Beddington, 1996). Null mutation of the *Hesx1/Rpx* gene leads to defects in forebrain development, eye abnormalities, and dysplasia of the pituitary gland which is presaged by the loss of *Six3*, *Pax6*, *Nkx2.1*, and *Shh* activity in the embryonic forebrain. *Hesx1* activity is therefore required for forebrain and pituitary differentiation but it is not clear how it is involved with anterior patterning since the primordia for anterior structures are properly specified in the absence of *Hesx1* activity (Dattani et al., 1998). Mutation of the *cerberus-1* (*Cerrl*) gene that is also localized to the anterior visceral endoderm like the *Hesx1/Rpx* gene does not result in any obvious abnormality in anterior patterning (Shawlot and Behringer, unpublished).

Organizing activity of tissues in the anterior region of the mouse gastrula has been tested by transplantation. Fragments of the anterior epiblast of early-streak and anterior ectoderm of late-streak embryo do not induce any changes in the body plan of the late-streak host embryo (Beddington, 1994; Tam et al., 1997b). Neither the graft of anterior visceral endoderm of the early-streak embryo alone nor that associated with the overlying epiblast tissue displays any inductive or organizing activity, suggesting that these tissues do not act as an organizer. Induction of anterior neural characteristics in the secondary axis can only be achieved by a combined graft of anterior visceral endoderm, anterior epiblast, and EGO (Tam, unpublished). Details of the morphogenetic interactions between these tissues during axis induction are not known, but this preliminary observation raises an intriguing possibility that anterior patterning requires the augmentation of the EGO activity by signals that regulate the competence of the prospective anterior tissues (Koshida et al., 1998). In zebrafish it has been shown that the anteroposterior patterning of the neuraxis also requires the action of anterior ectodermal cells (Houart et al., 1998). These anterior ectodermal cells may be equivalent to the anterior epiblast of the mouse gastrula. Whether the primitive endoderm, the anterior epiblast, and the EGO indeed have synergistic roles in head organization remains to be resolved.

IV. Tissue Patterning and the Specification of the Gastrula Organizer

A. Searching for the Nieuwkoop Signal

Mechanisms that specify the activity and localization of the organizer of vertebrate gastrulae are of fundamental importance to the establishment of the body plan. In lower vertebrates, molecules that specify the early steps in axis determination are mostly maternally supplied and locally sequestered within the egg cytoplasm (Kageura, 1997). In the *Xenopus* zygote, movement of the cortical cytoplasm presumably brings some yet to be identified dorsal determinant from the vegetal region to the prospective dorsal region of the embryo. It is believed that an inductive signal from dorsal vegetal tissues (the Nieuwkoop center) that has inherited these cytoplasmic factors will specify a set of dorsal blastomeres to become the organizer in the dorsal blastoporal lip of the gastrula (Nieuwkoop, 1985; Vodicka and Gerhart, 1995; Brannon and Kimelman, 1996).

Molecules of both the TGF-β and Wnt superfamily have been implicated in the establishment of the Nieuwkoop center in the amphibian embryo. Ectopic expression of various *Wnt* mRNAs in the ventral blastomeres leads to induction of dorsal mesoderm and axis duplication (McMahon and Moon, 1989; Sokol *et al.*, 1991; Fagotto *et al.*, 1997; reviewed by Cadigan and Nusse, 1997). Despite this positive effect of the *Wnt* molecule, no known *Wnt* gene has yet been found whose site and timing of expression are appropriate for its being an inductive factor of the Nieuwkoop center. There is, however, evidence that it is the downstream components of the Wnt signaling cascade that are essential for the formation of the organizer (Harland and Gerhart, 1997). The glycogen synthase kinase (GSK3) present in the *Xenopus* egg phosphorylates and enhances the proteolysis of β-catenin. An inhibition of GSK3 activity by *Dishevelled* (*Dsh*) leads to the stabilization of β-catenin and an increase in its cytosolic level. The cytoplasmic β-catenin forms complexes with members of the LEF1/TCF3 family of HMG box transcription factors and modifies their transcriptional activity. During the midblastula transition, TCF3 in conjunction with β-catenin activates *Siamois* on the dorsal tissue whereas the same factor when not accompanied by a high level of nuclear β-catenin represses *Siamois* expression in the ventral tissues of the embryo (Brannon *et al.*, 1997; Fan and Sokol, 1997), suggesting the localization of β-catenin has a direct effect on *Siamois* activity.

The paired-class homeobox gene *Siamois* is one of the earliest transcription factors to be expressed in the dorsal vegetal cells after the onset of zygotic transcription. It is a potential mediator of maternal dorsal signals and Spemann organizer formation, as revealed by the activation of *gsc* transcription (Lemaire *et al.*, 1995; Carnac *et al.*, 1996). Transcription of *gsc* is also induced by a *Siamois*-related homeobox gene, *Xtwn* (Laurent *et al.*, 1997). *Siamois* activity when expressed ectopically can induce a complete secondary body axis, suggesting that it may be associated with the formation of an ectopic organizer (Lemaire et al., 1995). Activation of *gsc* transcription can also be achieved by the combined action of *Wnt*, activin, and Vgl-like signals (Watabe et al., 1995). Vgl might act in concert with the Wnt signaling pathway to initiate and maintain the expression of genes of the Spemann organizer (Cui et al., 1996). In the chick embryo, ectopic expression of *cVg1* induces simultaneously the formation of a secondary primitive streak (Selerio et al., 1996) and the activation of organizer-specific genes (Shah et al., 1997). The Vgl factor is therefore involved not only with the formation and organization of the primitive streak but also the avian gastrula organizer. Whether separate mechanisms are involved in the formation of primitive streak and the organizer in the chick embryo is presently not known.

On the basis of the expression pattern of *cVg1* and *GSC* in the posterior blastoderm and the ability to induce a new axis by grafting tissues that express these genes, the extraembryonic posterior marginal zone and the Koller's sickle are reputed to be the avian counterpart of the Nieuwkoop center. In the mouse, the rapid growth and the extensive intermingling of cells that occur in the epiblast prior to the initiation of gastrulation seem incompatible with the stable registration of positional information within this tissue. In contrast, the visceral endoderm develops without dispersal of the descendants of individual cells, and this would permit positional information to be preserved during early postimplantation development (Lawson et al., 1986; Lawson and Pedersen, 1987; Gardner and Cockroft, 1988). If an analogy can be drawn between the mouse and the chick, the functional equivalent of the Nieuwkoop center in the mouse will be the extraembryonic ectoderm and the primitive endoderm that is associated with the posterior epiblast of the early gastrula.

B. Insight from Mutational and Transgenic Studies

Targeted mutations of a number of genes that are expressed in the peri-implantation mouse embryos have resulted in the arrest of embryonic development around the time of gastrulation (Table IV). A critical role of the primitive endoderm in the initiation or the progression of gastrulation is portrayed by the amelioration of the mutant phenotype in chimeras comprising wild type extraembryonic tissues and mutant embryonic tissues derived from ES cells lacking the activity of *Hnf4*, *Hdh*, *ActRIB*, *Smad2*, *Smad4*, *nodal*, *Otx2*, *Lim1*, and *Hnf3β* genes (Table IV; Duyao et al., 1995; Duncan et al., 1997; Dragatsis et al., 1998; Dufort et al., 1998; Sirard et al., 1998; Wang et al., 1998). Mutant embryos that have initiated gastrulation but show abnormal differentiation of the mesoderm and ectoderm also display ectopic expression of some organizer genes (Table IV). For instance, *nodal* mutant does not form a primitive streak, which is a predicted phenotype for the loss of a mouse Nieuwkoop center. Consistent with this prediction, *nodal* mutant fails to express *Gsc* (Conlon et al., 1994). In other mutant embryos, however,

3. The Organizer of the Gastrulating Mouse Embryo

the expression of many organizer-specific genes is ectopic rather than absent (Table IV). Further investigation is needed to find out if any organizer has indeed been specified in these mutant embryos to assess the impact of the mutation on the Nieuwkoop center.

The role of Wnt signaling has been studied by misexpressing *Wnt8C* (the chicken orthologue of *Wnt8* gene) in the mouse embryo. This leads to induction of a secondary axis with only posterior characteristics (Popperl *et al.*, 1997), thus implicating that Wnt signaling may influence body patterning. Interestingly, *Wnt8* expression is detected in the epiblast just when the primitive streak is formed (Bouillet *et al.*, 1996). *Wnt8* is therefore a potential candidate factor for the axis specification in the mouse but the impact of *Wnt8* mutation is not yet known. *Wnt3a*, *Wnt5a*, and *Wnt5b* genes are expressed in the mesoderm and ectoderm during, but not before, gastrulation (Moon *et al.*, 1997). In addition, none of the Wnt mutants produced so far have shown defects in the early events of axis formation: *Wnt1* plays a key role in specification of neural cell fate (McMahon and Bradley, 1990), disruption of *Wnt3a* perturbs somites and notochord formation (Takada *et al.*, 1994), and *Wnt7a* is implicated in the establishment of the normal polarity of the mouse limb (Parr and McMahon, 1995). Circumstantial evidence for a role of Wnt signaling in axis development has been provided by the finding of caudal axis duplication in the *fused* mutant. The *fused* gene encodes an anti-Wnt Axin factor which in the amphibian will negatively regulate the axis-inducing effect of Wnt signaling by binding to GSK3 and downregulating β-catenin activity (Hart *et al.*, 1998; Itoh *et al.*, 1998). The loss of Axin activity in the *fused* mutant therefore causes axis duplication similar to that produced by *Wnt* misexpression in *Xenopus* (Zeng *et al.*, 1997). In *Xenopus*, dorsal marginal blastomeres that have acquired organizer activity display nuclear localization of β-catenin (Brannon *et al.*, 1997; Fan and Sokol, 1997). An immunohistochemical study of peri-implantation mouse embryos has not revealed any tissues in which β-catenin is localized in the nucleus (Popperl *et al.*, 1997). This may suggest that either nuclear localization of β-catenin or the activity of this molecule is not a characteristic of the mouse organizer cells. *β-catenin* null mutant shows a disorganization of the epiblast and developmental arrest before gastrulation (Haegel *et al.*, 1995). Because of the early embryonic lethality, it is impossible to discern how the loss of β-catenin activity may affect the formation of the organizer. In *Drosophila*, *Wingless* activates *dishevelled* presumably via *Frizzled* (a putative cell surface Wnt receptor), which in turn inhibits the kinase activity *Zeste-white 3*, leading then to the activation of *Armadillo* (the *Drosophila* homologue of β-catenin). In the mouse, however, the $Dvl1-/-$ mutant does not display any developmental defects (Lijam *et al.*, 1997), probably due to the compensatory activity of other mouse *Dishevelled* genes. Genetic studies in the mouse so far do not implicate any significant role of Wnt signaling in the specification of the mouse gastrula organizer and have not identified any tissues as the source of Nieuwkoop signals.

C. The Role of Extraembryonic Tissues

The discovery of organizing activity in the early gastrula has pinpointed the latest time for specifying the organizer to be the pregastrulation stage of development. Since the emergence of the body plan is a spatial readout of the patterning signal, the anatomical organization of the interacting tissues must be relatively stable to retain the positional information conferred on the germ layers. Spatially defined information in a cell population is only developmentally relevant in the absence of extensive intermingling of cells. During the postimplantation period, extensive intermingling of cells occurs in the inner cell mass until the epithelialization of the epiblast (Gardner and Cockroft, 1998). In contrast, very little cell mixing takes place in the primitive endoderm and the cells maintain a relatively stable neighbor relationship even as they undergo morphogenetic movement during gastrulation (Lawson *et al.*, 1986; Lawson and Pedersen, 1987; Tam and Beddington, 1992; Thomas and Beddington, 1996; Thomas *et al.*, 1998). It is therefore likely that any patterning information that is established in the primitive endoderm will be retained at least until the onset of gastrulation when it will be replaced by epiblast-derived definitive endoderm. Of particular interest is that the primitive endoderm underneath the primitive streak is the most ancient population in the endodermal layer and it is not replaced by epiblast-derived definitive endoderm probably until hindgut formation (Lawson *et al.*, 1986; Lawson and Pedersen, 1987; Tam and Beddington, 1992). It was postulated that their presence may be important for maintaining the activity of the primitive streak.

Regionalized expression of genes in the primitive endoderm in the gastrulating embryo provides the most persuasive evidence for the patterning of this tissue layer (see Section IIIC2). In contrast to the multitude of genes that are expressed in the anterior visceral endoderm, other genes such as *Evx1* and *Fgf8* are expressed in the posterior visceral endoderm of the pre-streak and early-streak embryo (Dush and Martin, 1992; Crossley and Martin, 1995). Mutation of these two genes results in defective gastrulation. $Evx1-/-$ embryo fails to initiate primitive streak formation, whereas $Fgf8-/-$ embryo shows defective migration of the embryonic mesoderm (Spyropoulos and Capecchi, 1994; Meyers *et al.*, 1998). If the failure to form primitive streak can be taken as an indication of the absence of organizer, then activity of the *Evx1* gene may be required for the specification of the organizer. No direct evidence is available regarding the formation of organizer in the $Fgf8-/-$ mutant embryo, but the mutant phenotype suggests that *Fgf8* may be required primarily for the maintenance of primitive streak function. A recent study on the functional role of the *Hnf3β* gene has shown that chimeras made up of wild type extraembryonic tissues and homozygous $Hnf3β-/-$ epiblast cells form normal primitive streak but do not form any node (Dufort *et al.*, 1998). Therefore, the formation of the primitive streak and a morphologically definable node (and perhaps also the early gastrula organizer) may be separate events.

The rescue of the mutant chimeric embryo is not entirely due to the presence of the wild type primitive endoderm. The wild type cells of the host embryo also contribute to the extraembryonic ectoderm and the ectoplacental cone. At least three genes (*Bmp4*, *Sox2*, and *Sox17*) are now known to be expressed in the distal cells of the extraembryonic ectoderm that abuts the epiblast (Waldrip et al., 1998; Lovell-Badge, unpublished; Tam and Steiner, unpublished). The functional role of the SOX transcription factor is not known but a deficiency of BMP4 results in abnormal gastrulation (Winnier et al., 1995). The critical role of BMP signaling is further highlighted by the disruption of gastrulation when components of the signal transduction pathway such as SMAD2, SMAD4, and BMPR are missing (Mishina et al., 1995; Gu et al., 1998; Nomura and Li, 1998; Sirard et al., 1998; Waldrip et al., 1998; reviewed by Kretzschmar and Massague, 1998). In summary, these findings give strong support to the notion that the inductive signals emanating from the extraembryonic ectoderm may be important for patterning the epiblast of the peri-implantation embryo.

D. Emergence of Pattern Asymmetry and Formation of the Organizer

We propose that in the pregastrulation embryo a series of inductive and antagonistic signals are interacting dynamically to bring about the specification of the organizer. The following scenario of tissue interactions (Fig. 5; see color plate) may take place between the extraembryonic ectoderm, the primitive endoderm, and the epiblast and leads to the initial patterning of the epiblast and the formation of the organizer.

In the peri-implantation embryo, the induction by the inner cell mass triggers the proliferation of the polar trophectoderm (Copp, 1979). The inward growth of the polar trophectoderm provides the morphogenetic force that shapes the embryo into a cylindrical structure. Cavitation of the inner cell mass and formation of the epithelial epiblast are then accomplished under the inductive influence of the primitive endoderm (Coucouvanis and Martin, 1995). In the embryo, activities of the *T* and *Hex* genes are polarized respectively to proximal epiblast and distal visceral endoderm and mark the proximal–distal axis of the cylindrical embryo.

The first hint of asymmetry in the transverse plane of the embryo is the displacement of the *Hex* expression domain in the visceral endoderm to one side of the cylindrical embryo and the *T* expression domain to the opposite side. The shift in the *Hex* domain is likely to be due to the anterior displacement of the endoderm (Thomas et al., 1998), brought about by the more rapid growth of the cylindrical embryo on the prospective posterior side.

In the pre-streak embryo (6.0 day), expression of the *T* gene in the proximal epiblast mirrors the expression of the *Bmp4* genes in the distal part of the extraembryonic ectoderm. The expression of the *Cerrl* gene, which encodes a factor with anti-BMP activity (Bouwmeester et al., 1996; Belo et al., 1997; Glinka et al., 1997; Biben et al., 1998), in the anterior visceral endoderm may modulate the activity of the BMP from the extraembryonic ectoderm. This may predispose the anterior epiblast for an ectodermal fate (neural and surface ectoderm). On the posterior side of the embryo, expression of the *Evx1* and *Fgf8* genes in the visceral endoderm, the epiblast, and the distal part of the extraembryonic ectoderm confers a mesodermal fate to the epiblast.

At or just before the early-streak stage embryo (6.5 day), the extraembryonic ectoderm and the posterior visceral endoderm act as the mouse equivalent of the Nieuwkoop center. It is predicted that these tissues will express genes that are orthologous to the chick *Vg1* and the *Xenopus Siamois* genes. They induce the formation of the primitive streak (marked by *T* expression) and the early gastrula organizer which displays *Gsc* and *Hnf3β* activity (*Gsc*-expressing EGO, Fig. 1B) in the posterior epiblast.

V. An Organizer for All Seasons?

In the mouse gastrula, the organizer is first found in the posterior epiblast of the early-streak embryo and later as the node in the late-streak embryo. Both cell populations, like the organizer of other vertebrate gastrulae, can induce a secondary embryonic axis, differentiate to axial mesendoderm, and express a unique set of genes. Lineage analysis of the cells of the early gastrula organizer reveals a genealogical relationship with the cells in the anterior region of the primitive streak of the mid-streak stage embryo and the node of the late-streak embryo. A structure reminiscent of the node and expressing a subset of the organizer genes is still evident in the anterior part of the primitive streak of the early-organogenesis stage embryo (Sulik *et al.*, 1994; Wilson and Beddington, 1996), but its organizing property has not been determined. In the chick embryo, tissue fragments containing the tail bud are able to induce a secondary axis containing neural tissues that express genes characteristic of posterior neural tube (Knezevic *et al.*, 1998). The organizers that are found in embryos of successive developmental stages are therefore different guises of the gastrula organizer. The dynamic change in its cellular and molecular properties underlies the progressive potency to induce a neural axis of more posterior characteristics. Experimental evidence suggests there may be molecular and anatomical compartmentalization of cell populations within the organizer that may display divergent organizing potency in the avian and amphibian embryos. This has not been demonstrated in the mouse organizer, but findings of mutational analysis raise the possibility of separate head and trunk organizer activity. However, we should not disregard the patterning activity of tissues derived from the organizer such as the notochord, the prechordal plate, and the head process mesoderm (Lemaire *et al.*, 1997; Pera and Kessel, 1997). The apparent distinction between head and trunk organizing activity may reside not in the organizer per se but in its derivatives (Shimamura and Robenstein, 1997). Recently, it has been shown that the anterior neural ridge and the isthmus of the brain are capable of directing segment-specific patterning of the brain (Crossley *et al.*, 1996; Lee *et al.*, 1997; Ye *et al.*, 1998; reviewed by Tam *et al.*, 1997a). These secondary organizers are essential in the refinement of the global pattern established by the gastrula organizer in the early embryo.

Functional test of the organizing potency of the early gastrula organizer and the node reveals that neither cell population can induce an axis complete with anterior neural characteristics. However, preliminary results suggest that induction of anterior neural tissues can be achieved by transplanting the early gastrula organizer together with the anterior visceral endoderm and the anterior epiblast of the early-streak embryo. The last two tissues by themselves cannot induce any new axis in the host embryo and therefore do not act like an organizer. The ability to restore the head organizing activity highlights an important aspect of the mode of action of the organizer. The gastrula organizer is fully endowed with the capacity to pattern all parts of the anterior–posterior axis but can only do so by interacting with tissues of the appropriate competence or previously primed to respond to the patterning signal. The potency of the organizer is also constantly modulated by the surrounding germ layer tissues. This may involve the acquisition of permissive information from the anterior tissues in order to express the head organizing activity (Houart *et al.*, 1998; Koshida *et al.*, 1998). In the chick, the Hensen's node initially does not contain any patterning activity for left–right body asymmetry, but it acquires the information from the lateral tissues and then imposes an instructive control on the further elaboration of the laterality of the body plan (Pagán-Westphal and Tabin, 1998).

Acknowledgments

We thank Richard Behringer, Bruce Davidson, and Peter Rowe for reading the manuscript and Richard Behringer, Bill Shawlot, Kirsten Steiner, and Robin Lovell-Badge for allowing us to cite the unpublished results. Our work is supported by the National Health and Medical Research Council (NHMRC) of Australia, Human Frontier of Science Program (with Siew-lan Ang, Richard Behringer, and Hiroshi Sasaki), and Mr. James Fairfax. P.P.L.T. is a NHMRC Principal Research Fellow.

References

Acampora, D., Mazan, S., Lallemand, Y., Avantaggiato, V., Maury, M., Simeone, A., and Brulet, P. (1995). Forebrain and midbrain regions are deleted in *Otx2−/−* mutants due to a defective anterior neurectoderm specification during gastrulation. *Development* **121,** 3279–3290.

Ang, S.-L., and Rossant, J. (1994). *HNF-3β* is essential for node and notochord formation in mouse development. *Cell* **78,** 561–574.

Ang, S.-L., Jin, O., Rhinn, M., Daigle, N., Stevenson, L., and Rossant, J. (1996). A targeted mouse *Otx2* mutation leads to severe defects in gastrulation and formation of axial mesoderm and to deletion of rostral brain. *Development* **122,** 243–252.

Bally-Cuif, L., and Boncinelli, E. (1997). Transcription factors and head formation in vertebrates. *Bioessays* **19,** 127–135.

Barnes, J. D., Crosby, J. L., Jones, C. M., Wright, C. V., and Hogan, B. L. (1994). Embryonic expression of Lim-1, the mouse homolog of *Xenopus* Xlim-1, suggests a role in lateral mesoderm differentiation and neurogenesis. *Dev. Biol.* **161**, 168–178.

Beddington, R. S. P. (1981). An autoradiographic analysis of the potency of embryonic ectoderm in the 8th day postimplantation mouse embryo. *J. Embryol. Exp. Morphol.* **64**, 87–104.

Beddington, R. S. P. (1994). Induction of a second neural axis by the mouse node. *Development* **120**, 603–612.

Beddington, R. S. P., and Robertson, E. J. (1989). An assessment of the developmental potential of embryonic stem cells in the midgestation mouse embryo. *Development* **105**, 733–737.

Beddington, R. S. P., and Robertson, E. J. (1998). Anterior patterning in mouse. *Trends Genet.* **14**, 277–284.

Beddington, R. S. P., Rashbass, P., and Wilson, V. (1992). *Brachyury*—A gene affecting mouse gastrulation and early organogenesis. *Development Supp.* **1992**, 157–165.

Belo, J. A., Bouwmeester, T., Leyns, L., Kertesz, N., Gallo, M., Follettie, M., and De Robertis, E. M. (1997). *Cerberus-like* is a secreted factor with neuralizing activity expressed in the anterior primitive endoderm of the mouse gastrula. *Mech. Dev.* **68**, 45–57.

Biben, C., Stanley, E., Fabri, L., Kotecha, S., Rhinn, M., Drinkwater, C., Lah, M., Wang, C.-C., Nash, A., Hilton, D., Ang, S.-L., Mohun, T., and Harvey, R. P. (1998). Murine cerberus homologue *mCer-1*: A candidate anterior patterning molecule. *Dev. Biol.* **194**, 135–151.

Blum, M., Gaunt, S. J., Cho, K. W. Y., Steinbeisser, H., Blumberg, B., Bittner, D., and De Robertis, E. M. (1992). Gastrulation in the mouse: The role of the homeobox gene *goosecoid*. *Cell* **69**, 1097–1106.

Bouillet, P., Oulad-Abdelghani, M., Ward, S. J., Bronner, S., Chambon, P., and Dolle, P. (1996). A new member of the Wnt gene family, mWnt-8, is expressed during early embryogenesis and is ectopically induced by retinoic acid. *Mech. Dev.* **58**, 141–152.

Bouwmeester, T., and Leyns, L. (1997). Vertebrate head induction by anterior primitive endoderm. *Bioessays* **19**, 855–862.

Bouwmeester, T., Kim, S.-H., Sasai, Y., Lu, B., and De Robertis, E. M. (1996). Cerberus is a head-inducing secreted factor expressed in the anterior endoderm of Spemann's organizer. *Nature* **382**, 595–601.

Brannon, M., and Kimelman, D. (1996). Activation of *Siamois* by the Wnt pathway. *Dev. Biol.* **180**, 344–347.

Brannon, M., Gomperts, M., Sumoy, L., Moon, R. T., and Kimelman, D. (1997). A beta-catenin/XTcf-3 complex binds to the *siamois* promoter to regulate dorsal axis specification in *Xenopus*. *Genes Dev.* **11**, 2359–2370.

Cadigan, K. M., and Nusse, R. (1997). Wnt signaling: A common theme in animal development. *Genes Dev.* **11**, 3286–3305.

Carnac, G., Kodjabachian, L., Gurdon, J. B., and Lemaire, P. (1996). The homeobox gene *Siamois* is a target of the Wnt dorsalisation pathway and triggers organiser activity in the absence of mesoderm. *Development* **122**, 3055–3065.

Catala, M., Teillet, M. A., De Robertis, E. M., and Le Douarin, N. M. (1996). A spinal cord fate map in the avian embryo: While regressing, Hensen's node lays down the notochord and floor plate thus joining the spinal cord lateral walls. *Development* **122**, 2599–2610.

Chiang, C., Litingtung, Y., Lee, E., Young, K. E., Corden, J. L., Westphal, H., and Beachy, P. A. (1996). Cyclopia and defective axial patterning in mice lacking Sonic hedgehog gene function. *Nature* **383**, 407–413.

Cho, K. W. Y., Blumberg, B., Steinbeisser, H., and De Robertis, E. M. (1991). Molecular nature of Spemann's organizer: The role of the *Xenopus* homeobox gene *goosecoid*. *Cell* **67**, 1111–1120.

Collignon, J., Varlet, I., and Robertson, E. J. (1996). Relationship between asymmetric nodal expression and the direction of embryonic turning. *Nature* **381**, 155–158.

Conlon, F. L., Barth, K. S., and Robertson, E. J. (1991). A novel retrovirally induced embryonic lethal mutation in the mouse: Assessment of the developmental fate of embryonic stem cells homozygous for the 413.d proviral integration. *Development* **111,** 969–981.

Conlon, F. L., Lyons, K. M., Takaesu, N., Barth, K. S., Kispert, A., Herrmann, B., and Robertson, E. J. (1994). A primary requirement for *nodal* in the formation and maintenance of the primitive streak in the mouse. *Development* **120,** 1919–1928.

Copp, A. J. (1979). Interactions between inner cell mass and trophectoderm of the mouse blastocyst. II. The fate of the polar trophectoderm. *J. Embryol. Exp. Morphol.* **51,** 109–120.

Coucouvanis, E., and Martin, G. R. (1995). Signals for death and survival: A two-step mechanism for cavitation in the vertebrate embryo. *Cell* **83,** 279–287.

Crossley, P., and Martin, G. (1995). The mouse *Fgf8* gene encodes a family of polypeptides that is expressed in regions that direct outgrowth and patterning in the developing embryo. *Development* **121,** 439–451.

Crossley, P., Martinez, S., and Martin, G. (1996). Midbrain development induced by FGF8 in the chick embryo. *Nature* **380,** 66–68.

Cui, Y. Z., Trian, Q., and Christian, J. L. (1996). Synergistic effects of Vg1 and Wnt signals in the specification of dorsal mesoderm and endoderm. *Dev. Biol.* **180,** 22–34.

Dattani, M. T., Martinez-Barbera, J.-P., Thomas, P. Q., Brickman, J. M., Gupta, R., Martensson, I.-L., Toresson, H., Fox, M., Wales, J. K. H., Hindmarsh, P. C., Krauss, S., Beddington, R. S. P., and Robinson, I. C. A. F. (1998). Mutations in the homeobox gene *HESX1/Hesx1* associated with septo-optic dysplasia in human and mouse. *Nat. Genet.* **19,** 125–133.

Davidson, B. P., Camus, A., and Tam, P. P. L. (1999). Cell fate and lineage specification in the gastrulating mouse embryo. In "Cell Lineage and Fate Determination" (S. A. Moody, Ed.), pp. 491–504. Academic Press, San Diego.

Dias, M. S., and Schoenwolf, G. C. (1990). Formation of ectopic neuroepithelium in chick blastoderms: Age-related capacities for induction and self-differentiation following transplantation of quail Hensen's nodes. *Anat. Rec.* **229,** 437–448.

Doniach, T. (1993). Planar and vertical induction of anteroposterior pattern during the development of the amphibian central nervous system. *J. Neurobiol.* **24,** 1256–1275.

Downs, K. M., and Davies, T. (1993). Staging of gastrulating mouse embryos by morphological landmarks in the dissecting microscope. *Development* **118,** 1255–1266.

Dragatsis, I., Efstratiadis, A., and Zeitlin, S. (1998). Mouse mutant embryos lacking huntingtin are rescued from lethality by wild-type extraembryonic tissues. *Development* **125,** 1529–1539.

Dufort, D., Schwartz, L., Harpal, K., and Rossant, J. (1998). The transcription factor HNF3β is required in visceral endoderm for normal primitive streak morphogenesis. *Development* **125,** 3015–3025.

Duncan, S. A., Nagy, A., and Chan, W. (1997). Murine gastrulation requires *HNF-4* regulated gene expression in the visceral endoderm: Tetraploid rescue of *Hnf-4−/−* embryos. *Development* **124,** 279–287.

Dush, M. K., and Martin, G. R. (1992). Analysis of mouse *Evx* genes: *Evx-1* displays graded expression in the primitive streak. *Dev. Biol.* **151,** 273–287.

Duyao, M. P., Auerbach, A. B., Ryan, A., Persichetti, F., Barnes, G. T., McNeil, S. M., Ge, P., Vonsattel, J. P., Gusella, J. F., Joyner, A. L., and MacDonald, M. E. (1995). Inactivation of the mouse Huntington's disease gene homolog Hdh. *Science* **269,** 407–410.

Fagotto, F., Guger, K., and Grumbiner, B. M. (1997). Induction of the primary dorsalizing center in *Xenopus* by the Wnt/Gsk/beta-catenin signaling pathway, but not by Vg1, activin or noggin. *Development* **124,** 453–460.

Fainsod, A., Deissler, K., Yelin, R., Mrom, K., Epstein, M., Pillemer, G., Steinbesser, H., and Blum, M. (1997). The dorsalizing and neural inducing gene follistatin is an antagonist of BMP-4. *Mech. Dev.* **63,** 39–50.

Fan, M. J., and Sokol, S. Y. (1997). A role for *Siamois* in Spemann organizer formation. *Development* **124,** 2581–2589.

Faust, C., Schumacher, A., Holdener, B., and Magnuson, T. (1995). The *eed* mutation disrupts anterior mesoderm production in mice. *Development* **121,** 273-285.

Filosa, S., Rivera-Pérez, J. A., Gómez, A. P., Gansmuller, A., Sasaki, H., Behringer, R. R., and Ang, S.-L. (1997). *goosecoid* and *HNF-3β* genetically interact to regulate neural tube patterning during mouse embryogenesis. *Development* **124,** 2843-2854.

Gardner, R. L., and Cockroft, D. L. (1998). Complete dissipation of coherent clonal growth occurs before gastrulation in mouse epiblast. *Development* **125,** 2397-2402.

Gasca, S., Hill, D. P., Klingensmith, J., and Rossant, J. (1995). Characterization of a gene trap insertion into a novel gene, *cordon-bleu*, expressed in axial structures of the gastrulating mouse embryo. *Dev. Genet.* **17,** 141-154.

Gerhart, J., Doniach, T., and Stewart, R. (1991). Organizing the *Xenopus* organizer. *In* "Gastrulation: Movements, Patterns, and Molecules" (R. Keller, W. H. Clark, Jr., and F. Griffin, Eds.), pp. 57-77. Plenum, New York.

Gilbert, S., and Saxen, L. (1993). Spemann organizer: Models and molecules. *Mech. Dev.* **41,** 73-89.

Gimlich, R. L., and Cooke, J. (1983). Cell lineage and the induction of second nervous system in amphibian development. *Nature* **306,** 471-473.

Glinka, A., Wu, W., Onichtchouk, D., Blumenstock, C., and Niehrs, C. (1997). Head induction by simultaneous repression of Bmp and Wnt signaling in *Xenopus*. *Nature* **389,** 517-519.

Glinka, A., Wu, W., Delius, H., Monaghan, A. P., Blumenstock, C., and Niehrs, C. (1998). Dickkopf-1 is a member of a new family of secreted proteins and functions in head induction. *Nature* **391,** 357-362.

Greco, T. L., Takada, S., Newhouse, M. M., McMahon, J. A., McMahon, A. P., and Camper, S. A. (1996). Analysis of the *vestigial tail* mutation demonstrates that *Wnt3a* gene dosage regulates mouse axial development. *Genes Dev.* **10,** 313-324.

Gu, Z., Nomura, M., Simpson, B. B., Lei, H., Feijen, A., Van den Eijnden-van Raaij, J., Donahoe, P. K., and Li, E. (1998). The type I activin receptor ActRIB is required for egg cylinder organization and gastrulation in the mouse. *Genes Dev.* **12,** 844-857.

Haegel, H., Larue, L., Ohsugi, M., Fedorov, L., Herrenknecht, K., and Kemler, R. (1995). Lack of β-catenin affects mouse development at gastrulation. *Development* **121,** 3529-3537.

Hansen, C. S., Marion, C. D., Steele, K., George, S., and Smith, W. C. (1997). Direct neural induction and selective inhibition of mesoderm and epidermis inducers by Xnr3. *Development* **124,** 483-492.

Harland, R., and Gerhart, J. (1997). Formation and function of Spemann's organizer. *Annu. Rev. Cell Dev. Biol.* **13,** 611-667.

Harrison, S. M., Dunwoodie, S. L., Arkell, R. M., Lehrach, H., and Beddington, R. S. P. (1995). Isolation of novel tissue-specific genes from cDNA libraries representing the individual tissue constituents of the gastrulating mouse embryo. *Development* **121,** 2479-2489.

Hart, M. J., de los Santos, R., Albert, I. N., Rubinfeld, B., and Polakis, P. (1998). Downregulation of β-catenin by human axin and its association with the APC tumor suppressor, β-catenin and GSK3β. *Curr. Biol.* **8,** 573-581.

He, X., Saint-Jeannet, J. P., Wang, Y., Nathans, J., Dawid, L., and Varmus, H. (1997). A member of the Frizzled protein family mediating axis induction by Wnt-5A. *Science* **275,** 1652-1654.

Heasman, J. (1997). Patterning the *Xenopus* blastula. *Development* **124,** 4179-4191.

Hermesz, E., Mackem, S., and Mahon, K. A. (1996). *Rpx*: A novel anterior-restricted homeobox gene progressively activated in the prechordal plate, anterior neural plate and Rathke's pouch of the mouse embryo. *Development* **122,** 41-52.

Herrmann, B. G. (1991). Expression pattern of the *Brachyury* gene in whole-mount T^{Wis}/T^{Wis} mutant embryos. *Development* **113,** 913-917.

Hogan, B. L. M., Thaller, C., and Eichele, G. (1992). Evidence that Hensen's node is a site of retinoic acid synthesis. *Nature* **359,** 237-241.

Houart, C., Westerfield, M., and Wilson, S. W. (1998). A small population of anterior cells patterns the forebrain during zebrafish gastrulation. *Nature* **291,** 788–792.

Hsu, D. R., Economides, A. N., Wang, X., Eimon, P. M., and Harland, R. (1998). The *Xenopus* dorsalizing factor gremlin identifies a novel family of secreted proteins that antagonize BMP activities. *Mol. Cell* **1,** 673–683.

Itoh, K., Krupnik, V. E., and Sokol, S. Y. (1998). Axis determination in *Xenopus* involves biochemical interactions of axin, glycogen synthase kinase 3 and β-catenin. *Curr. Biol.* **8,** 591–594.

Izpisua-Belmonte, J.-C., De Robertis, E. M., Storey, K. G., and Stern, C. D. (1993). The homeobox gene *goosecoid* and the origin of organizer cells in the early chick blastoderm. *Cell* **74,** 645–659.

Jacobson, A. G., and Tam, P. P. L. (1982). Cephalic neurulation in the mouse embryo analyzed by SEM and morphometry. *Anat. Rec.* **203,** 375–396.

Jones, C. M., and Smith, J. C. (1998). Establishment of a BMP-4 morphogene gradient by long-range inhibition. *Dev. Biol.* **194,** 12–17.

Kageura, H. (1997). Activation of dorsal development by contact between the cortical dorsal determinant and the equatorial core cytoplasm in eggs of *Xenopus laevis*. *Development* **124,** 1543–1551.

Knezevic, V., De Santo, R., and Mackem, S. (1998). Continuing organizer function during chick tail development. *Development* **128,** 1791–1801.

Koshida, S., Shinya, M., Mizuno, T., Kuroiwa, A., and Takeda, H. (1998). Initial anteroposterior pattern of the zebrafish central nervous system is determined by differential competence of the epiblast. *Development* **125,** 1957–1966.

Kretzschmar M, and Massague, J. (1998). SMADs: Mediators and regulators of TGF-β signaling. *Curr. Opin. Genet. Dev.* **8,** 103–111.

Lane, M. C., and Keller, R. E. (1997). Microtubule disruption reveals that Spemann's organizer is subdivided into two domains by a vegetal alignment zone. *Development* **124,** 895–906.

Laurent, M. N., Blitz, I. L., Hashimoto, C., Rothbächer, U., and Cho, W.-Y. (1997). The *Xenopus* homeobox gene *Twin* mediates Wnt induction of *Goosecoid* in establishment of Spemann's organizer. *Development* **124,** 4905–4916.

Lawson, K. A., and Pedersen, R. A. (1987). Cell fate, morphogenetic movement and population kinetics of embryonic endoderm at the time of germ layer formation in the mouse. *Development* **101,** 627–652.

Lawson, K. A., Meneses, J. J., and Pedersen, R. A. (1986). Cell fate and cell lineage in the endoderm of the presomite mouse embryo, studied with an intracellular tracer. *Dev. Biol.* **115,** 325–339.

Lawson, K. A., Meneses, J. J., and Pedersen, R. A. (1991). Clonal analysis of epiblast fate during germ layer formation in the mouse embryo. *Development* **113,** 891–911.

Lee, S. M., Danielian, P. S., Fritzsch, B., and McMahon, A. P. (1997). Evidence that FGF8 signaling from the midbrain–hindbrain junction regulates growth and polarity in the developing midbrain. *Development* **124,** 959–969.

Lemaire, L., and Kessel, M. (1997). Gastrulation and homeobox genes in chick embryos. *Mech. Dev.* **67,** 3–16.

Lemaire, P., Garrent, N., and Gurdon, J. B. (1995). Expression cloning of *Siamois*, a *Xenopus* homeobox gene expressed in dorsal-vegetal cells of blastulae and able to induce a complete secondary axis. *Cell* **81,** 85–94.

Lemaire, L., Röeser, T., Izpisúa-Belmonte, J. C., and Kessel, M. (1997). Segregating expression domains of two *goosecoid* genes during the transition from gastrulation to neurulation in chick embryos. *Development* **124,** 1443–1452.

Leyns, L., Bouwmeester, T., Kim, S.-H., Piccolo, S., and De Robertis, E. M. (1997). Frzb-1 is a secreted antagonist of Wnt signaling expressed in the Spemann organizer. *Cell* **88,** 747–756.

Lijam, N., Paylor, R., McDonald, M. P., Crawley, J. N., Deng, C.-X., Herrup, K., Stevens, K. E., Maccaferri, G., McBain, C. J., Sussman, D. J., and Wynshaw-Boris, A. (1997). Social interaction and sensorimotor gating abnormalities in mice lacking *Dvl1*. *Cell* **90,** 895–905.

Matsuo, I., Kuratani, S., Kimura, C., Takeda, N., and Aizawa, S. (1995). Mouse *Otx2* functions in the formation and patterning of rostral head. *Genes Dev.* **9,** 2646–2658.

Matzuk, M. M., Lu, N., Vogel, H., Sellheyer, K., Roop, D. R., and Bradley, A. (1995). Multiple defects and perinatal death in mice deficient in follistatin. *Nature* **374,** 360–363.

McMahon, A. P., and Bradley, A. (1990). The Wnt-1 (int-1) proto-oncogene is required for development of a large region of the mouse brain. *Cell* **62,** 1073–1085.

McMahon, A. P., and Moon, R. T. (1989). int-1—A protooncogene involved in cell signaling. *Development Suppl.* **107,** 161–167.

McMahon, J. A., Takada, S., Zimmerman, L. B., Fan, C.-M., Harland, R. M., and McMahon, A. P. (1998). Noggin-mediated antagonism of BMP signaling is required for growth and patterning of the neural tube and somite. *Genes Dev.* **12,** 1438–1452.

Melby, A. E., Warga, R. M., and Kimmel, C. B. (1996). Specification of cell fates at the dorsal margin of the zebrafish gastrula. *Development* **122,** 2225–2237.

Meyers, E. N., Lewandoski, M., and Martin, G. R. (1998). An *Fgf8* mutant allelic series generated by Cre- and Flp-mediated recombination. *Nat. Genet.* **18,** 136–141.

Mishina, Y., Suzuki, A., Ueno, N., and Behringer, R. R. (1995). *Bmpr* encodes a type I bone morphogenetic protein receptor that is essential for gastrulation during mouse embryogenesis. *Genes Dev.* **9,** 3027–3037.

Moon, R. T., Brown, J. D., and Torres, M. (1997). WNTs modulate cell fate and behavior during vertebrate development. *Trends Genet.* **13,** 157–162.

Moos, M., Jr., Wang, S., and Krinks, M. (1995). Anti-dorsalizing morphogenetic protein is a novel TGF-beta homolog expressed in the Spemann organizer. *Development* **121,** 4293–4301.

Nagy, A., Gocza, E., Diaz, E. M., Prideaux, V. R., Ivanyi, E., Markkula, M., and Rossant, J. (1990). Embryonic stem cells alone are able to support fetal development in the mouse. *Development* **110,** 815–821.

Nagy, A., Rossant, J., Nagy, R., Abramow-Newerly, W., and Roder, J. C. (1993). Derivation of completely cell-culture-derived mice from early-passage embryonic stem cells. *Proc. Natl. Acad. Sci. USA* **90,** 8424–8428.

Nieuwkoop, P. D. (1985). Inductive interactions in early amphibian development and their general nature. *J. Embryol. Exp. Morphol.* **89,** 333–347.

Nomura, M., and Li, E. (1998). Smad2 role in mesoderm formation, left–right patterning and craniofacial development. *Nature* **393,** 786–790.

Pagán-Westphal, S. M., and Tabin, C. J. (1998). The transfer of left–right positional information during chick embryogenesis. *Cell* **93,** 25–35.

Parr, B. A., and McMahon, A. P. (1995). Dorsalizing signal Wnt-7a required for normal polarity of D–V and A–P axes of mouse limb. *Nature* **374,** 350–353.

Pera, E. M., and Kessel, M. (1997). Patterning of the chick forebrain anlage by the prechordal plate. *Development* **124,** 4153–4162.

Perry, W. L., III, Vasicek, T. J., Lee, J. J., Rossi, J. M., Zeng, L., Zhang, T., Tilghman, S. M., and Costantini, F. (1995). Phenotypic and molecular analysis of a transgenic insertional allele of the mouse *Fused* locus. *Genetics* **141,** 321–332.

Piccolo, S., Sasai, Y., Lu, B., and De Robertis, E. M. (1996). Dorsoventral patterning in *Xenopus*: Inhibition of ventral signals by direct binding of chordin to BMP-4. *Cell* **86,** 589–598.

Poelmann, R. E. (1981). The head process and the formation of the definitive endoderm in the mouse embryo. *Anat. Embryol.* **162,** 41–49.

Pöpperl, H., Schmidt, C., Wilson, V., Hume, C. R., Dodd, J., Krumlauf, R., and Beddington, R. S. P. (1997). Misexpression of Cwnt8C in the mouse induces an ectopic embryonic axis and causes a truncation of the anterior neuroectoderm. *Development* **124,** 2997–3005.

Psychoyos, D., and Stern, C. D. (1996). Restoration of the organizer after radical ablation of Hensen's node and the anterior primitive streak in the chick embryo. *Development* **122,** 3263–3273.

Quinlan, G. A., Williams, E. A., Tan, S.-S., and Tam, P. P. L. (1995). Neuroectodermal fate of epiblast cells in the distal region of the mouse egg cylinder: Implication for body plan organization during early embryogenesis. *Development* **121,** 87–98.

Rashbass, P., Cooke, L. A., Herrmann, B. G., and Beddington, R. S. (1991). A cell autonomous function of Brachyury in T/T embryonic stem cell chimaeras. *Nature* **353,** 348–351.

Rashbass, P., Wilson, V., Rosen, B., and Beddington, R. S. (1994). Alterations in gene expression during mesoderm formation and axial patterning in Brachyury (T) embryos. *Int. J. Dev. Biol.* **38,** 35–44.

Rhinn, M., Dierich, A., Shawlot, W., Behringer, R. R., Le Meur, M., and Ang, S.-L. (1998). Sequential roles for *Otx2* in visceral endoderm and neuroectoderm for forebrain and midbrain induction and specification. *Development* **125,** 845–856.

Sasai, Y., Lu, B., Steinbesser, H., Geissert, D., Gont, L. K., and De Robertis, E. M. (1994). *Xenopus* chordin: A novel dorsalising factor activated by organizer-specific homeobox genes. *Cell* **79,** 779–790.

Sasai, Y., Lu, B., Piccolo, S., and De Robertis, E. M. (1996). Endoderm induction by the organizer-secreted factors chordin and noggin in *Xenopus* animal capps. *EMBO J.* **15,** 4547–4555.

Sasaki, H., and Hogan, B. L. (1993). Differential expression of multiple fork head related genes during gastrulation and axial pattern formation in the mouse embryo. *Development* **118,** 47–59.

Schumacher, A., Faust, C., and Magnuson, T. (1996). Positional cloning of a global regulator of anterior–posterior patterning in mice. *Nature* **383,** 250–253.

Selerio, E. A. P., Connolly, D. J., and Cooke, J. (1996). Early developmental expression and experimental axis determination by the chicken *Vg1* gene. *Curr. Biol.* **6,** 1476–1486.

Selleck, M. A. J., and Stern, C. D. (1991). Fate mapping and cell lineage analysis of Hensen's node in the chick embryo. *Development* **112,** 615–626.

Selleck, M. A., and Stern, C. D. (1992). Commitment of mesoderm cells in Hensen's mode to the chick embryo to notochord and somite. *Development* **114,** 403–415.

Shah, S. B., Skromne, I., Hume, C. R., Kessler, D. S., Lee, K. J., Stern, C. D., and Dodd, J. (1997). Misexpression of chick Vg1 in the marginal zone induces primitive streak formation. *Development* **124,** 5127–5138.

Shawlot, W., and Behringer, R. R. (1995). Requirement for *Lim1* in head-organizing function. *Nature* **374,** 425–430.

Shawlot, W., Min Deng, J., and Behringer, R. R. (1998). Expression of the mouse *cerberus*-related gene, *Cerrl*, suggests a role in anterior neural induction and somitogenesis. *Proc. Natl. Acad. Sci. USA* **95,** 6198–6203.

Shih, J., and Fraser, S. E. (1995). Distribution of tissue progenitors within the shield region of the zebrafish gastrula. *Development* **121,** 2755–2765.

Shih, J., and Fraser, S. E. (1996). Characterizing the zebrafish organizer: Microsurgical analysis at the early-shield stage. *Development* **122,** 1313–1322.

Shimamura, K., and Rubenstein, J. L. R. (1997). Inductive interactions direct early regionalization of the mouse forebrain. *Development* **124,** 2709–2718.

Simeone, A., Acampora, D., Mallamaci, A., Stornaiuolo, A., D'Apice, M. R., Nigro, V., and Boncinelli, E. (1993). A vertebrate gene related to *orthodenticle* contains a homeodomain of the *bicoid* class and demarcates anterior neuroectoderm in the gastrulating mouse embryo. *EMBO J.* **12,** 2735–2747.

Sirard, C., de la Pompa, J. L., Elia, A., Itie, A., Mirtsos, C., Cheung, A., Hahn, S., Wakeham, A., Schwartz, L., Kern, S. E., Rossant, J., and Mak, T. W. (1998). The tumor suppressor gene *Smad4/Dpc4* is required for gastrulation and later for anterior development of the mouse embryo. *Genes Dev.* **12,** 107–119.

Smith, J. C., and Slack, J. M. W. (1983). Dorsalisation and neural induction: Properties of the organizer in *Xenopus laevis*. *J. Embryol. Exp. Morphol.* **78,** 291–298.

Sokol, S., Christian, J. L., Moon, R. T., and Melton, D. A. (1991). Injected Wnt RNA induces a complete body axis in *Xenopus* embryos. *Cell* **67,** 741–752.

Spyropoulos, D., and Capecchi, M. R. (1994). Targeted disruption of the *even-skipped* gene, *evx1*, causes early postimplantation lethality of the mouse conceptus. *Genes Dev.* **8**, 1949–1961.

Storey, K. G., Crossley, J. M., De Robertis, E. M., Norris, W. E., and Stern, C. D. (1992). Neural induction and regionalisation in the chick embryo. *Development* **114**, 729–741.

Storey, K. G., Selleck, M. A. J., and Stern, C. D. (1995). Neural induction and regionalisation by different subpopulations of cells in Hensen's node. *Development* **121**, 417–428.

Streit, A., Lee, K. J., Woo, I., Roberts, C., Jessell, T. M., and Stern, C. D. (1998). Chordin regulates primitive streak development and the stability of induced neural cells, but is not sufficient for neural induction in the chick embryo. *Development* **125**, 507–519.

Sulik, K., DeHart, D. B., Inagaki, T., Carson, J. L. M., Vrablic, T., Gesteland, K., and Schoenwolf, G. C. (1994). Morphogenesis of the murine node and notochordal plate. *Dev. Dyn.* **201**, 260–278.

Taira, M., Jamrich, M., Good, P. J., and Dawid, I. B. (1992). The LIM domain-containing homeobox gene *Xlim-1* is expressed specifically in the organizer region of *Xenopus* gastrula embryos. *Genes Dev.* **6**, 356–366.

Takada, S., Stark, K. L., Shea, M. J., Vassileva, G., McMahon, J. A., and McMahon, A. P. (1994). Wnt-3a regulates somite and tail bud development in the mouse embryo. *Genes Dev.* **8**, 174–189.

Tam, P. P. L. (1989). Regionalisation of the mouse embryonic ectoderm: Allocation of prospective ectodermal tissues during gastrulation. *Development* **101**, 55–67.

Tam, P. P. L., and Beddington, R. S. P. (1987). The formation of mesodermal tissues in the mouse embryo during gastrulation and early organogenesis. *Development* **99**, 109–126.

Tam, P. P. L., and Beddington, R. S. P. (1992). Establishment and organization of germ layers in the gastrulating mouse embryo. In "Postimplantation Development in the Mouse," Ciba Found. Symp. Vol. 165, pp. 27–49. Wiley, Chichester.

Tam, P. P. L., and Behringer, R. R. (1997). Mouse gastrulation: The formation of a mammalian body plan. *Mech. Dev.* **68**, 3–25.

Tam, P. P. L., and Quinlan, G. A. (1996). Mapping vertebrate embryos. *Curr. Biol.* **16**, 106–108.

Tam, P. P. L., Quinlan, G. A., and Trainor, P. A. (1997a). The patterning of progenitor tissues for the cranial region of the mouse embryo during gastrulation and early organogenesis. *Adv. Dev. Biol.* **5**, 137–200.

Tam, P. P. L., Steiner, K. A., Zhou, S. X., and Quinlan, G. A. (1997b). Lineage and functional analyses of the mouse organizer. *Cold Spring Harbor Symp. Quant. Biol.* **62**, 135–144.

Thomas, P., and Beddington, R. S. P. (1996). Anterior primitive endoderm may be responsible for patterning the anterior neural plate in the mouse embryo. *Curr. Biol.* **6**, 1487–1496.

Thomas, P. Q., and Rathjen, P. D. (1992). *HES-1*, a novel homeobox gene expressed by murine embryonic stem cells, identifies a new class of homeobox genes. *Nucleic Acids Res.* **20**, 5840.

Thomas, P. Q., Johnson, B. V., Rathjen, J., and Rathjen, P. D. (1995). Sequence, genomic organization, and expression of the novel homeobox gene, *Hesx1*. *J. Biol. Chem.* **270**, 3869–3875.

Thomas, P. Q., Brown, A., and Beddington, R. S. P. (1998). *Hex*: A homeobox gene revealing peri-implantation asymmetry in the mouse embryo and an early transient marker of endothelial cell precursors. *Development* **125**, 85–94.

Thomsen, G. H. (1997). Antagonism within and around the organizer: BMP inhibitors in vertebrate body patterning. *Trends Genet.* **13**, 209–211.

Tuckett, F., and Morriss-Kay, G. M. (1985). The kinetic behaviour of the cranial neural epithelium during neurulation in the rat. *J. Embryol. Exp. Morphol.* **85**, 111–119.

Varlet, I., Collingnon, J., and Robertson, E. J. (1997). *nodal* expression in the primitive endoderm is required for specification of the anterior axis during mouse gastrulation. *Development* **124**, 1033–1044.

Vodicka, M. A., and Gerhart, J. C. (1995). Blastomere derivation and domains of gene expression in the Spemann Organizer of *Xenopus laevis*. *Development* **121**, 3505–3518.

Waldrip, W. R., Bikoff, E. K., Hoodless, P. A., Wrana, J. L., and Robertson, E. J. (1998). *Smad2* sig-

naling in extraembryonic tissues determines anterior–posterior polarity of the early mouse embryo. *Cell* **92,** 797–808.
Wang, S., Krinks, M., Lin, K., Luyten, F. P., and Moos, M. (1997). *Frzb,* a secreted protein expressed in the Spemann organizer, binds and inhibits wnt-8. *Cell* **88,** 757–766.
Wang, X., Li, C., Xu, X., and Deng, C. (1998). The tumor suppressor SMAD4/DPC4 is essential for epiblast proliferation and mesoderm induction in mice. *Proc. Natl. Acad. Sci. USA* **95,** 3667–3672.
Watabe, T., Kim, S., Candia, A., Rothbächer, U., Hashimoto, C., Inoue, K., and Cho, K. W. Y. (1995). Molecular mechanisms of Spemann's organizer formation: Conserved growth factor synergy between *Xenopus* and mouse. *Genes Dev.* **9,** 3038–3050.
Weinstein, D. C., Ruiz i Altaba, A., Chen, W. S., Hoodless, P., Prezioso, V. R., Jessell, T. M., and Darnell, J. E., Jr. (1994). The winged-helix transcription factor HNF-3 beta is required for notochord development in the mouse embryo. *Cell* **78,** 575–588.
Wilson, V., and Beddington, R. S. P. (1996). Cell fate and morphogenetic movement in the late mouse primitive streak. *Mech. Dev.* **55,** 79–90.
Wilson, V., and Beddington, R. S. P. (1997). Expression of T protein in the primitive streak is necessary and sufficient for posterior mesoderm movement and somite differentiation. *Dev. Biol.* **192,** 45–58.
Wilson, V., Manson, L., Skarnes, W. C., and Beddington, R. S. P. (1995). The *T* gene is necessary for normal mesodermal morphogenetic cell movements during gastrulation. *Development* **121,** 877–886.
Winnier, G., Blessing, M., Labosky, P. A., and Hogan, B. L. M. (1995). Bone morphogenetic protein-4 is required for mesoderm formation and patterning in the mouse. *Genes Dev.* **9,** 2105–2116.
Ye, W., Shimamura, K., Rubenstein, J. L. R., Hynes, M. A., and Rosenthal, A. (1998). FGF and Shh signals control dopaminergic and sertonergic cell fate in the avian neural plate. *Cell* **93,** 755–766.
Yuan, S., and Schoenwolf, G. C. (1998). *De novo* induction of the organizer and formation of the primitive streak in an experimental model of notochord reconstitution in avian embryos. *Development* **125,** 201–213.
Yuan, S., Darnell, D. K., and Schoenwolf, G. C. (1995a). Mesodermal patterning during avian gastrulation and neurulation: Experimental induction of notochord from non-notochordal precursor cells. *Dev. Genet.* **17,** 38–54.
Yuan, S., Darnell, D. K., and Schoenwolf, G. C. (1995b). Identification of inducing, responding, and suppressing regions in an experimental model of notochord formation in avian embryos. *Dev. Biol.* **172,** 567–584.
Zeng, L., Fagotto, F., Zhang, T., Hsu, W., Vasicek, T. J., Perry, W. L., III, Lee, J. L., Tilghman, S. M., Gumbiner, B. M., and Costantini, F. (1997). The mouse *Fused* locus encodes axin, an inhibitor of the Wnt signaling pathway that regulates embryonic axis formation. *Cell* **90,** 181–192.
Zhou, X., Sasaki, H., Lowe, L., Hogan, B. L., and Kuehn, M. R. (1993). Nodal is a novel TGF-beta-like gene expressed in the mouse node during gastrulation. *Nature* **361,** 543–547.
Zimmerman, L. B., De Jesús-Escobar, J. M., and Harland, R. M. (1996). The Spemann organizer signal noggin binds and inactivates bone morphogenetic protein 4. *Cell* **86,** 599–606.
Zoltewicz, J. S., and Gerhart, J. C. (1997). The Spemann organizer of *Xenopus* is patterned along its anterior–posterior axis at the earliest gastrula stage. *Dev. Biol.* **192,** 482–491.

4

Molecular Genetics of Gynoecium Development in *Arabidopsis*

John L. Bowman,* Stuart F. Baum,* Yuval Eshed,*
Joanna Putterill,[†] and John Alvarez[‡]
*Section of Plant Biology
University of California, Davis
Davis, California 95616

[†]School of Biological Sciences
University of Auckland
Auckland, New Zealand

[‡]Department of Biological Sciences
Monash University
Clayton, Melbourne
Victoria 3168, Australia

I. Introduction
  A. Carpel Functions: Protection, Pollination, and Dispersion
  B. Evolutionary Origins of the Carpel
  C. Aims of This Review
II. Gynoecium Structure and Development in *Arabidopsis*
  A. Pattern Elements and Tissue Types of the Carpel
  B. Gynoecium Development
  C. Origins of Carpel Tissue Types
III. Molecular Genetics of Carpel Development
  A. The Floral Homeotic Mutants of *Arabidopsis*
  B. *AGAMOUS* Is a Key Regulatory Gene Specifying Carpel Identity
  C. *SPATULA* Directs Differentiation of Carpel Tissues Independently of *AGAMOUS*
  D. *CRABS CLAW* Controls Several Aspects of Carpel Differentiation
  E. *ETTIN* May Specify or Respond to Regional Identity in the Carpel Primordium
  F. *FRUITFULL* Controls Valve Maturation
  G. *AGL1* and *AGL5* Are Involved in Defining Tissues at the Valve–Replum Boundary
  H. Genes with Pleiotropic Mutant Phenotypes
IV. Lessons from Analyses of Gene Expression Patterns
  A. How Many Genes?
  B. Gene Expression Patterns May Delineate Compartments in Developing Carpels
V. Conclusions
  A. Pattern Formation in the Carpel
  B. Carpel Fusion
  C. Evolutionary Origins of the Carpel
  References

Carpels are the ovule-bearing structural units in angiosperms. In *Arabidopsis,* the specification of carpel identity is achieved by at least two separate pathways: a pathway mediated by the C class gene *AG* and an *AG*-independent pathway. Both pathways are negatively regulated by A class genes. Two genes, *SPT* and *CRC,* can promote differentiation of carpel tissue independently of *AG* and are thus components of the *AG*-independent pathway. *CRC* and *SPT* appear to act in a redundant manner to promote the differentiation of subsets of carpel tissues. The carpel primordium is subdivided into regional domains, both medial versus lateral and abaxial versus adaxial. Based on morphological and gene expression analyses, it appears likely that these domains define developmental compartments. The medial domain appears fated to differentiate into the marginal tissue types of the carpel (septum with transmitting tract and placenta with ovules), whereas the lateral domain gives rise to the ovary walls. The expression of *ETT* defines the abaxial domain, and this gene is involved in the abaxial–adaxial and, possibly, the apical–basal patterning of tissues in the carpel. Once regional domains have been established, the differentiation of tissue and cell types occurs. The MADS-box genes *FUL* and *AGL1/5* are involved in the differentiation of specific tissue types in the valves and valve margins. Thus, the genes identified can be arranged in a functional hierarchy: specification of carpel identity, patterning of the carpel primordium and directing the differentiation of the specialized tissues of the carpel. © 1999 Academic Press.

I. Introduction

One of the key defining features of angiosperms is the carpel, whose primary function is to house the ovules. Indeed the term angiosperm is from the Greek, "seeds within a capsule." Carpels also mediate the interactions between the male and female gametes, providing opportunity for discrimination between potential male gametes. In most species carpels develop directly into fruits which provide protection for the enclosed developing seeds and effect seed dispersal by a variety of mechanisms. It can be argued that these features of the carpel have been a major driving force in evolution, allowing angiosperms to dominate the vegetation in most ecosystems today.

A. Carpel Functions: Protection, Pollination, and Dispersion

The overall structure of the carpel reflects its primary functions: to enclose and protect the developing ovules, to mediate pollination, and to develop into a fruit to be used as a vector for seed dispersal.

In angiosperms carpels are the ovule-bearing structural units, and they are found in the center of the flower, internally to the sepals, petals, and stamens, if they are present. In some families of flowering plants there is more than one carpel per flower, whereas in other families they occur singly. The term gynoecium refers to the carpels of the flower collectively. In species with multiple carpels, they may be united into a single structure (syncarpous gynoecium) or occur as numerous

4. Arabidopsis Carpel Development

solitary structures (apocarpous gynoecium). In this review we will use the term carpel when referring to a single ovule bearing structural unit, and the term gynoecium when referring to the carpels of the flower collectively.

Two types of fusion are recognized in the ontogeny of carpels (Verbeke, 1992). Congenital, or phylogenetic, fusion refers to a compound structure that is formed from a single primordium but is thought to be composed of multiple parts in a phylogenetic sense. Postgenital, or ontogenetic, fusion involves surfaces that have already developed as individual entities, with the contacting epidermal cells often redifferentiating into parenchyma cells. In plants, where cell migrations do not occur during development, inner morphological spaces, or cavities, must be formed either by a postgenital fusion event or by programmed cell death of subepidermal cells. In the case of carpels, the cavity in which the ovules develop is formed by a postgenital fusion event. In a phylogenetic sense, varying degrees of fusion are observed, with most angiosperms exhibiting complete postgenital fusion of the carpels but with some of the lower dicots having ovaries in which only partial fusion occurs (e.g., Igersheim and Endress, 1997). In this sense, carpels are unique among plant organs for forming, via fusion, a cavity protected from the environment. The carpels comprising syncarpous gynoecia may be united by congenital, postgenital, or a combination of congenital and postgenital fusion events. In this case, the carpels are usually fused along their edges, or margins, and the margins of fusion representing the boundaries between the fused carpels may be obscured by subsequent evolutionary modifications in many extant species.

Due to their enclosure of the ovules, carpels provide a selective barrier to fertilization at the interspecific and, in some cases, the intraspecific levels. The distal tips of the carpels, or in some species the margins of the fused carpels, are specialized for the reception of pollen grains and the conduction of growing pollen tubes to the ovary. The receptive surface at the distal tips of the carpels is the stigma. The stigma, which is often composed of papillate cells, usually secretes substances that stimulate the germination of intraspecific pollen grains. The region between the stigma and the ovary is defined as the style, the length of which varies enormously between species (from essentially nonexistent to many centimeters). The interior regions of the style are often filled with a tissue specialized for the growth of pollen tubes. This tissue, called the transmitting tract, begins at the stigmatic surface, extends the length of the style, and often continues into the ovary, thus providing a conducting path for pollen tubes from the site of pollen germination to the ovules. Species-specific (and in some cases self versus nonself) discrimination of pollen is usually effected either at the level of the stigma or in the transmitting tract tissue (Nasrallah and Nasrallah, 1993; Newbigin *et al.*, 1993).

Another function of the carpel is to differentiate as a fruit, whose primary functions are to protect the developing seed(s) and to aid in their dispersal via wind, water, and animal vectors. In many cases, after fertilization, the ovaries of the carpels differentiate directly into the fruit, whereas in others, additional parts of the flower such as the receptacle also contribute to the formation of the fruit. For the purposes of this review, we will consider carpel development to end at fertilization, at which point fruit development commences, and for the most part, our discussion will be confined to those developmental events that occur during carpel development.

B. Evolutionary Origins of the Carpel

The evolutionary origin of the carpel, and its homologies with the reproductive structures of other seed plants, are presently enigmatic. Several evolutionary scenarios have been proposed and many of these invoke the idea that the carpel has its origins in an ancestral foliar organ (for reviews, see Friis and Endress (1990) and Doyle (1994, 1996)). In one scenario, an ancestral leaflike sporophyll, bearing ovules on the adaxial surface of the leaflike organ itself, would fold conduplicately such that its margins would be adjacent to one another. Subsequent fusion along the apposed margins would result in a structure similar to a solitary conduplicate carpel. Conduplicate carpels are considered to be primitive and have been hypothesized to be ancestral to at least some of the more advanced carpels of modern angiosperms (Bailey and Swamy, 1951; Cronquist, 1988). An alternative view holds that ascidate carpel morphology is basal in the angiosperms and that the carpel walls are derived from an ancestral bractlike organ which subtended a reduced shoot system bearing ovules (Taylor, 1991). In either case, the carpels and most of their associated tissues can be considered to be foliar in nature.

C. Aims of This Review

Despite the enormous variety in carpel morphology (for a review, see Weberling (1989)), most carpels share the same structural features: an apical stigmatic surface, a basal ovary enclosing ovules, and a style with transmitting tract tissue connecting the stigma to the ovary. Since the carpels of *Arabidopsis thaliana* (Brassicaceae) are typical of the angiosperms as a whole, this review will focus on the structure and development of the *Arabidopsis* carpel and those genes that have been shown to be controlling aspects of its specification, growth, and differentiation. In considering what can be addressed in this review on carpel development, a few concepts must be kept in mind based on the premise that carpels have their evolutionary origins in an ancestral leaflike sporophyll.

First, what makes carpels different from leaves? Given their likely evolutionary relationship, there should exist a set of genes whose purpose is to modify organs from a leaflike ground state into a carpel. Loss of the function of such genes would result in leaflike organs developing in the place of carpels, or given the likely partial redundancy of such genes, organs with characteristics of both leaves and carpels developing in the place of carpels.

Second, is regional specification within developing organs governed by common mechanisms? Given their evolutionary relationship, fundamental events of leaf and carpel development are likely to be conserved or at least highly similar. For instance, leaves and carpels, and indeed most organs produced on the flanks of the apical or floral meristems, exhibit conspicuous polarities along both their proximal–distal (basal–apical) and abaxial–adaxial axes (Fig. 1; see color plate). Thus, how an organ primordium is partitioned into regions with distinct developmental fates is a common problem, and assuming carpels evolved from leaflike organs, they are likely to share much of the genetic machinery that controls these basic pattern elements. However, because carpels are more complex than leaves, there may also be carpel-specific developmental pathways that augment the common ones. Mutations in genes of common pathways would likely affect the development of most organs produced by the apical and flower meristems, whereas mutations in organ-specific pathways would be, by nature, more limited in their phenotypic defects.

Third, once a primordium has been specified to develop into a carpel, how is this information translated into the differentiation of the complex tissue types of the carpel? This question could be rephrased in the context of the first two questions: once a primordium acquires a fate of carpel identity and is partitioned into regional domains, what directs the differentiation of specific tissue types? In a simplistic scenario, mutations in genes controlling such aspects of carpel differentiation would either lack specific tissue types, exhibit homeotic conversions of tissue types, or exhibit rather more nebulous mutant phenotypes. In summarizing, it must be noted that these developmental processes need not be independent of one another and that a single gene could be involved in multiple processes.

II. Gynoecium Structure and Development in *Arabidopsis*

The superior gynoecium of *Arabidopsis* is typical of the Brassicaceae; it is composed of two congenitally fused carpels creating a single ovary, which is topped with a postgenitally fused short solid style and stigmatic papillae (Figs. 1 and 2, see color plate; Hill and Lord, 1989; Smyth *et al.*, 1990). Below the ovary is a short internode, referred to as the gynophore, that connects the gynoecium to the floral receptacle. A postgenitally fused septum divides the ovary into two locules, with parietal placentae developing at the adaxial margins of fusion of the two carpels (Figs. 1, 2, and 3E; see color plate). Thus, the *Arabidopsis* gynoecium provides a system to examine the specification of tissue types within an organ as well as congenital and postgenital fusion events.

A. Pattern Elements and Tissue Types of the Carpel

The gynoecium is composed of a number of apical–basal and abaxial–adaxial pattern elements (Figs. 1–3; see color plate). By a pattern element we really mean a tissue type or group of associated tissue types that develop in a stereotypical position within the organ. In the apical–basal axis, the mature gynoecium is composed of four distinct pattern elements: from apex to base, the stigma, the style, the ovary, and the gynophore (Figs. 1 and 2; Sessions and Zambryski, 1995). The stigma, style, and gynophore are essentially radially symmetrical with respect to tissue types. The stigmatic surface is composed of a uniform single cell layer and as such does not display any readily apparent polarity. The style is solid, with the central region composed of transmitting tract tissue, giving the style an abaxial–adaxial polarity of tissue types (Figs. 1 and 3D). The gynophore is a stemlike structure connecting the base of the gynoecium to the floral receptacle; it displays radial symmetry similar to that of other stems (Figs. 1 and 3F; see color plate).

The ovary is not radially symmetrical but instead displays bilateral symmetry in addition to an adaxial–abaxial polarity (Figs. 1 and 3E; see color plate). The ovary is a bisected cylinder, with the septum being the plane of bilateral symmetry, and, for simplicity, can be subdivided into four major tissues: the placentae and ovules, the septum, the valves, and a region we have designated the abaxial replum (Figs. 1 and 3E; see color plate). The replum, as defined by Weberling (1989), is the robust housing of the septum and placenta to which the valves are attached. (In the Brassicaceae, fruits dehisce along the margin of fusion between the valves and the replum, with the valves detaching from the replum, allowing seed dispersal.) In terms of tissue types presented in Fig. 1, the abaxial replum, septum, and placentae with ovules comprise the replum as defined by Weberling (1989). In the mature gynoecium, the four strands of placentae are positioned axially along the false septum in an adaxial position near the margins of fusion of the two carpels. Both the septum and parietal placentae with accompanying ovules can be considered adaxial tissues as they develop from the internal margins of the carpels. The ovary wall is divided into two distinct domains: the valves which comprise most of the ovary wall and the abaxial replum at the margins of fusion of the two carpels [Figs. 1, 3E (see color plate), and 4B]. The abaxial replum develops along the external margins of fusion and is thus an abaxial tissue type. The valves, which are derived from the central regions of the carpels, are composed of both adaxial and abaxial tissues. Thus, the mature gynoecium exhibits apical–basal, abaxial–adaxial, and radial pattern elements which are reflected in the tissue and cell types which develop in a position-dependent manner. To provide a baseline for the analysis of mutant phenotypes, the morphology and development of each of these pattern elements and tissue types are described in more detail in the following subsections based on previous observations as well as our own observations (Bowman *et al.*, 1989; Hill and Lord, 1989; Okada *et al.*, 1989; Smyth *et al.*, 1990; Sessions and Zambryski, 1995; Sessions, 1997; Gu *et al.*, 1998; Lennon *et al.*, 1998). We begin with a description of the tissue and cell types observed in the mature carpel and then discuss the differentiation of the tissues during gynoecium development.

1. Stigma

In *Arabidopsis* the stigma consists of a single layer of epidermal cells at the apical tip of the gynoecium (Figs. 2 and 4A; Elleman *et al.*, 1992; Kandasamy *et al.*, 1994; Webb, 1994; Sessions and Zambryski, 1995). It is composed of about 150 elongate bulbous cells approximately 50 μm in length, which are swollen toward their base. These cells are specialized for pollen germination and hydration, possibly by virtue of their unique extracellular cuticle. The stigmatic surface is dry, which is typical for species of the Brassicaceae. Pollen tubes of germinated pollen grains penetrate the cuticle of the stigmatic papilla cells and grow inside the cellulose–pectin layer of the cell wall. The pollen tubes exit the base of the papillar cells and enter the central transmitting tract tissue of the underlying style. Thus, the stigmatic papillae can be considered as the beginning of the transmitting tract.

2. Style

The style is defined as that region between the stigmatic papillae and the ovary. In the *Arabidopsis* gynoecium, the style is a couple hundred micrometers in height (Fig. 4A; Sessions and Zambryski, 1995). It is a radially symmetric structure composed of a central region of transmitting tract tissue, surrounded by chlorenchyma tissue, and a distinctive epidermis [Figs. 3D (see color plate), 4A, and 4D]. The central transmitting tissue occupies approximately one third of the style diameter and is composed of axially elongated cells (Figs. 2 and 3D; see color plate). Pollen tubes grow intercellularly through this region, the cells of which are characterized by an abundance of extracellular polysaccharides that can be visualized by alcian blue staining (Figs. 3D and 3E; see color plate). In the mature gynoecium, there is no conspicuous evidence of a suture reflecting the site of the postgenital fusion that occurs at the center of the style. Xylem and phloem tissues are embedded in the chlorenchyma tissue and are connected to the medial vascular bundles (Figs. 5A and 5B; see color plate). The epidermis of the style consists of rectangular cells which have distinctive epicuticular thickenings, and these cells are interspersed with stomata (Fig. 4D).

3. Ovary Wall

The ovary wall at maturity can be divided into two distinct regions, the valves and the abaxial regions of the replum (Figs. 1 and 3E; see color plate). We will treat the valves and abaxial replum separately here because they appear to be distinct in terms of gene expression from the earliest stages of carpel development (see later).

At maturity, the valves consist of six cell layers comprising four distinct cell types (Figs. 3A, 3C, and 3E; see color plate; Sessions and Zambryski, 1995; Gu *et al.*, 1998). The six cell layers are, from the outside to the inside, the abaxial epi-

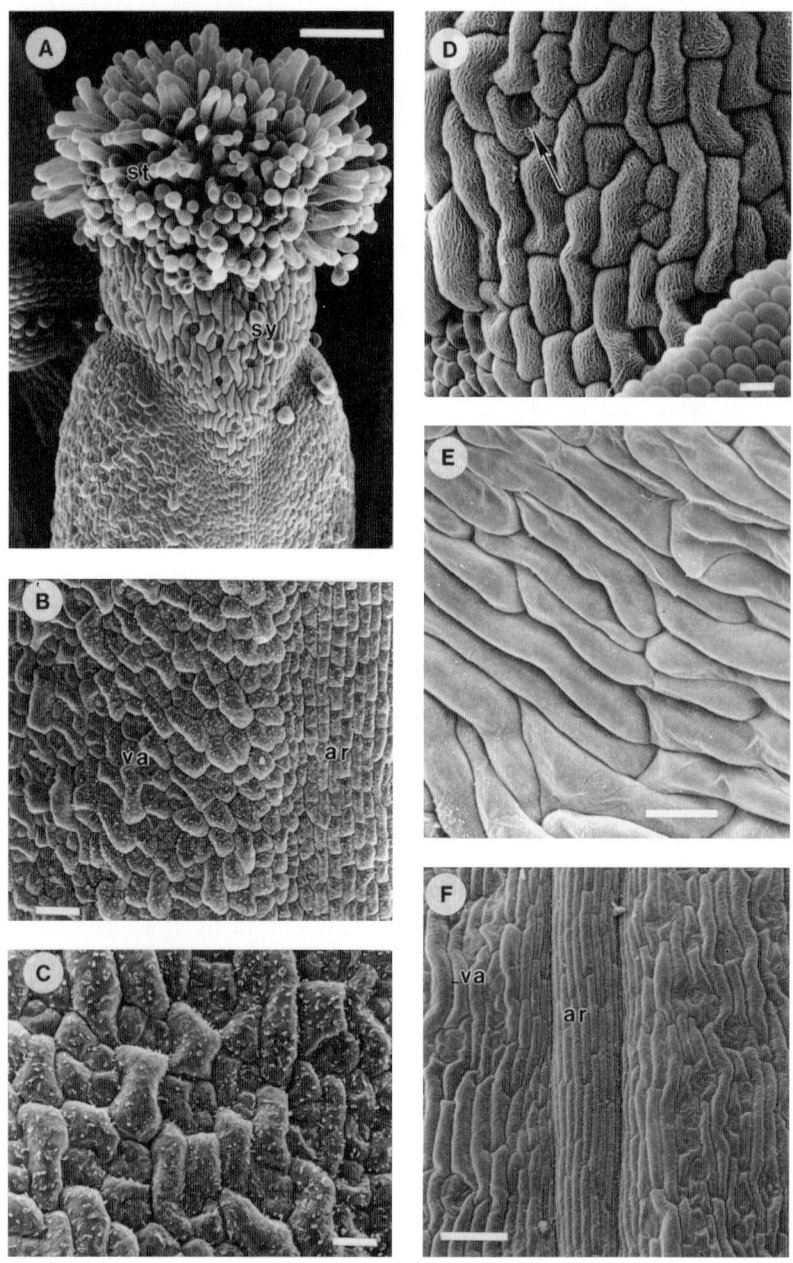

Fig. 4 Epidermal cell morphology of mature wild-type carpels: (A) apical portion of carpel; (B) detail of valve and abaxial replum; (C) close-up of B; (D) detail of style epidermis; (E) septum epidermis postfertilization; (F) detail of valve and abaxial replum postfertilization. ar, Abaxial replum; st, stigmatic papillae; sy, style; va, valve; arrow, stomate. Bars = 100 μm in A, 50 μm in F, 20 μm in B and E, and 10 μm in C and D. From Bowman, 1994.

dermis, three layers of chlorenchyma, a distinct subepidermal adaxial layer, and the adaxial epidermis. The only exception to this pattern occurs at the positions where the lateral vascular bundles develop. The lateral vascular bundles develop near the position of the innermost chlorenchyma layer (Fig. 3E; see color plate). At maturity, the abaxial epidermis consists of vertical files of cells (Figs. 3A and 3C; see color plate) whose outer surface is coated with a characteristic wax layer (Fig. 4B). Interspersed in the abaxial epidermis are stomata, with one stomate for every 4–5 surrounding cells (Fig. 4C). The epidermal cells markedly elongate axially during fruit development (Fig. 4F). The chlorenchyma tissue is composed of cells that are relatively small and isodiametric (Figs. 3A, C, and E; see color plate). The cells of the subepidermal adaxial layer are elongated in the axial dimension, while being relatively small in cross section (Figs. 3A, C, and E; see color plate). The adaxial epidermis is composed of cells that are elongate in the lateral plane but are relatively short in the axial plane, the converse of the subepidermal layer (Figs. 3A, C, and E; see color plate). One interesting question is whether this pattern of cell layers is developmentally related to the pattern observed in *Arabidopsis* leaves, which also consist of six cell layers: abaxial epidermis, three layers of spongy mesophyll, palisade mesophyll, and adaxial epidermis. This is in contrast to sepals, which do not have differentiated mesophyll layers.

Throughout most of their development the cells comprising the abaxial region of the replum are morphologically similar to those of the valves; however, at maturity, the cells of the abaxial replum are much smaller than those of the corresponding layer of the valve, likely due to a lack of cell expansion (Fig. 5D; see color plate). The abaxial epidermis of the replum consists of small rectangular block shaped cells (with no interspersed stomata) whose morphology is easily distinguished from that of the adjacent cells of the valves (Fig. 4B). The mesophyll cells of the abaxial replum are also much smaller than those of the valve (Figs. 3E and 5D; see color plate). The medial vascular bundle develops in a similar radial position as the lateral vascular bundle. Cells interior to the medial vascular bundle are considered to comprise the septum.

4. Septum

A false septum divides the mature ovary into two lateral locules (Figs. 2 and 3E; see color plate). The septum is considered to be false because it does not correspond to the lateral walls of the fused carpels but rather develops as an extension of their margins. The septum is formed by a postgenital fusion event (Figs. 6C and D, see color plate; Hill and Lord, 1989; Sessions and Zambryski, 1995). As in the case of the style, at maturity little evidence of a suture is visible at the site of fusion (Figs. 3E and 5D; see color plate), although it is likely to be similar to other Brassicaceae (e.g., *Capsella bursa-pastoris*), in which portions of the suture could be detected by transmission electron microscopy (Boeke, 1971). The epidermis of the septum consists of cuboidal cells which elongate axially postfertilization (Fig.

4E). Situated at the point of fusion between adjoining adaxial epidermis layers is the transmitting tract. At carpel maturity, the transmitting tract is composed of 4–5 cell layers interior to the point of fusion, including the fused layers of epidermis (Fig. 5C; see color plate). The transmitting tract tissue provides the pathway for the pollen tubes after they exit the style, and growth of the pollen tubes through this region is again extracellular (Lennon *et al.*, 1998). Cues, possibly emanating from unfertilized ovules, cause the pollen tubes to leave the transmitting tract and grow intercellularly between the epidermal cells of the septum (Hülskamp *et al.*, 1995). Abaxial to the transmitting tract tissue is a large lacuna in which linear arrays of cells oriented parallel, perpendicular, and oblique to the plane of the septum are present.

5. Placenta and Ovules

Four strands of parietal placentae develop along the intersections of the septum and the ovary walls (Figs. 2 and 3E; see color plate). Each strand of placental tissue gives rise to a row of 10–15 equally spaced ovules. The number and spacing of ovule primordia represent another intriguing pattern element within the carpel. At maturity, the placenta appears as a small ridge of tissue on the flanks of the septum. For the purposes of this review, the placenta with associated ovules will be considered a single tissue.

6. Vasculature

Entering into the basal end of the gynophore are two bundles, each of which bifurcates proximal to the ovary base to produce one medial and one lateral bundle. In addition, within the ovary of a mature gynoecium, isolated strands of xylem tissue are observed in the funiculi of the ovules. The style contains two arrays of tracheary elements whose cell shape resembles mature parenchyma cells that have produced a reticulate pattern of lignified secondary walls. Each array is connected to a medial vascular bundle (Figs. 5A and B; see color plate). There are a number of anatomical and developmental differences between the medial and lateral bundles: (1) The medial bundles have a larger cross-sectional area and contain more xylem and phloem tissue than the lateral bundles. (2) The medial bundles produce a contiguous column of vessel elements that extends into the style whereas the lateral bundles do not. (3) The maturation of a contiguous column of vessel elements is completed first in the medial bundles followed by the lateral bundles (Fig. 5A). Within a mature carpel, residual procambium is evident and located between the mature xylem and phloem elements. The vessel elements exhibit a helical pattern of secondary wall thickenings that is characteristic of protoxylem vessel members. At the distal end of the lateral bundles, in the region where the ovary wall tapers along the side of the style, tracheary elements are connected to the lateral bundle and follow the curved outer surface of the valve. During the later phase of fruit development, a reticulate pattern of tracheary elements is observed within the valve (Sessions and Zambryski, 1995). These elements are connected to the aforementioned tracheary elements that are located at the distal end of the lateral bundles. The maturation of the reticulate pattern is basipetal, starting at the distal end of the valves.

Associated with the lateral vascular bundles are myrosin cells, which are idioblasts that contain myrosinase (β-thioglucosidase). Myrosinase hydrolyzes thioglucosides into mustard oil (Isothiocyanate), glucose, and sulfate (Fahn, 1979). Sessions and Zambryski (1995) observed myrosin cells only in vascular bundles of the valves in an invariant position next to phloem sieve tube members. Metcalfe and Chalk (1979) describe myrosin cells as being the only distinctive cell type found in the Brassicaceae.

7. Gynophore

The gynophore is, in effect, an internode and as such has all the usual attributes of a stemlike structure (Fig. 3F; see color plate). It is radially symmetric with a ring of vascular bundles. The abaxial epidermis of the gynophore resembles that of the pedicel and as such is distinct from the abaxial epidermises of the replum and valves.

B. Gynoecium Development

Arabidopsis flower development has been divided into stages using morphological landmarks (Smyth *et al.*, 1990) and these have been used as a framework on which to place developmental events during carpel development (Hill and Lord, 1989; Sessions, 1997). The gynoecium begins its development as an ovate primordium composed of morphologically uniform cells (stages 5 and 6). The perimeter of the primordium subsequently grows upward, producing a cylinder with a deep invagination which will form the internal surfaces of the fused gynoecium (stages 7 and 8). The placenta, ovules, and septum develop from a ridge of tissue that originates on the adaxial side of the cylinder (stages 8–10). Postgenital fusion of the septum and style completes the formation of the two-loculed gynoecium (stages 9–12).

By early stage 5, floral meristems have initiated the outer three whorls of floral organs, and the remaining floral meristem that will give rise to the gynoecium primordium is relatively flat and symmetric. By late stage 5, the remaining floral meristem becomes dome-shaped and starts to expand in the lateral dimension. These trends continue into stage 6, when the carpel primordium becomes distinct and is wider in the lateral dimension as compared to the medial [Figs. 6A, 6B (see color plate), 7A]. The cells comprising the primordium are uniform in appearance and are densely cytoplasmic (Fig. 6B; see color plate). During these stages, both

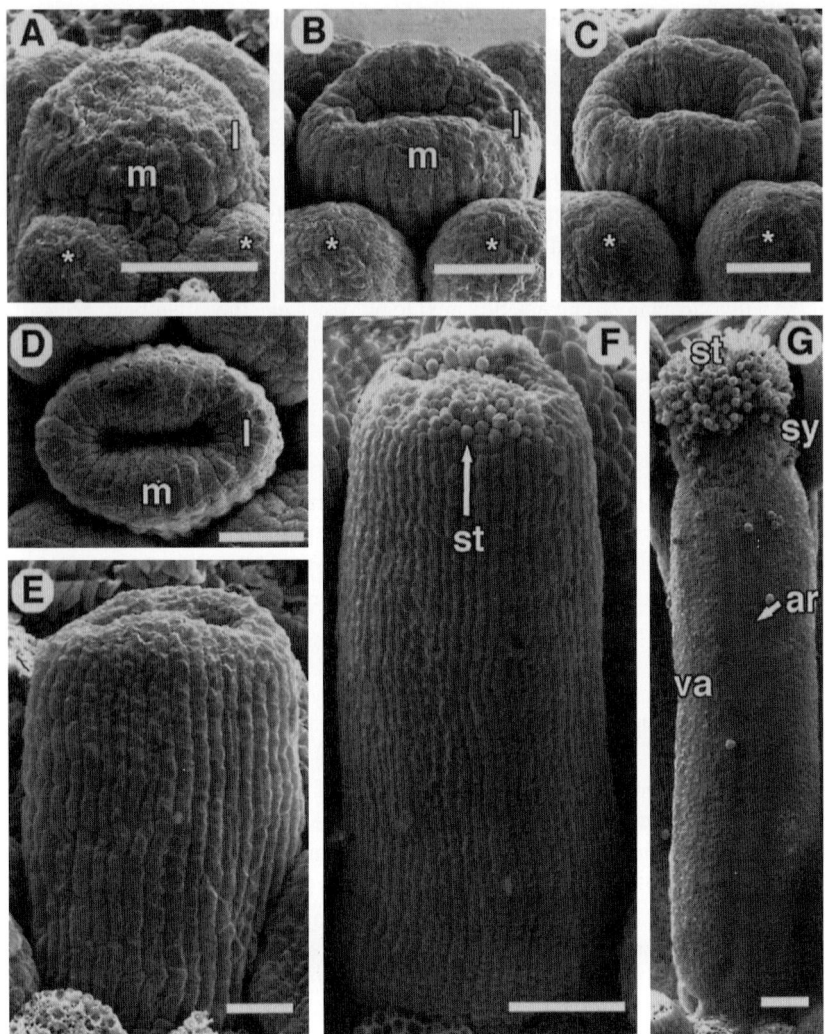

Fig. 7 SEMs of developing wild-type gynoecia: (A) stage 6 gynoecium primordium; (B) early stage 7 gynoecium; (C) stage 7 gynoecium; (D) apical view of stage 7 gynoecium; (E) stage 8 gynoecium; (F) stage 10 gynoecium; (G) stage 13 gynoecium. ar, Abaxial replum; l, lateral; m, medial; st, stigmatic papillae; sy, style; va, valve; *, medial stamen. Bar = 100 μm in G, 50 μm in F, and 20 μm in A–E.

the epidermal (L1) and subepidermal (L2) cell layers undergo predominantly anticlinal divisions, giving the primordium an ordered appearance (Hill and Lord, 1989). During late stage 6 and early stage 7, the initial indication of the gynoecial cylinder becomes apparent (Fig. 7B). Due to the periclinal divisions in the L2 at the circumference of the gynoecial primordium, an indentation, or crease, in the lateral plane becomes evident in the center of the gynoecial primordium (Figs. 7B, 7C, and 7E). During stage 7, apical growth of the margins of the carpel primordium continues, and the formation of the gynoecial cylinder becomes conspicuous [Figs. 6E (see color plate), 7B, 7C, and 7E]. At this point in development the cylinder is already 5–6 cell layers thick (Fig. 6E; see color plate), and its medial internal surfaces are tightly appressed to one another [Figs. 6E (see color plate) and 7D].

During stages 7 and 8, regional subdivision of the gynoecium into valve versus replum becomes morphologically evident. During late stage 7, a ridge of tissue, the medial ridge, develops in the medial plane from the adaxial side of the cylinder (Fig. 6C, see color plate; Sessions, 1997). The medial ridge gives rise to both the septum and the placenta (Fig. 6D; see color plate). It is initially formed by subepidermal periclinal divisions in the medial adaxial region of the cylinder. Stage 8 sees a continuation of the trends seen in late stage 7; the medial ridge continues to grow by subepidermal periclinal divisions such that the medial ridges emanating from opposite sides of the gynoecial cylinder contact each other in the center of the cylinder (Fig. 6D; see color plate). During these stages, the abaxial epidermis consists of vertical cell files, whose appearance is uniform around the circumference of the cylinder (Figs. 7C and 7D). The lateral regions of the cylinder consist of 5–6 cell layers, suggesting that all cell layers of the valve have formed by this time (Figs. 6C and 6D; see color plate). During these stages, much of the vertical growth of the gynoecium occurs through cell divisions, producing conspicuous cell files [Fig. 6E, 6F (see color plate), and 7B–F]. By the end of stage 7, the gynoecial cylinder is approximately 80 μm high, and by the end of stage 8, the height reaches approximately 200 μm (Sessions, 1997). Radial expansion of the cylinder is also due largely to cell division, as evidenced by the increase from 30 vertical cell files in the gynoecium in Fig. 7C to the 45 vertical cell files in the gynoecium in Fig. 7E. Throughout the stages in which the gynoecial cylinder increases in height, cell divisions occur along the entire length of the cylinder, again with cell files being conspicuous (Fig. 5E; see color plate). This is in contrast to the cell division patterns observed in developing *Arabidopsis* leaves, where most cell divisions are localized at the base of the leaf (Pyke *et al.*, 1991). Stage 8 also sees the initiation of vascular bundles in the medial and lateral positions (Sessions, 1997).

During stage 9, the regions of valve, placenta, septum, style, and stigma become morphologically distinct from one another (Fig. 6D; see color plate). By stage 9, the abaxial epidermis consists of 60–70 vertical cell files. Within the developing valves, the characteristic cell layers of the valve become evident (Figs. 6C and 6D; see color plate). The medial ridge undergoes dramatic changes during stage 9, with a high level of mitotic activity occurring throughout the ridge (Fig. 5E; see color plate). At this stage, differences are apparent in the cells of the central and marginal regions of the medial ridge (Figs. 6C and 6D; see color plate). Ovule primordia originate from placentae that arise from the marginal regions of the medial ridge (Fig. 6D; see color plate). The placenta arises from periclinal divisions in the L2 of the flanks of the medial ridge. Likewise, periclinal divisions in the L2 in the central portion of the medial ridge contribute to the formation of the septum; continued cell divisions in the subepidermal regions of the central medial ridge result in the developing septum appearing as a pair of prominent ridges which fuse in the middle of the hollow cylinder (Figs. 5C and 6D; see color plate). At the distal tip of the medial ridge, ovule primordia are not initiated; instead periclinal divisions within the medial ridge result in the formation of the solid style (Sessions, 1997). Thus, the adaxial regions of style are derived from the medial ridge, whereas the abaxial regions are derived from the surrounding cylinder. Finally, in late stage 9, stigmatic papillae begin their differentiation in the medial regions of the distal tip of the gynoecium (Fig. 7F).

Cellular differentiation of the tissue types within the gynoecium occurs during stages 10–12. During stage 10, the abaxial epidermal vertical cell files begin to show differences in morphology, corresponding to the valve and abaxial replum regions of the ovary wall, and by stage 12, the characteristic abaxial epidermal cells of the valves and abaxial replum become clearly apparent (Figs. 4B and 7G). From stage 8, the central regions of the medial ridge contact each other, an event which might mark the beginning of the postgenital fusion of the septum. From stages 10 through early stage 12, it appears that the epidermal cells of the tips of the septal outgrowths are fused, but there is not any histological sign of dedifferentiation into transmitting tract cells until mid-stage 12 (Fig. 5C; see color plate). The postgenital fusion of the solid style also occurs at this time, with continued divisions of cells within the medial ridge resulting in the complete filling of the cylinder. During stages 11 and 12, the stigmatic papillae become conspicuous, elongating significantly during stage 12, and continuing to elongate until just prior to fertilization (Figs. 7F and 7G). Finally, the style becomes distinct from the underlying ovary, which grows laterally such that the style appears as a tapered extension above the ovary (Fig. 7G).

C. Origins of Carpel Tissue Types

1. Cell Lineage

Does cell lineage play any role in the origin of specific tissue types? Due to the nature of cell division patterns in angiosperm meristems which are organized in a tunica-corpus fashion (Schmidt, 1924; Satina *et al.*, 1940), the cells of the L1 are clonally related, as are the cells of the L2. Cell divisions in these layers are predominantly anticlinal. The cells internal to these layers divide both anticlinally and periclinally and form a third group of cells (L3) of distinct cell lineage origin. Periclinal cell divisions occur very rarely in the L1. When they do, they result in invasion of the L2 layer by cells derived from the L1 layer. Likewise, L2 cells occasionally divide periclinally (more frequently than L1 cells) and invade the L1 or

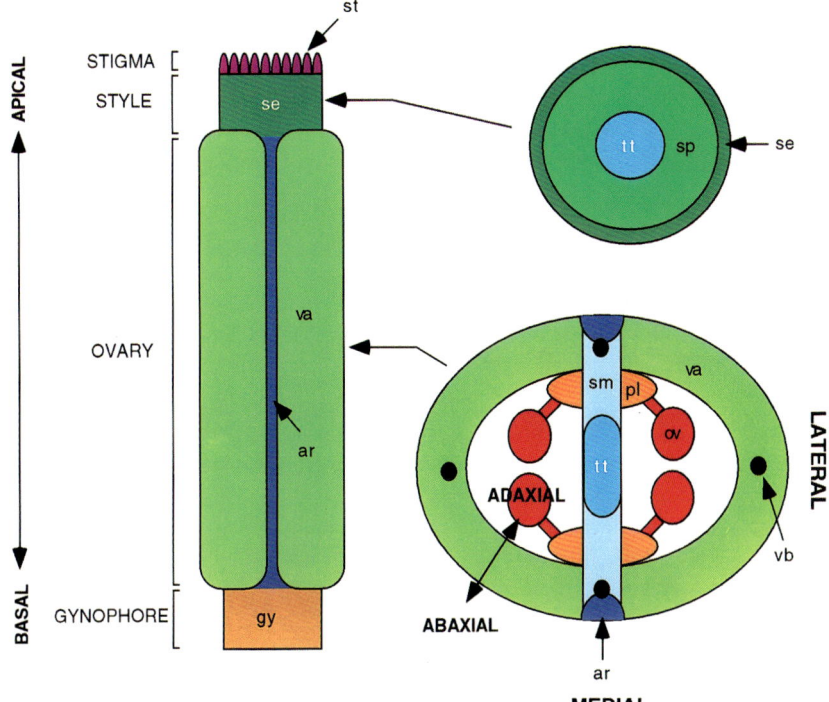

Ch. 4, Fig. 1 Cartoon of the major tissue types of the *Arabidopsis* carpel. st, Stigmatic papillae (pink); se, style epidermis (dark green); sp, style parenchyma (green); tt, transmitting tract (blue); va, valve (light green); ar, abaxial replum (dark blue); sm, septum (light blue); pl, placenta (orange); ov, ovules (red); vb, vascular bundle (black); gy, gynophore (tan).

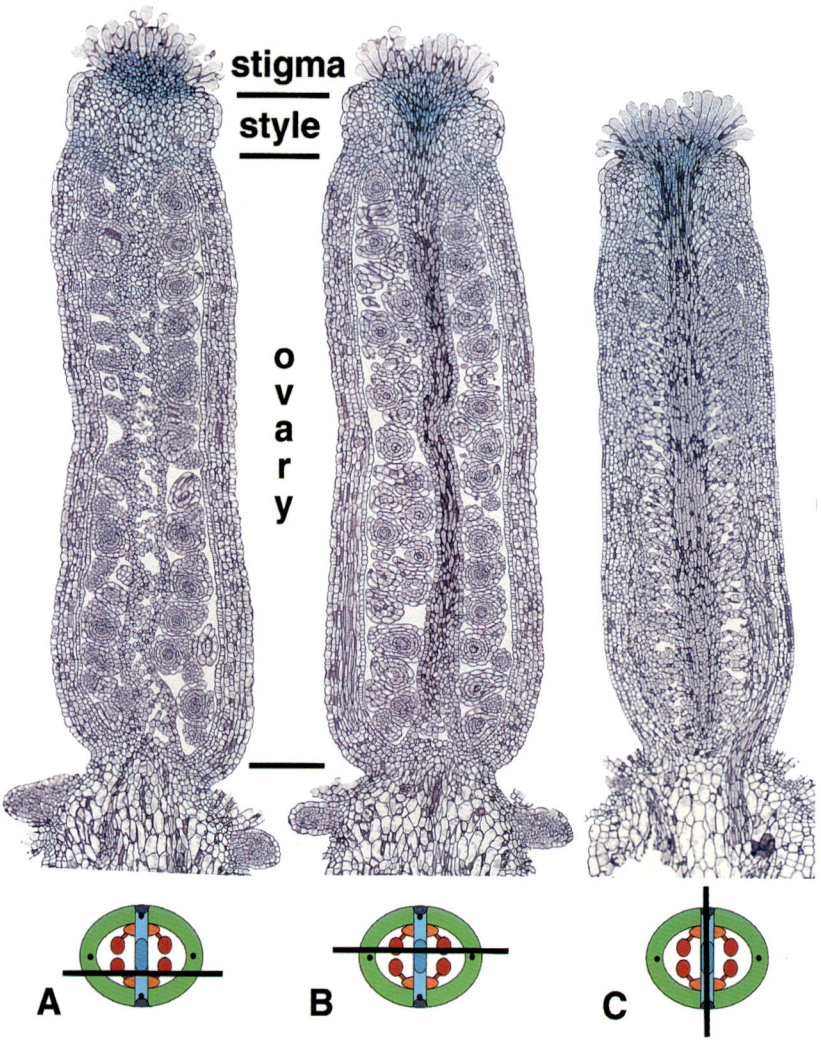

Ch. 4, Fig. 2 Longitudinal sections through mature (stage 12) wild-type carpels. The planes of section with respect to the ovary tissues are indicated.

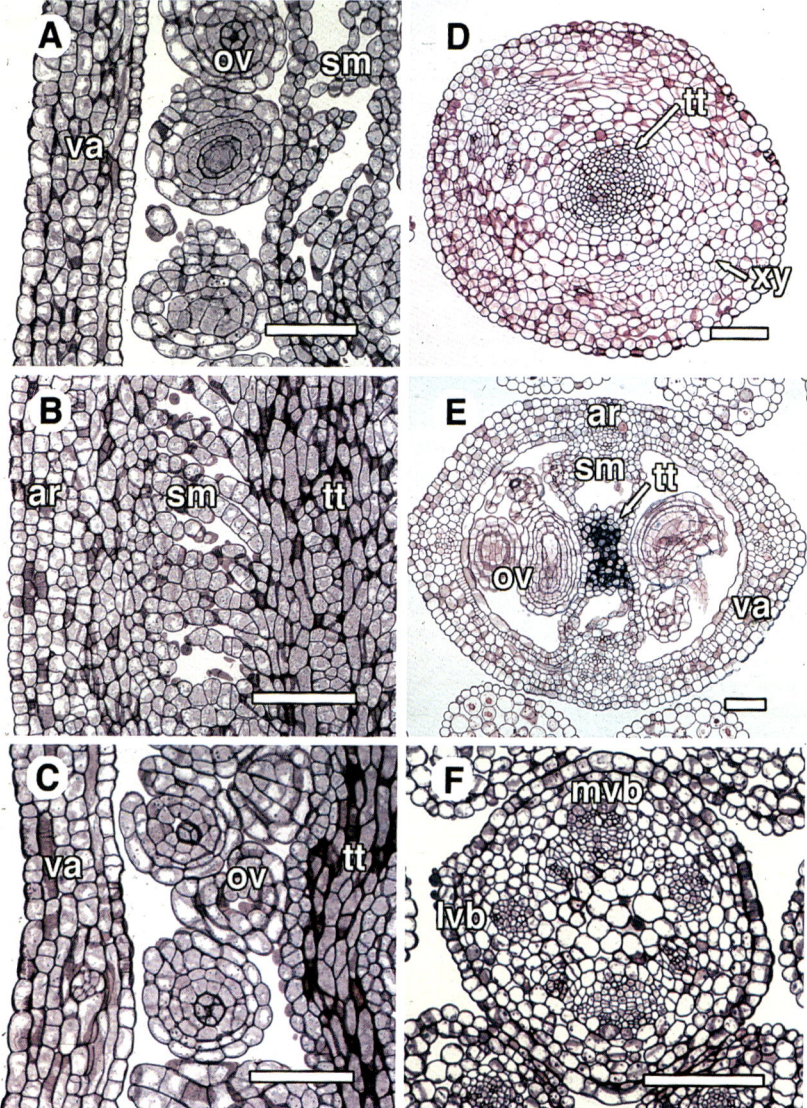

Ch. 4, Fig. 3 Histology of stage 12 wild-type carpel tissue types: (A) close-up of 2A; (B) close-up of 2C; (C) close-up of 2B; (D) transverse section through the lower portion of the style; (E) transverse section through the ovary; (F) transverse section through the gynophore. In D–F, medial positions are top and bottom; lateral positions are left and right. ar, Abaxial replum; lvb, lateral vascular bundle; mvb, medial vascular bundle; ov, ovule; sm, septum; tt, transmitting tract; va, valve; xy, xylem elements. Bars = 50 μm.

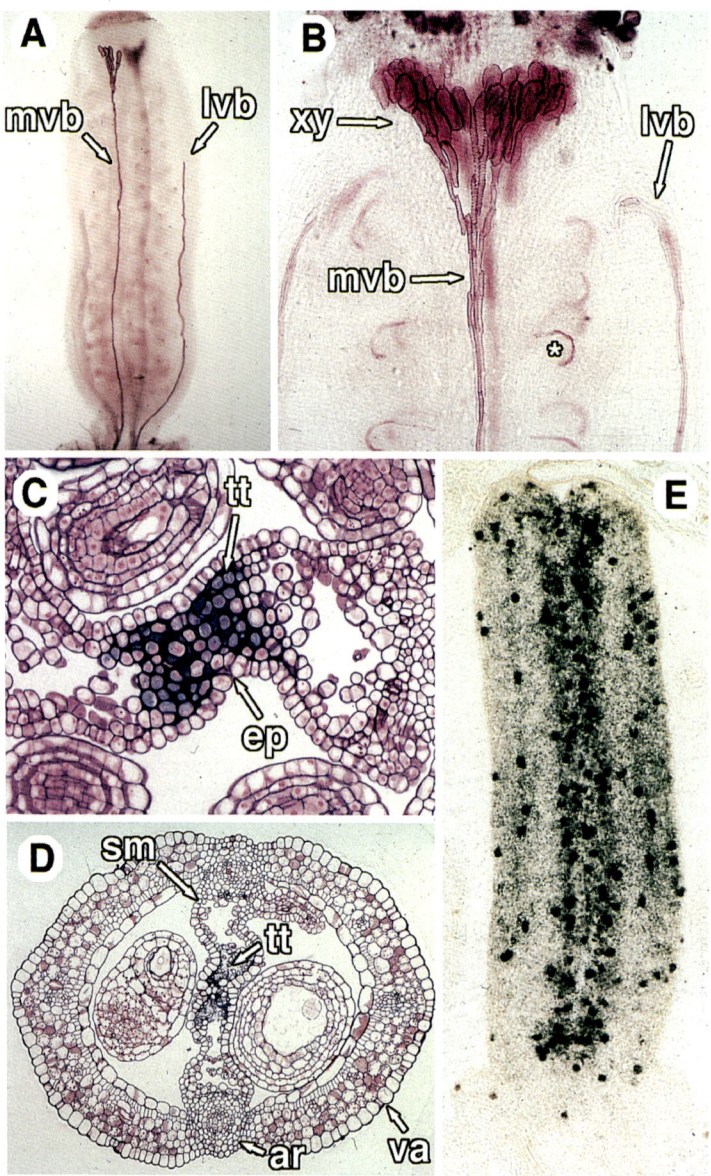

Ch. 4, Fig. 5 Wild-type gynoecia: (A) xylem elements in developing gynoecium; (B) detail of A; (C) detail of developing transmitting tract tissue; (D) transverse section of postfertilization ovary; (E) longitudinal section of stage 9 gynoecium labeled with [^{3}H]thymidine for 4 h. ar, Abaxial replum; ep, septum epidermis; lvb, lateral vascular bundle; mvb, medial vascular bundle; sm, septum; tt, transmitting tract; va, valve; xy, stylar xylem array; *, funicular xylem elements.

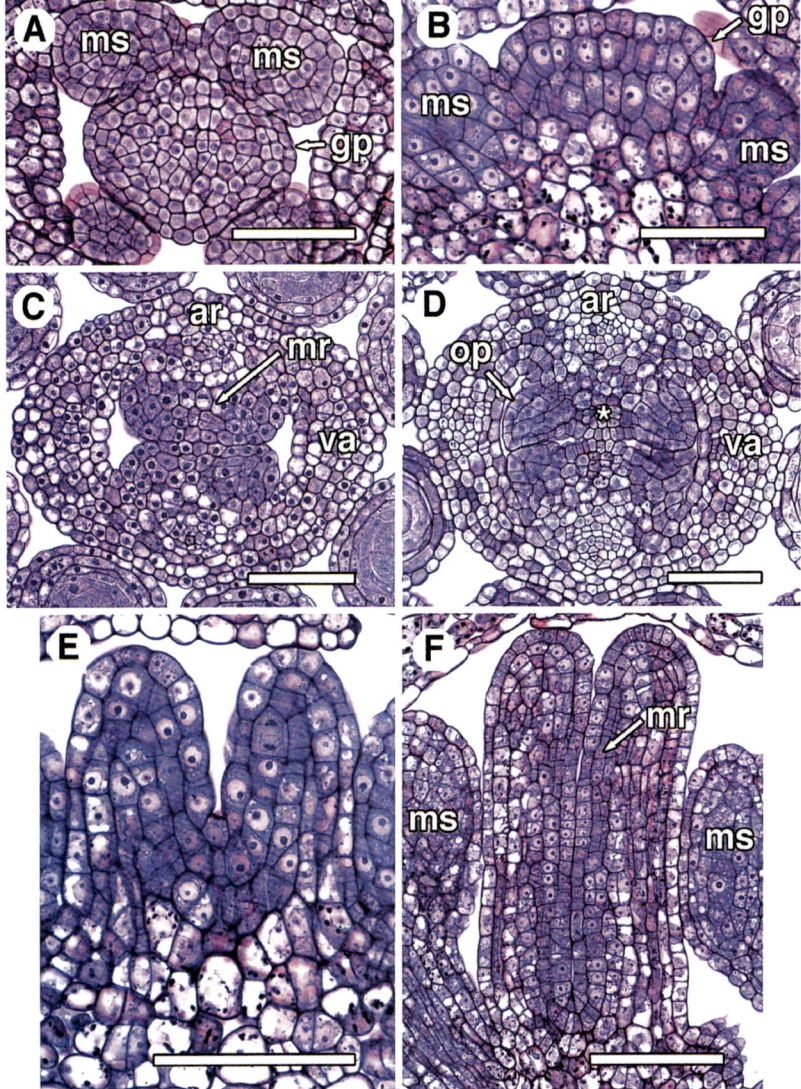

Ch. 4, Fig. 6 Histology of developing wild-type gynoecia: (A) transverse section of stage 6 primordium; (B) longitudinal section of stage 6 primordium; (C) transverse section of stage 8 ovary; (D) transverse section of stage 9 ovary; (E) longitudinal section of stage 7 gynoecium; (F) longitudinal section of stage 8 gynoecium. ar, Abaxial replum; gp, gynoecium primordium; mr, medial ridge; ms, medial stamen; op, ovule primordium; va, valve; *, septum primordium. Bars = 50 μm.

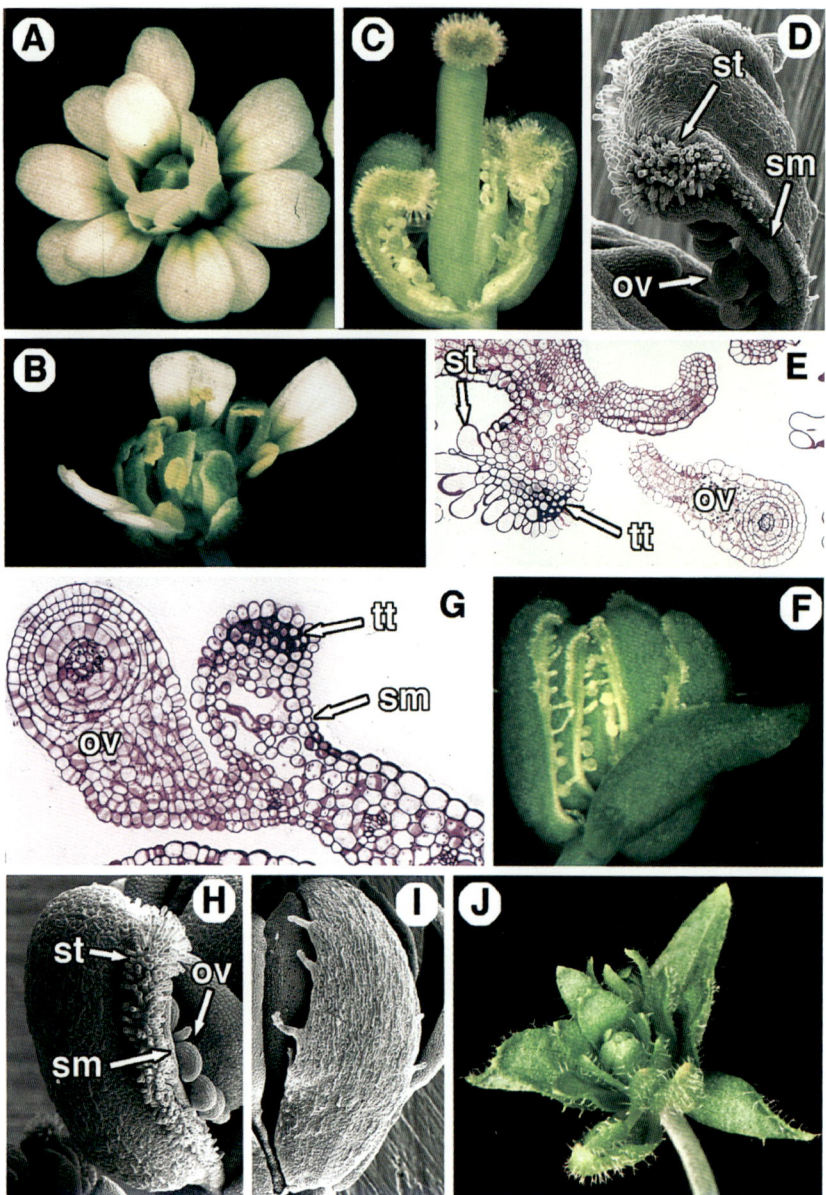

Ch. 4, Fig. 8 A and C class mutant phenotypes: (A) *ag-1* flower; (B) *ag-4* flower; (C) *ap2-2* flower; (D) detail of *ap2-2* first-whorl carpel; (E) transverse section of *ap2-2* first-whorl carpel; (F) *ap2-2 pi-2 ag-1* flower; (G) transverse section through carpelloid organ of *ap2-2 pi-1 ag-1* flower; (H) detail of first-whorl organ of *ap2-2 ag-1* flower; (I) detail of first-whorl organ of *spt-2 ap2-2 ag-1* flower; (J) *spt-2 ap2-2 pi-1 ag-1* flower. ov, Ovule; sm, septum; st, stigmatic papillae; tt, transmitting tract.

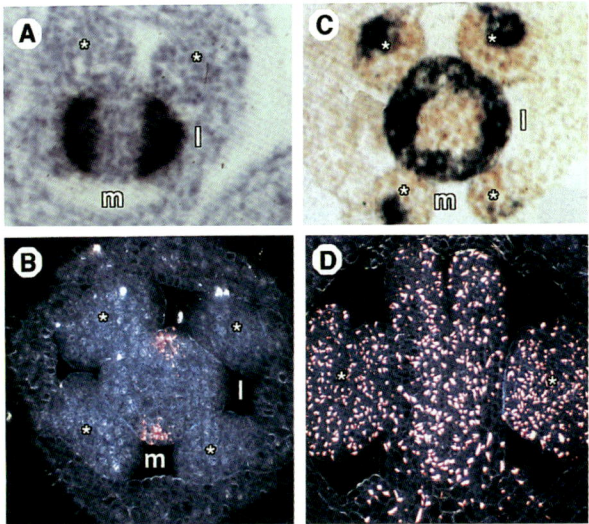

Ch. 4, Fig. 10 Gene expression patterns in gynoecium primordia: (A) *CRC* expression in stage 6 primordium; (B) *AGL5* expression in stage 6 primordium; (C) *ETT* expression in stage 6 primordium; (D) *AG* expression in stage 8 gynoecium. l, Lateral; m, medial; *, medial stamen. (C) courtesy of Allen Sessions, Jennifer Nemhauser, and Patti Zambryski.

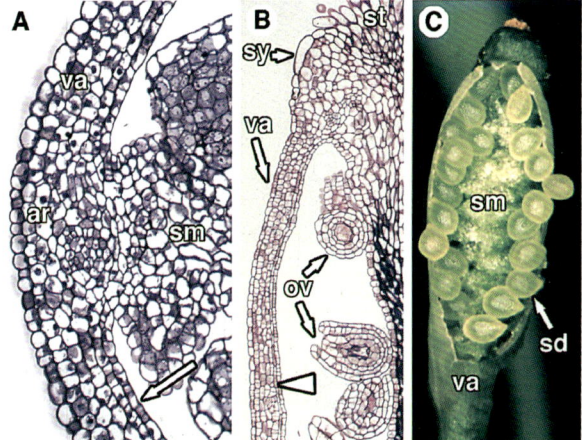

Ch. 4, Fig. 11 *ful* mutant phenotypes: (A) transverse section of ovary of mature *ful* gynoecium; (B) longitudinal section of ovary of mature *ful* gynoecium; (C) fruit of *ful* mutant split in the middle of the valve. ar, Abaxial replum; ov, ovule; sd, developing seed; sm, septum; st, stigma; sy, style; va, valve; arrow, adaxial epidermal layer; arrowhead, adaxial subepidermal layer.

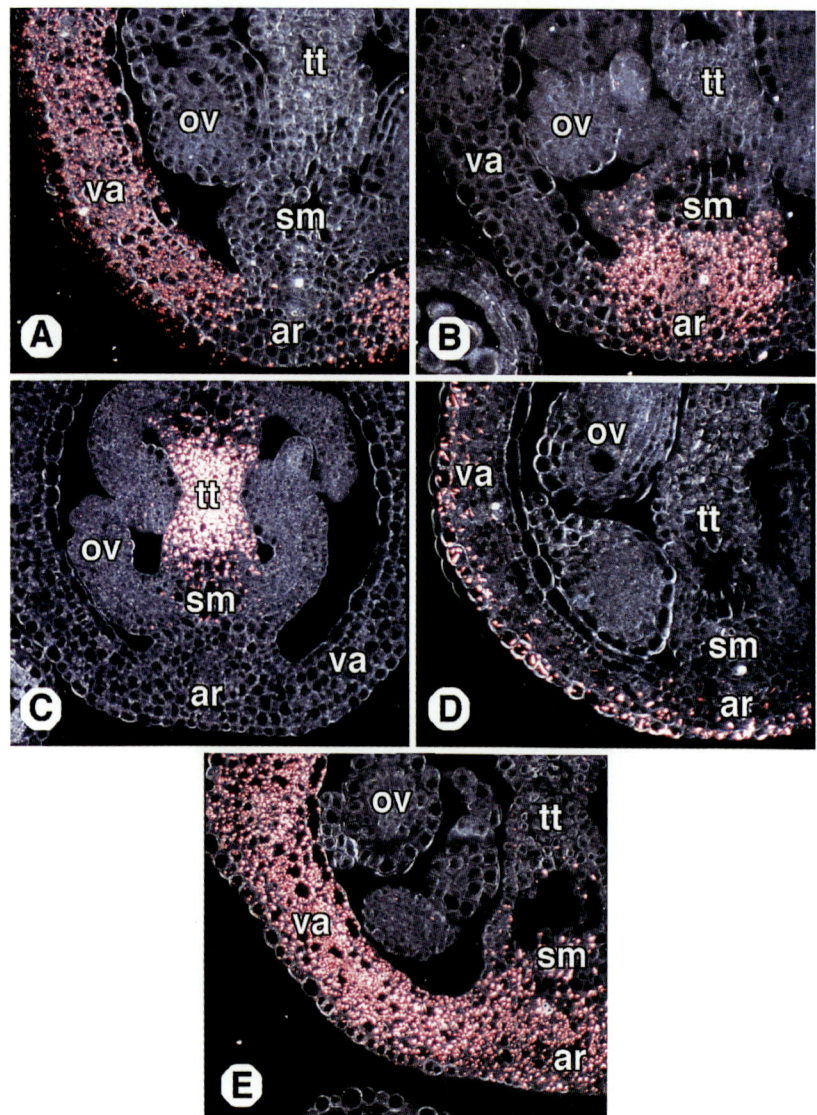

Ch. 4, Fig. 12 Gene expression patterns in stage 12 gynoecia: (A) *FUL*; (B) *KNAT2*; (C) enhancer trap line TJ3878; (D) enhancer trap line TJ283; (E) enhancer trap line YJ2. ar, Abaxial replum; ov, ovule; sm, septum; tt, transmitting tissue; va, valve.

L3 cell layers. Cell divisions in the L3 occur both anticlinally and periclinally. To coordinate both cell division frequencies and tissue differentiation, extensive communication pathways must exist between these clonally distinct layers.

Genetic mosaic and sector boundary analyses in shoots have shown that in most cases cell lineage is not correlated with cell fate in plants (for review, see Dawe and Freeling, 1991). This has been shown specifically for carpels in both dicotyledonous (Satina and Blakeslee, 1943; Derman and Stewart, 1973) and monocotyledonous (Dellaporta *et al.*, 1991) plants. Using periclinal chimeras in *Prunus persica*, Derman and Stewart showed that all three apical cell layers give rise to tissue in the carpels but that there was no strict correlation between cell lineage and fate. The style and ovary walls were shown to have contributions from all three cell layers, whereas the ovules and stigma consisted of derivatives primarily from the L1 and L2 layers. Similar conclusions were drawn for *Datura stramonium*, although in this case the L3 appears to have contributed to a significant portion of the septum and carpel walls (Satina and Blakeslee, 1943). In the case of *Datura*, the stigma and transmitting tract were found to be of L1 origin and the style primarily of L2 origin (Satina, 1944).

Cell lineage studies in *Arabidopsis* have been limited to analyses of the shoot apical meristem in the seed and, more recently, sector boundary analysis of the early stages of flower development (Shevchenko *et al.*, 1976; Furner and Pumfrey, 1992; Irish and Sussex, 1992; Bossinger and Smyth, 1996). Using sector boundary analyses, Bossinger and Smyth (1996) deduced that each carpel of the *Arabidopsis* gynoecium is derived from eight ancestral cells in the floral meristem. That these initial cells are in a linear arrangement supports the notion that the carpel is foliar in nature (Bossinger and Smyth, 1996). Lineage analyses to date in *Arabidopsis* have not addressed cell lineage patterns during carpel differentiation; however, it is likely that *Arabidopsis* will be similar in this regard to other angiosperms. Based on studies of other species and cell division patterns observed in thin sections of *Arabidopsis* carpels (Hill and Lord, 1989; Sessions, 1997), it is likely that the majority of the carpels are derived from the L1 and L2 layers. For example, the stigma appears to be entirely of L1 origin, whereas the style, transmitting tract tissue, and ovules have their origins in both the L1 and L2 cell lineages. The ovary walls are composed of at least L1 and L2 tissues, with the contribution of the L3 presently unknown. Thus, it is likely that in a broad sense there are few, if any, lineage restrictions in the *Arabidopsis* carpel. However, present studies do not rule out cell lineage patterns having a role late in the development of certain tissue or cell types.

2. Developmental Compartments

Another way to consider the origin of the tissue types is to ask whether there exist developmental compartments in carpels. Many of the tissue types we associate with carpels are derived from the marginal regions of the organs. When we refer to the margins of the carpel in the syncarpous congenitally fused *Arabidopsis* gynoecium, we mean those tissues that would constitute the margins of the carpels if the two carpels of the gynoecium were not fused. The regions that are not considered marginal will be referred to as central. In general, the stigma, septum, and placenta are all derived from the marginal regions of the carpel, whereas the majority of the ovary wall is derived from the central regions of the carpel primordium. These origins are somewhat obscured by the congenital fusion of the two carpel primordia in *Arabidopsis*. However, one interpretation is that the medial domains of the stage 6 carpel primordium represent the marginal regions of the congenitally fused primordia, whereas the lateral domains represent the central regions. In this scenario the septum, placenta, and abaxial replum would be derived from the medial domains, whereas the valves would have their derivation in the lateral domains. Support for this interpretation comes from morphological analyses of gynoecium development. One striking feature when comparing stage 8 (Fig. 6C), stage 9 (Fig. 6D), and stage 12 (Fig. 3E) (see color plate) gynoecia is the relatively constant size of the abaxial replum in terms of cell numbers. The epidermis of the abaxial replum consists of approximately 12–14 cells at each of these stages. This contrasts sharply with the valves, in which the cell number increases from approximately 20 at stage 8 to approximately 55 by stage 12 in the abaxial epidermis. The adaxial epidermis also undergoes an increase in cell number from approximately 8 at stage 8 to a little over 20 by stage 12. Thus, although the replum occupies a small fraction of the circumference late in development, it occupies a portion nearly equal to the valves early in development. Prior to the emergence of the medial ridge, there are no morphological markers distinguishing the valve and replum regions; however, gene expression patterns (see later) indicate that the medial and lateral domains of the stage 6 gynoecium primordium are distinct and are of approximately equal size. It is plausible to suggest that the cells occupying the medial domain of the stage 6 gynoecium primordium are fated to give rise to the tissues of the replum, whereas those of the lateral domain give rise to the valves.

3. Evolutionary Origins of Carpel Tissues

Examination of fossil angiosperms and potential basal angiosperms has provided a picture, although incomplete, of the ancestral angiosperm carpel (Bailey and Swamy, 1951; van Heel, 1981; Friis and Crepet, 1987; Cronquist, 1988; Friis and Endress, 1990; Taylor, 1991; Endress and Igersheim, 1997; Igersheim and Endress, 1997). The earliest angiosperm carpels were likely apocarpous and either conduplicate or ascidate, lacking a distinct style. Early fossil carpels have a decurrent stigma; the stigmatic surface extended along the length of the margins of fusion of the carpels. This distribution of stigma is similar to that of several conduplicate and ascidate carpels from taxa that are considered basal within the extant angiosperms (Bailey and Swamy, 1951; Endress and Igersheim, 1997; Igersheim and Endress, 1997). In many of these extant species the stigmatic papillae and their secretions constitute a type of transmitting tract through which pollen tubes grow to effect fertilization of the ovules. It is plausible that the ancestral carpels possessed this type of transmitting tract, with the well-differentiated transmitting tract tissue of most modern angiosperms being derived from the originally decurrent stigma. This would suggest that the stigma and transmitting tract tissue have a common origin in the marginal tissues of the ancestral carpel. Thus, in an evolutionary sense, the stigmatic and transmitting tract tissue would have a marginal origin, consistent with its present developmental origin. Subsequent evolutionary modifications of the stigma would be its apicalization and the development of a style on which the apical stigma is perched for better exposition. Ovules are found in all seed plants, and therefore they predate the origin of carpels, which are found only in the angiosperms. It follows that ovules, and possibly the placental tissue from which they arise, have a different, more ancient, evolutionary origin than carpels, and their development is likely to be independent of the development of the other tissues of the carpel. For the purposes of this review, the ovules, which themselves represent an entire organ system, and the placental tissue will be considered as a single independent lineage and will not be discussed in detail (for reviews of ovule development, see Chaudhury *et al.*, 1998; Drews *et al.*, 1998; Gasser *et al.*, 1998; and Grossniklaus and Schneitz, 1998). Different hypotheses concerning the origin of the carpel place the placental tissue in different places in the ancestral carpel; however, in any interpretation subsequent evolutionary modifications of placental position are required to account for all of the extant forms (Taylor, 1991; Doyle, 1994). Thus, the relationship between the ancestral condition and the present position of the placental tissue and ovules in the adaxial marginal position in *Arabidopsis* is equivocal.

III. Molecular Genetics of Carpel Development

A primary technique for gene identification has been to isolate mutants that specifically disrupt the process of interest. Despite extensive screens, very few mutations have been isolated whose mutant phenotype is restricted to the carpel. This might sound unlikely considering there are an estimated 21,000 genes in the *Arabidopsis* genome (Ecker, 1998), and a reasonable fraction of these might be expected to be expressed in the carpel. Several reasons can account for the paucity of carpel-specific mutants. First, since the carpel is thought to be evolutionarily derived from a leaflike progenitor, a large number of genes are expected to be involved in the development of both leaves and carpels. Second, analyses of enhancer trap collections suggest that, other than in very specialized tissues, the expression of most genes is not limited to a particular organ type. Thus, it is likely that most genes are involved in multiple developmental processes. An extension of this is that since the carpel is formed late in the development of the plant, observing a mutant phenotype in the carpel can be precluded if such mutations result in lethality (it has been estimated that mutations in approximately 5000 *Arabidopsis* genes result in embryo lethality; Jürgens *et al.*, 1991), result in apical meristem disruption, or otherwise indirectly prevent the development of carpels. Third, it is becoming apparent from genome projects that there is a significant amount of genetic redundancy built into most organisms (e.g., Mewes *et al.*, 1997). This may be particularly prevalent in plants, where the incidence of polyploidy is common (Leitch and Bennett, 1997). This theme of partial redundancy will resurface with the description of many of the genes known to be involved in carpel differentiation.

What follows is a discussion of several genes for which functionality in carpel development has been demonstrated. We will start with those genes whose mutant phenotype suggests that they are involved in specifying carpel identity and then proceed to those genes that may be involved in establishing or interpreting regional positional information within the carpel primordium. Finally, we will cover those genes whose mutant phenotype suggests that they might be responsible for directing the differentiation of specific tissue or cell types. In the course of these vignettes it is our hope to cover what is known about some of the genes directing the growth and differentiation of the carpels and offer speculations on possible mechanisms of gene function.

A. The Floral Homeotic Mutants of *Arabidopsis*

The recent application of developmental genetics has led to progress in understanding how organ type is specified in the flower. The advances are based on both genetic analyses of mutants which specifically disrupt floral development and molecular cloning of the corresponding genes to reveal the nature of their biochemical function. A model, now known widely as the ABC model, based on genetic data was proposed to explain how a limited set of genes acting alone and in combination could specify the identity of the floral organs (Bowman *et al.,* 1989; Carpenter and Coen, 1990; Schwarz-Sommer *et al.*, 1990; Bowman *et al.*, 1991a; reviewed in Coen and Meyerowitz, 1991; and Weigel and Meyerowitz, 1994). Subsequent molecular analyses of the relevant genes and ectopic expression studies in *Arabidopsis* have largely supported the genetic model and led to numerous refinements and further insights (e.g., Mandel *et al.*, 1992; Jack *et al.*, 1994; Sakai *et al.*, 1995; Krizek and Meyerowitz, 1996; reviewed in Weigel and Meyerowitz, 1994). A brief description of the floral homeotic mutants and the ABC model follows, with those aspects relevant to carpel development highlighted.

Studies have focused on a set of floral homeotic genes, mutations in which result in the misinterpretation of positional information in the developing flower, and subsequent homeotic transformation of floral organ types. These floral homeotic mutants fall into three classes, designated A, B, and C, and each of the mutations results in organ identity defects in two adjacent whorls. The *Arabidopsis* flower

4. Arabidopsis Carpel Development

consists of four concentric whorls of organs: sepals, petals, stamens, and carpels. Class A mutants [represented by *apetala2* (Fig. 8C; see color plate) and *apetala1*] have homeotic conversions in the first two whorls, with, in the case of *ap2*, the first-whorl organs developing as carpels rather than sepals, and stamens, when present, arising in the second and third whorls (Komaki *et al.*, 1988; Bowman *et al.*, 1989; Kunst *et al.*, 1989; Irish and Sussex, 1990; Bowman *et al.*, 1991a, 1993; Gustafson-Brown *et al.*, 1994; Jofuku *et al.*, 1994). Class B mutants (represented by *pistillata* and *apetala3*) have alterations in the middle two whorls, with sepals instead of petals developing in the second whorl and carpels instead of stamens in the third whorl (Bowman *et al.*, 1989; Hill and Lord, 1990; Bowman *et al.*, 1991a; Jack *et al.*, 1992; Goto and Meyerowitz, 1994). The inner two whorls are affected in class C mutants (represented by *agamous*, Fig. 8A), with petals developing in place of stamens and another flower replacing the carpels (Bowman *et al.*, 1989; Yanofsky *et al.*, 1990; Bowman *et al.*, 1991a).

The basic tenets of the ABC model (Bowman *et al.*, 1991a) are that (1) each of the classes of homeotic gene function acts in a field composed of two adjacent whorls, the particular whorls being those that are altered when the corresponding genes are in mutant form; (2) the combination of floral homeotic gene activities specifies the type of organ that develops (e.g., class A alone specifies sepals, classes A + B specify petals, classes B + C specify stamens, and class C alone specifies carpels); and (3) the class A and class C activities are mutually antagonistic such that loss of A results in C activity in all four whorls and vice versa. For most of the genes it has been shown that the first and third tenets are satisfied at the transcriptional level; however, for some class A genes, the restriction of their organ identity activity to the outer two whorls must occur posttranscriptionally (Drews *et al.*, 1991; Jofuku *et al.*, 1994; Liu and Meyerowitz, 1995).

B. AGAMOUS Is a Key Regulatory Gene Specifying Carpel Identity

A key regulatory gene whose action is central in the specification of carpel identity in *Arabidopsis* flowers is the C class gene *AGAMOUS* (*AG*). Loss-of-function mutations in *AG* result in homeotic conversions in the third and fourth whorls of the flower (Bowman *et al.*, 1989). Petals develop in the third whorl rather than the wild-type stamens while those cells that would normally develop into the two-carpeled gynoecium instead behave as if they constituted another flower meristem (Fig. 8A; see color plate). The result of these alterations is an indeterminate flower with repeating sets of organs: (sepals, petals, petals)$_n$. Based on this mutant phenotype, four distinct functions of *AG* are identified: (1) specification of stamen identity, (2) specification of carpel identity, (3) control of flower meristem determinacy, and (4) negative regulation of A class activity. That these roles are, at least in part, separable is illustrated by the partial loss-of-function allele *ag-4*, in which the third-whorl organs can be nearly normal stamens, the fourth-whorl organs are carpelloid sepals (sepals with marginal placental and septal tissue), and the flower meristem is indeterminate, producing extra whorls of stamens and carpelloid sepals interior to the fourth whorl (Fig. 8B; Sieburth *et al.*, 1995). The roles of *AG* in specifying carpel identity and the repression of A class activity will be discussed in this section. The roles of *AG* in the specification of stamen identity and in the control of floral meristem determinacy will not be discussed further.

1. *AG* and the Specification of Carpel Identity

In an otherwise wild-type flower, *AG* is absolutely required for the cells of the fourth whorl to acquire carpel identity. Flowers of plants carrying a null loss-of-function allele such as *ag-2* fail to develop any carpelloid tissue (Fig. 8A; see color plate). *AG* encodes a member of the large MADS-box containing transcription factor gene family (Yanofsky *et al.*, 1990). The functions of this gene family in plant development and how their biological specificity is attained have been well reviewed by Riechmann and Meyerowitz (1997) and will not be addressed here. Consistent with its proposed functions based on loss-of-function alleles, *AG* is expressed in the third and fourth whorls of developing flowers (Yanofsky *et al.*, 1990; Drews *et al.*, 1991). Expression commences at stage 3, well before the third- and fourth-whorl organ primordia are initiated from the floral meristem (Drews *et al.*, 1991). During stages 6–8, during which time the gynoecial primordium forms and develops as a cylinder, *AG* is expressed uniformly throughout the developing carpel (Fig. 10D; see color plate). At later stages of development, *AG* expression is restricted to specific cell types within the carpels as cellular differentiation occurs (Bowman *et al.*, 1991b). Thus, *AG* expression is detected in stigmatic papillae and ovules but is excluded from those cells that will give rise to the gametes. While the early expression pattern reflects the function of *AG* to specify carpel identity, without conditional mutant alleles, it is unclear if the later *AG* expression pattern relates to functions in those specific cell types in which it is expressed.

Ectopic expression of *AG* in transgenic *Arabidopsis* plants has shown that *AG* is sufficient for the specification of carpels in some contexts within the flower. When *AG* is constitutively expressed under the control of the Cauliflower Mosaic Virus 35S promoter, the first-whorl organs develop as carpels rather than sepals, phenocopying *ap2* mutants (Mizukami and Ma, 1992). Thus, it seems that ectopic *AG* expression can override the function of A class in the outer whorl of the flower. It appears that under some circumstances, ectopic *AG* expression can also override B class activity. This proposal is based on the observation that when the *Brassica napus AG* ortholog, *BAG*, is constitutively expressed in transgenic tobacco plants, all four whorls of the flower can develop into carpelloid organs (Mandel *et al.*, 1992). In this case it is unclear whether the transgene *BAG*, the endogenous tobacco *AG* gene (*NAG1*; Kempin *et al.*, 1993), or a combination of both is responsible for the ubiquitous development of carpels in the flower. Further support for the hypothesis that *AG* expression can override both the A and B class activities comes from experiments in which a constitutively activated form of *LEAFY* (*LFY*) is expressed. *LFY* has been proposed to be an upstream activator of *AG* (Weigel and Meyerowitz, 1993), and expression of a constitutively activated form of *LFY* (*LFY:VP16*) is sufficient to activate *AG* throughout the floral meristem (Parcy *et al.*, 1998). In these plants, all floral organs develop as carpels, and their development as such requires *AG* activity since *ag* mutations are epistatic to the effects of the *LFY:VP16* transgene (Parcy *et al.*, 1998). This suggests that if *AG* activity is at a sufficiently high level, all cells of the flower meristem can be fated to develop as carpels, regardless of the activity of the other ABC homeotic genes. Thus, expression of *AG* is sufficient in some contexts to confer carpel identity on developing floral organ primordia. Notably, the leaves of transgenic plants constitutively expressing *AG* or *BAG* do not develop carpelloid features, even in those plants in which all the floral organs are carpelloid (Mandel *et al.*, 1992; Mizukami and Ma, 1992), indicating that there must be other floral-specific factors that allow cells to become competent to respond to ectopic *AG* expression. Likewise, *AG* is ectopically expressed in the leaves of *curly leaf* (*clf*) mutants, and this is insufficient for these organs to develop as carpels (Goodrich *et al.*, 1997). Thus, ectopic *AG* expression is not sufficient to transform leaves into carpels.

2. The Interaction of A and C Class Homeotic Genes

One of the basic tenets of the ABC model is the mutual antagonism between the A and C class activities. As with C class activity, A class activity has a number of components: specification of sepal identity, specification of petal identity, and repression of C class activity. In *Arabidopsis*, C class activity is defined by *AG* activity. A class activity is somewhat more complicated, with the function being shared by several players, three of which are *APETALA1* (*AP1*), *APETALA2* (*AP2*), and *LEUNIG* (*LUG*) (Komaki *et al.*, 1988; Bowman *et al.*, 1989; Kunst *et al.*, 1989; Irish and Sussex, 1990; Bowman *et al.*, 1991a, 1993; Gustafson-Brown *et al.*, 1994; Jofuku *et al.*, 1994; Liu and Meyerowitz, 1995). *AP2* participates in all three functions, whereas *AP1* participates only in the organ identity functions and not the repression of C class activity. Conversely, *LUG* is involved in repression of C class activity but does not appear to act in specification of organ identity. The restriction of C class activity (*AG* expression) to the inner two whorls of the flowers is mediated at the transcriptional level by *LUG* and *AP2* (Drews *et al.*, 1991; Liu and Meyerowitz, 1995). Likewise, the accumulation of *AP1* mRNA appears to be negatively regulated by *AG* in the inner two whorls (Gustafson-Brown *et al.*, 1994). In contrast, *AP2* mRNA is present in the stamens and carpels, and thus, the restriction of A class activity to the outer two whorls of the flower has posttranscriptional components (Jofuku *et al.*, 1994). Thus, the mutual antagonism between the A and C class activities involves several genes whose products interact at both the transcriptional and posttranscriptional levels.

3. *AG*-Independent Carpel Development

Due to the mutual antagonism of the A and C class activities, in wild-type flowers A class activity is restricted to the outer two whorls, and C class activity to the inner two whorls. In *ag* mutants, A class activity is in all whorls, whereas in *ap2* mutants, C class activity is present in all four whorls. In each of these cases, either A class or C class is present in each of the whorls. Thus, the lack of carpel development in *ag* mutants could be due to either an absolute requirement for *AG* in the specification of carpel identity or a repression of other genes involved in carpel development by the ectopic A class activity. This has been tested by constructing lines that lack both A and C class activities: *ap2 ag* double mutants and *ap2 pi ag* triple mutants (for simplicity, the triple mutant lacking A, B, and C class activities is often used since all of its organs are essentially identical; Bowman *et al.*, 1991a). Organs with features of both carpels (stigmatic tissue, fusion of organs along their margins, and ovules) and leaves (stellate trichomes and stipules) develop in the first whorl of *ap2 ag* double-mutant flowers and all whorls of *ap2 pi ag* triple-mutant flowers (Figs. 8F and 8H; see color plate). This leads to the conclusion that there must exist genes that can specify carpel development in the absence of *AG* gene function and that these genes are also negatively regulated by A class activity (Bowman *et al.*, 1991a). Stated another way, there exists a second pathway to carpel development that can act independently of, and parallel to, the program specified by *AG*.

Intriguingly, the carpelloid leaves that develop in the ABC triple mutant still possess most of those tissues that are usually thought to define carpels (Alvarez and Smyth, 1999). These organs may develop as solitary organs, or they may be fused with adjacent organs. Carpel tissues such as stigmatic papillae, style, septum, transmitting tract, placenta, and ovules can be identified, although when the carpels develop in a solitary fashion, their relative positions may be somewhat modified from that of wild type (Figs. 8F–H, see color plate; Alvarez and Smyth, 1999). Rather than being confined to the apex as they are in fused carpels, the stigmatic papillae and style tissue can be found all along the edges of the unfused organs. Adaxial to the marginal rim of stigmatic papillae are two parallel ridges of tissue: the one adjacent to the stigma is septal tissue containing transmitting tract, and the one most adaxial is placental tissue from which ovules develop (Figs. 8G and 8H; see color plate). The primary carpelloid components lacking in the organs of the ABC triple mutant are the specialized cell layers of the valve walls. Rather than consisting of the six distinct layers seen in wild type, the walls of these organs are sepal-like, being composed of spongy mesophyll cells internal to abaxial and adaxial epidermises reminiscent of sepals or leaves (Figs. 8G and 8H, see color plate; Alvarez and Smyth, 1999). The overall shape of these organs still resembles that of wild-type carpels: they have parallel margins and the "ovary" wall has a curvature similar to that of wild-type ovary walls. Thus, one carpel tissue that appears to absolutely require *AG* activity for appropriate differentiation is the valve.

4. Arabidopsis Carpel Development

A similar displacement of stigmatic and style tissues is also observed in the solitary carpels that develop in the first whorl of *ap2* flowers (Figs. 8C–E, see color plate; Alvarez and Smyth, 1999). In this regard, it is noteworthy that the other carpel tissue that appears to be lacking in solitary carpels, in either *ap2* or *ap2 pi ag* flowers, is the abaxial replum. However, in the fused carpelloid organs of the triple mutant (and in *ap2* single mutants where the first-whorl carpels are fused to one another), a tissue resembling that of wild-type replum tissue develops at the margin of fusion between the organs. One hypothesis is that the formation of the abaxial replum as a distinct tissue is induced during the process of organ fusion. In the absence of fusion, tissue that would normally differentiate into abaxial replum instead differentiates into style and stigma. These differences in the development of fused versus unfused carpels underscore the notion that there is communication between the fused carpels, a concept that has been noted previously (for review, see Verbeke, 1992).

Since there exists a second pathway to carpel development that can act independently of *AG*, it would be of interest to examine whether *LFY:VP16* is acting solely through *AG* to effect the differentiation of carpels. In light of the antagonism between the A and C class activities, the epistasis of *ag* mutations to the *LFY:VP16* transgene with respect to organ identity in the flower can be interpreted in two ways. Either the *LFY:VP16* transgene is acting solely through *AG* or, alternatively, in the *ag LFY:VP16* plants the second pathway is still inhibited by the action of the A class genes. Examination of *ap2 ag* double mutants which carry the *LFY:VP16* transgene could address this question. In this regard it is of interest to note that there is *AG*-independent differentiation of carpel tissues in *ap1 lfy* double mutants (Weigel and Meyerowitz, 1993). Thus, even if *LFY* does act through *AG*-independent pathways, the converse, that the alternate pathways require *LFY*, is not the case.

C. *SPATULA* Directs Differentiation of Carpel Tissues Independently of *AGAMOUS*

Plants carrying *spatula* (*spt*) mutations produce fruits that are flattened in the medial plane, primarily in the distal region (Alvarez and Smyth, 1999). The most conspicuous defect in *spt* carpels is the lack of transmitting tissue in both the style and the septum and an accompanying loss of the postgenital fusion of the septum (Fig. 9C). However, several other defects in carpel growth are evident, suggesting that *SPT* does not merely control the production of transmitting tract tissue. For example, mature (stage 12) *spt-2* carpels are sometimes unfused at the distal region, through the style and uppermost portion of the ovary (Fig. 9D). In *spt* carpels, the development of stigmatic papillae at the apex of the gynoecium is delayed in time and reduced in amount. That there are pleiotropic defects in carpel development is supported by the observation that the gynoecial cylinder displays altered

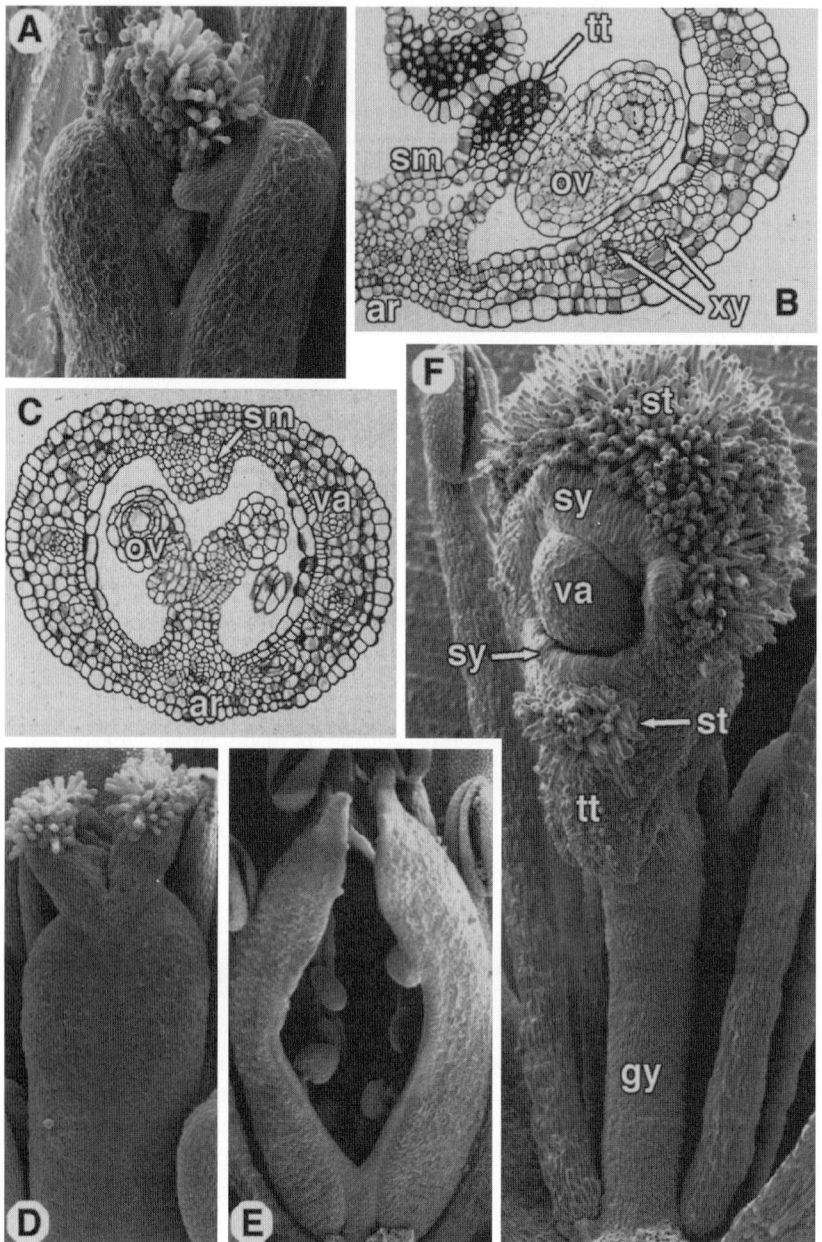

Fig. 9 *crc*, *spt*, and *ett* mutant phenotypes: (A) upper portion of *crc-1* gynoecium; (B) transverse section of *crc-1* ovary; (C) transverse section of *spt-2* ovary; (D) *spt-2* gynoecium; (E) *crc-1 spt-2* gynoecium; (F) *ett-3* gynoecium. ar, Abaxial replum; gy, "gynophore"; ov, ovule; sm, septum; st, stigmatic papillae; sy, style; tt, transmitting tract; va, valve; xy, xylem elements.

growth as early as stages 7 to 8; in *spt* gynoecia, the lateral regions of the tip of the gynoecial cylinder are higher than the medial, opposite of what is observed in wild type (Alvarez and Smyth, 1999). In the weaker mutant allele, *spt-1*, the postgenital fusion of the septum is more similar to that of wild type, even though transmitting tissue fails to develop. Thus, the development of these two tissues can occur independently. Despite the absence of transmitting tract tissue, the uppermost ovules of the carpel may be fertilized, indicating that transmitting tract tissue is not absolutely required for fertilization.

Mutations in *ag* are epistatic to mutations in *spt* (Alvarez and Smyth, 1999). To test whether this is due to *SPT* being downstream of *AG* or a result of ectopic A class function repressing *SPT* in *ag* mutants, *spt ap2 ag* triple and *spt ap2 pi ag* quadruple mutants were constructed. Analysis of these genotypes has provided a broader understanding of the role that *SPT* plays in carpel development (Alvarez and Smyth, 1999). The first-whorl organs of *spt ap2 ag* triple mutants and all floral organs of the quadruple mutant exhibit reduced carpellody, with no signs of stigmatic papillae, style tissue, septum, or transmitting tract evident (Figs. 8I and 8J; see color plate). The only marginal structures of the floral organs are small outgrowths which display some ovulelike characteristics (Fig. 8I; see color plate). That there still exist outgrowths suggests that additional genes promoting carpelloid growth are still active in this background. These organs often have stellate trichomes on their abaxial surfaces, an indication that these organs are leaflike (Fig. 8J; see color plate). *SPT* must be active in promoting the development of most of the tissues associated with carpel development, and this function is, at least in some contexts, redundant with the action of *AG*. *SPT* also influences the growth pattern of the carpels. The organs of the *spt ap2 pi ag* quadruple mutants are narrow and flattened, with margins that are not parallel, but are more leaf- or sepal-like in overall shape (Fig. 8J; see color plate). Thus, in the case of *SPT*, the relative simplicity of its single mutant phenotype belies the role *SPT* plays directing several aspects of carpel growth and differentiation.

Assuming that the *spt-2* allele is close to null, two possible interpretations of *SPT* function are that (1) *SPT* acts to impart carpel identity on primordial cells or (2) the primary function of *SPT* is to promote the growth of the marginal tissues of the carpel. These two hypotheses are not necessarily mutually exclusive. In support of the first hypothesis, *SPT* appears to be a primary player in the development of *AG*-independent carpel differentiation, as evidenced by the *spt ap2 pi ag* quadruple mutant phenotype. In the quadruple mutants, not only are the marginal tissues affected but also the overall shape of the organ is altered, suggesting either that *SPT* is active throughout the carpel or that differentiation of marginal tissues can influence growth patterns of the central regions of the organ. The first hypothesis would be further supported if ectopic expression of *SPT* can impart a fate of carpel development on other floral organs. The second interpretation is based on the observation that most of the tissues that are affected in *spt* single and multiple mutant combinations are those derived from the marginal regions of the carpels. If the expression of *SPT* is localized to the marginal regions of the carpels, then the second interpretation may gain some credence. Regardless of whether it imparts carpel or regional specificity, *SPT* is critical for the development of several carpel tissues, and that it can act independently of *AG* to direct differentiation of carpel tissues.

D. *CRABS CLAW* Controls Several Aspects of Carpel Differentiation

Mutations in *CRABS CLAW* (*CRC*) result in defects in several aspects of carpel development (Alvarez and Smyth, 1999). First, the gynoecium is shorter and broader than that of wild type. Analysis of developing *crc-1* (a putative null allele) carpels reveals that the primordium at stage 6 is larger in the lateral and medial dimensions than that of the wild type. Second, the apical region fails to fuse properly, with a concomitant loss of style tissue (Fig. 9A). It is unclear whether this is a reflection of earlier developmental defects such as the uneven growth of the gynoecial cylinder at its apex or a failure to develop style tissue later in development. Third, the number of ovules produced is reduced to less than half of that of wild type. Fourth, the abaxial epidermis of the carpel and the vasculature of the valves differentiate precociously, and ovules may occasionally develop from the abaxial epidermis. The development of a reticulate pattern of tracheary elements between the major vascular bundles in developing gynoecia, an event that in wild type occurs primarily during fruit development (Sessions and Zambryski, 1995), is an indication of precocious differentiation of the valves (Fig. 9B). Fifth, there is a slight loss of floral meristem determinacy, which is manifested by the development of a third carpel in a fraction of gynoecia of *crc* flowers. This rather nebulous mutant phenotype could be interpreted to mean either that *CRC* plays a number of different and separable roles in carpel development or that it encodes a protein that acts as a general growth regulator in carpel development. In support of the former interpretation, the phenotypic defects of *crc* mutants appear, at least in part, to be genetically separable, and the mRNA expression pattern of *CRC* consists of distinct and separable domains.

CRC encodes a protein containing a putative zinc finger and a helix–loop–helix domain, which has sequence similarity to the HMG-1 box (Landsman and Bustin, 1993), suggesting that it may be part of the transcriptional machinery (Bowman and Smyth, 1999). Consistent with its mutant phenotype, *CRC* mRNA is detected at stage 6 in two lateral hemispherical domains of the carpel primordium (Fig. 10A; see color plate). Subsequently, the expression pattern is dynamic and ephemeral, with two distinct expression domains in the carpel: the epidermal and internal domains. The epidermal carpel domain consists primarily of the abaxial epidermal cell layer and this domain may be a continuation of the early *CRC* expression in the lateral regions of the stage 6 carpel primordium. The second domain in the carpel is restricted to four internal groups of cells adjacent to the pla-

4. *Arabidopsis* Carpel Development

cental tissue. One explanation of the complex pattern of *CRC* expression is that the different domains of expression could be responsible for distinct phenotypic effects. To some extent we can correlate these utilizing the *crc-2* allele, a promoter mutation in which the internal expression is intact but the epidermal expression domain is largely missing. The *crc-2* mutant phenotype includes both a partial loss of floral meristem determinacy and a lack of fusion at the apex of the carpels, although the loss of fusion (and the reduction in style tissue) is not as severe as that observed in *crc-1* carpels. In contrast, ovule number in *crc-2* mutant carpels is similar to that of wild type. This suggests that the internal expression domain of the carpel is related to the number of ovules produced by the placental tissue and contributes to the formation of style tissue and the apical fusion of the carpels. Likewise, the epidermal carpel expression domain can be correlated with the control of floral meristem determinacy, and again, to at least some extent, the development of the style and the apical fusion of the carpels.

The multifunctional nature of *CRC* is also revealed in its genetic interactions with other carpel regulators. For example, flowers of the *crc ap2 pi ag* quadruple mutants still display development of stigma, septum, and placental tissue along the margins of the floral organs, but the development of these tissues is irregular and greatly reduced as compared to that of the *ap2 pi ag* triple mutant (Alvarez and Smyth, 1999). One tissue that does appear to be largely missing in the quadruple mutants is the style; this is consistent with the *crc* single mutant phenotype, where the style is also largely absent. The shape of the floral organs of *crc ap2 pi ag* quadruple mutants is affected; they are shorter, broader, and flatter relative to those of the *ap2 pi ag* triple mutant, again consistent with the *crc* single mutant. Thus, *CRC* plays a role along with *AG* in the development of some carpel tissues, particularly the style. *CRC* is one of the genes active in the *spt ap2 pi ag* quadruple mutants, as the *crc spt ap2 pi ag* pentuple mutant exhibits a reduction in carpelloid outgrowths as compared to the quadruple mutant (Alvarez and Smyth, 1999). However, there must exist additional genes active in this background since some carpelloid outgrowths still develop in the pentuple mutant. Partial redundancy of *CRC* and *SPT* function is revealed by examining carpels of *crc spt* double mutants (Fig. 9E; Alvarez and Smyth, 1999). In the double mutants, carpels develop that are strikingly different in several aspects as compared to carpels of each of the single mutants. The two carpels of the gynoecium are unfused, except at their base, and exhibit a reduction of other carpel tissues, including stigmatic papillae and style tissue, a reduction in ovule number, and a complete loss of septal tissue (Fig. 9E). These data suggest that both *CRC* and *SPT* share roles in many aspects of carpel development.

Genetic interactions between *crc* and *ag* reinforce the idea that *CRC* regulates floral meristem determinacy (Alvarez and Smyth, 1999). Although *ag* mutations are epistatic to *crc* mutations with respect to defects in the carpel (in *ag* mutants, no carpels ever develop), plants homozygous for *crc* and heterozygous for *ag* exhibit a unique phenotype. The floral meristem in *crc ag/+* flowers is often indeterminate, with successive whorls of stamens and carpels developing internal to the fourth-whorl carpels, thus confirming a role of *CRC* in controlling floral meristem determinacy. The phenotype of these plants also raises a more general observation: although most mutations affecting flower development that have been isolated are recessive in a wild-type background, many display semidominance in genetic backgrounds in which other genes are also compromised in their function. For example, plants of the genotypes *sup ag/+* (Schultz *et al.*, 1991), *ap1 cal/+* (Bowman *et al.*, 1993), *ag ap2/+* (Bowman, 1994), *clv1/+ clv3/+* (Clark *et al.*, 1995), and *crc spt/+* (Alvarez and Smyth, 1999) all have phenotypes distinguishable from their respective single mutant phenotypes.

Screens for genetic enhancers of *crc* have been enlightening in elucidating functions of *CRC* during carpel development, with three distinct classes of genetic enhancers being identified to date (Y.E. and J.L.B., unpublished). The first class of mutants is characterized by a loss of fusion of the carpels in the gynoecium, such as the interaction previously observed in *crc spt* flowers (known genes identified that fall into this class include *LEUNIG, ETTIN*, and *PERIANTHIA*, some of which are described in more detail below). A second class consists of mutants whose interactions with *crc* result in a loss of floral meristem determinacy, such as the interaction previously observed in *crc ag/+* flowers. The third class displayed a novel phenotype: the external (abaxial) development of tissues normally only found internally (adaxially) in the carpel (e.g., transmitting tract tissue, septum, placenta, and ovules). That the single mutant phenotypes of most of the mutations of the second and third classes are indistinguishable from wild type again suggests that genetic redundancy is prevalent in many developmental processes and that such screens in other genetic backgrounds may prove informative. In the case of *crc*, the isolation of enhancers that exhibit defects in cell fate specification indicates that one of the roles of *CRC* is to help specify the identity of tissues in the abaxial regions of the carpel, a role that was not readily apparent from analyzing its single mutant phenotype.

E. *ETTIN* May Specify or Respond to Regional Identity in the Carpel Primordium

A remarkable feature of loss-of-function *ettin* (*ett*) alleles is the major redistribution of tissue types found in the gynoecium, suggesting that *ETT* plays a role in establishing or interpreting regional information in the gynoecial primordium (Fig. 9F; Sessions and Zambryski, 1995; Sessions, 1997; Sessions *et al.*, 1997). The redistributions of tissues are distinct in the medial versus the lateral domains. Gynoecia of plants carrying a null *ett* allele have greatly reduced valve tissue (and in many cases lack valve tissue altogether), with the basal boundary of the valve shifted apically (Fig. 9F). Rings of style tissue and stigmatic papillae often surround each of the reduced patches of valve tissue when these patches are less than approximately one third of their normal length (Fig. 9F). In cases of larger patches of valve tissue, the style and stigma do not develop at the base of the valve. In the lateral plane, the lost valve tissue is replaced by a tissue that has characteristics of both the abaxial style (i.e., style epidermis) and the gynophore, resulting in a structure that superficially resembles an extended internode, or gynophore (Fig. 9F). However, this structure often contains ovary-like adaxial tissues, such as placentae and ovules, suggesting it is not a true gynophore. In contrast, the medial domain, which would ordinarily differentiate as abaxial replum, develops as adaxial style tissue, with copious transmitting tract tissue developing from both the epidermal and subepidermal cell layers. Where valves occur, forming an ovary, placental tissue with accompanying ovules and an incomplete septum develop from the adaxial region of the reduced cylinder. Each of these phenotypes occurs in an allele strength dependent manner, such that partial loss-of-function alleles progressively retain more valve tissue and the ectopic growth of stylar transmitting tract tissue is confined to more apical regions in the medial plane. The rather complex phenotype of *ett* can be simplified if one considers that in *ett* gynoecia, all of the tissues normally identified with carpels are present, but there are alterations with respect to their apical–basal and abaxial–adaxial distributions. Thus, *ETT* provides the clearest example yet of a gene involved early in specifying or responding to regional domains in the gynoecial primordium.

A model proposed to explain the function of *ETT* in the gynoecium is based on the idea that *ETT* defines boundaries in the gynoecial primordium, thus providing regional positional information (Sessions, 1997). In this model, *ETT* activity establishes two boundaries, one at the abaxial base of the gynoecial primordium (the basal boundary) and a second boundary defining the border between the abaxial and adaxial tissues of the primordium (the apical boundary). These boundaries would define compartments of the gynoecial primordium, with the different compartments fated to differentiate into different tissue types. During the development of wild-type gynoecia, the basal boundary would remain at the base of the carpel, while the apical boundary would remain at the distal rim of the developing cylinder, delimiting the abaxial and adaxial tissues of the carpels. In *ett* mutants, the apical displacement of the basal boundary results in the tissues of the carpel being shifted apically in both the lateral and medial domains, while basal displacement of the apical boundary results in a basal displacement of the abaxial–adaxial boundary in the medial plane. To account for the development of concentric rings of valve, style, and stigmatic tissues in intermediate *ett* alleles, Sessions (1997) also postulates that the apical boundary may be duplicated or split below the basal boundary in some instances. Interestingly, the expression pattern of *ETT* is consistent with its proposed role in defining the boundaries described earlier (Sessions *et al.*, 1997). *ETT* is expressed in stage 5–6 flower meristems in a ring of cells in a pattern that would be expected if the *ETT*-expressing cells are fated to become the abaxial tissues of the carpel primordium (Fig. 10C; see color plate). Expression persists in the abaxial cells of the carpel primordium from stages 6 through 8 and then ceases prior to differentiation of tissue types from the developing cylinder (expression is later seen in the developing vasculature traces of the carpel). In this regard, analysis of the expression in *ett* gynoecia of other early markers of abaxial or adaxial cell fate would be informative to determine where in the hierarchy of specifying regional information *ETT* may be acting. Thus, the expression pattern of *ETT* is consistent with a role in establishing or responding to boundaries. A functional test to determine whether *ETT* is involved in establishing boundaries would be to examine whether ectopic expression of *ETT* is sufficient to induce ectopic boundaries in certain contexts by using promoters that drive expression in different subsets of cells in the developing gynoecium (e.g., Lawrence and Struhl, 1996).

Although the model successfully describes the *ett* single mutant phenotypes, several questions remain concerning the phenotypes of double mutants involving *ett*. In *ap2-2 ett* double mutants, only the margins of the first-whorl carpels are affected by the *ett* mutation, with the valves developing in a near-normal fashion, indicating that *ETT* is not required for valve differentiation per se (Sessions, *et al.*, 1995; Sessions *et al.*, 1997). The difference between the first- and fourth-whorl carpels is not likely due to a whorl-specific effect, since *ETT* is expressed in the first-whorl carpels in manner similar to those of the fourth whorl (Sessions *et al.*, 1997). In support of this, *crc ett* double mutants have unfused fourth-whorl carpels where the phenotypic effects of *ett* are mitigated such that near-normal valves develop on most carpels (J. L. B. and J. A., unpublished observations). However, in *crc ett* gynoecia, placenta and other adaxial tissues develop ectopically along the abaxial margins of fusion of the carpels (J. A. and D. R. Smyth, unpublished results). These results suggest that the medial (margins) and lateral domains are responding in a different manner to lack of *ETT* activity. Perhaps *ETT* has distinct functions in each of these domains, by interacting, or being redundant, with different factors in these two domains. Alternatively, *ETT* might be a component of a communication pathway connecting fused carpels. In this regard, an important observation is that *spt* mutations are largely epistatic to *ett* mutations (Alvarez and Smyth, 1998). Fused carpels with near-normal valve and abaxial replum tissues develop in the *spt ett* double mutants, reinforcing the idea that *ETT* is not absolutely required for the differentiation of any particular tissue type within the carpel. The epistasis of *spt* mutations over *ett* mutations suggests that the *ett* mutant phenotype may be due to ectopic *spt* activity and that *ETT* is involved in restricting the growth and differentiation of marginal tissue types to the appropriate locations (Alvarez and Smyth, 1998). Functional tests of this hypothesis require the ectopic expression of *SPT* and examination of its expression pattern in wild-type and *ett* mutant gynoecia.

Intriguingly, *ETT* encodes a putative transcription factor with sequence similarity to the DNA-binding domains of *ARF1* and *IAA24*, which has been shown to encode *MONOPTEROS* (*MP*), a gene involved in formation of vascular strands (Kim *et al.*, 1997; Ulmasov *et al.*, 1997; Hardtke and Berleth, 1998). *ARF1* and

IAA24/MP belong to a large family (the ARF family) of genes encoding proteins that are implicated in binding auxin response elements in the promoters of early auxin response genes (Abel and Theologis, 1996; Ulmasov *et al.*, 1997). The proteins encoded by the *ARF1* and *IAA24/MP* genes share a second region of similarity that has been postulated to be a protein–protein interaction domain, facilitating interactions between family members. In addition, several of the early auxin response genes encode members of the ARF family, or the related Aux/IAA family, whose members encode proteins with the protein–protein interaction domain, but not the DNA-binding domain (Abel and Theologis, 1996; Guilfoyle, 1998). Interactions between members of these large gene families provide a myriad of possibilities in terms of the transcriptional regulation of target genes in response to auxin, and at least one member has been shown to be directly involved in auxin signal transduction (Rouse *et al.*, 1998).

Although *ETT* lacks the protein–protein interaction domain found in many of the other ARF family members, an attractive hypothesis is that *ETT* is somehow involved in interpreting positional information ultimately derived from auxin signals. That auxin may be involved in patterning the gynoecium is supported by the phenotypes of *pinoid* (*pid*) and *pin-formed* (*pin*) mutants in which most of the adaxial tissues of the gynoecium are lacking (Goto *et al.*, 1991; Okada *et al.*, 1991; Bennett *et al.*, 1995). Gynoecia of *pid* and *pin* flowers usually consist of a solid stemlike structure, capped with stigmatic papillae; in the cases where valves form, they are shifted apically, as in *ett* mutants. The link between the *PID* and *PIN* genes and auxin is circumstantial: inhibitors of polar auxin transport can phenocopy their mutant phenotypes (Okada *et al.*, 1991) and polar auxin transport has been reported to be greatly reduced in *pin* (Okada *et al.*, 1991). However, this reduction is conditional upon developmental age, and close to wild-type levels of polar transport can be observed in rapidly growing stems of both *pin* and *pid* mutant plants (Bennett *et al.*, 1995). Perhaps more convincing evidence for these genes' involvement in auxin signaling comes from the analysis of *pid axr1* double mutants, which display a severe synergistic interaction (Bennett *et al.*, 1995); *AXR1* encodes a protein involved in an auxin signaling pathway (Leyser, 1998). One attractive model to account for these observations is that auxin, or an auxin mediated signal produced in the gynoecial primordium, establishes polarity or, alternatively, induces appropriate patterns of tissue growth. Without more data, further speculations on molecular mechanisms are unconstrained; however, such a model immediately suggests a plethora of relevant experimental approaches.

F. *FRUITFULL* Controls Valve Maturation

Mutations in *FRUITFULL* (*FUL*; also known as *AGL8*) results in inappropriate maturation of the cells of the valves (Gu *et al.*, 1998). Other carpel tissues, such as the replum, septum, and placental tissue, appear largely unaffected. Defects in valve differentiation are detected before anthesis but are most obvious postfertilization during fruit development, when the valves fail to elongate appropriately in *ful* mutants. Despite the failure of the valve cells to elongate during fruit development, the cells of the replum and septum apparently elongate normally (Gu *et al.*, 1998). The lack of valve expansion results in fruits that often split open due to the expansion of the relatively normal seeds developing within (Fig. 11C; see color plate). Examination of stage 12 *ful* carpels reveals that the six cell layers of the valve form and differentiate from one another appropriately but that each of the layers fails to mature properly (Figs. 11A and 11B, see color plate; Gu *et al.*, 1998). In addition, the adaxial epidermis of *ful* mutants consists of more than twice as many cells than the wild type, indicating a failure in control of cell division in this layer (Fig. 11A, see color plate; Gu *et al.*, 1998). The adaxial subepidermal layer also appears to be much reduced in its axial elongation as compared to wild type (Fig. 11B; see color plate). That the valves of *ful* mutants fail to differentiate and elongate appropriately even though the rest of the tissues of the carpel develop properly indicates that the differentiation of these tissues can occur independently, perhaps reflecting the hypothesized early division between marginal (replum/septum) and central (valve) domains within the developing carpel.

The *FUL* gene was shown to encode *AGL8*, a gene previously identified on the basis of sequence similarity with *AG* (Mandel and Yanofsky, 1995). *FUL* is expressed in all cell layers of the developing valves (Fig. 12A, see color plate; Mandel and Yanofsky, 1995; Gu *et al.*, 1998); however, it is not clear when its expression becomes restricted to the valves. Ectopic expression experiments in which *FUL* is constitutively expressed suggest that its expression may be sufficient to confer valve identity on cells that would normally differentiate as abaxial replum tissue and cells normally fated to become style tissue (Liljegren *et al.*, 1998). However, ectopic expression of *FUL* is not sufficient to confer valve identity to other tissues of the carpel. Thus, *FUL* is sufficient only in certain contexts within the carpel to confer valve identity, and conversely, loss-of-function mutants do not lack valve identity but rather have valves that do not mature properly. This suggests that *FUL* interacts with other factors to confer valve identity and that only where these other factors are also present is its expression sufficient to specify valve identity. An interesting question is the relationship between *FUL* and *AG*: since proper differentiation of the valve requires *AG* activity, it may be that either the expression or the function of *FUL* in the carpel valve is downstream of *AG*. Consistent with this hypothesis, the expression pattern of *FUL* from stage 3 through stage 6 is a subset of that of *AG*. Alternatively, *FUL* may not be downstream of *AG*, but rather *FUL* could act in conjunction with *AG* to specify valve fate. Analysis of *FUL* expression in *ap2 pi ag* triple mutants and construction of *ful ap2 pi ag* quadruple mutants could help clarify the relationship between *FUL* and *AG*.

FUL is also expressed in the inflorescence meristem, the vasculature of the developing inflorescence stem, and other regions of the plant (Mandel and Yanofsky,

1995). Its function in these other regions is likely to be redundant with other genes since the *ful* single mutant phenotype only exhibits some cauline leaf defects in addition to its carpel defects (Gu *et al.*, 1998). Thus, *FUL* is another example of a gene that is utilized multiple times during shoot development, and only its role in valve differentiation is revealed by its single mutant phenotype.

G. *AGL1* and *AGL5* Are Involved in Defining Tissues at the Valve–Replum Boundary

The *AGL1* and *AGL5* genes were also originally identified by their sequence similarity with *AG* (Ma *et al.*, 1991). The expression patterns of *AGL1* and *AGL5* are indistinguishable (Savidge *et al.*, 1995; Flanagan *et al.*, 1996; Y.E. and J.L.B., unpublished) and sequence analysis suggests that they reflect a recent gene duplication event. In mature carpels they are expressed in a few cell layers at the valve–replum boundary, in the epidermis of the septum, in the placental tissue, and in ovules. The expression at the valve–replum boundary persists well past fertilization into fruit development. Their expression in carpels commences during stage 6, when they are expressed in two medial domains of the carpel primordium (Fig. 10B; see color plate). It is likely that the later expression pattern reflects a continuation and further refinement of the initial medial expression domains in the primordium. Given their similar sequence and expression patterns, it is not surprising that they are functionally redundant. Single mutants in either gene do not exhibit a phenotype distinguishable from the wild type, but in *agl1 agl5* double mutants, the valve–replum boundary fails to differentiate appropriately, resulting in fruits that do not dehisce (Liljegren *et al.*, 1998). Thus, these genes are required for the proper differentiation of the valve–replum boundary within the carpel and perhaps the continued differentiation of those cells in the dehiscence zone in the developing fruit. No phenotype is observed in the other tissues in which *AGL1* and *AGL5* are expressed, suggesting that their function in these tissues may be redundant with other genes. A candidate for such a gene is another MADS-box gene, *AGL11*, which is expressed only in a subset of the tissues expressing *AGL1* and *AGL5*: the epidermis of the septum, the placental tissue, and developing ovules (Rounsley *et al.*, 1995). Thus, triple mutants, at the minimum, will have to be analyzed before an assessment can be made as to the function of *AGL1*, *AGL5*, and *AGL11* in these tissues.

In contrast to most of the other genes discussed in this review, *AGL1* and *AGL5* (and possibly *AGL11*) are likely to function downstream of *AG* and may represent direct targets of *AG*. This has been hypothesized for *AGL5* based on the ability of ectopic *AG* expression to activate an *AGL5-GUS* reporter gene (Savidge *et al.*, 1995). Consistent with this proposal, neither *AGL1* nor *AGL5* is expressed in *ag* mutants, and the lack of *AGL5* expression in *ap2 ag* double mutants suggests that *AG* may be absolutely required for its expression. The expression patterns of

AGL1/AGL5 are only a subset of the domain of *AG*, both spatially and temporally, indicating that *AG* acts in conjunction with other more spatially restricted factors to activate *AGL1/AGL5*. Based on their mutant phenotypes and restricted expression patterns, *AGL1* and *AGL5* may represent genes involved in specifying tissue or cell types within the developing carpel. Although they are expressed in the medial domains of stage 6 carpel primordia, they are unlikely to be involved in specifying regional identity within the primordia as their ectopic expression throughout the carpel has little phenotypic effect on carpel development (Y. E. and J. L. B., unpublished). In this interpretation, *AGL1/AGL5* would be responding to earlier acting genes which specify carpel identity (e.g., *AG*) and to as yet unidentified genes which specify regional identity within the carpel primordium and subsequently acting to direct the differentiation of specific tissue and cell types.

Phylogenetic analyses indicate that the four genes *AG, AGL1, AGL5,* and *AGL11* form a monophyletic clade within the *Arabidopsis* MADS-box gene family (Purugganan *et al.*, 1995; Münster *et al.*, 1997; Liljegren *et al.*, 1998). *AGL1* and *AGL5* are more closely related to each other than they are to either *AG* or *AGL11*. Thus, an attractive evolutionary scenario is that the ancestral *AG* gene underwent two successive gene duplications to give rise to *AGL11* and the *AGL1/AGL5* progenitor, which subsequently also experienced a gene duplication. Each of the genes *AGL1, AGL5,* and *AGL11* either assumed a subset of the functions of the ancestral *AG* gene or acquired new functions associated with carpel development, and the genes appear to have fallen under the control of *AG* activity. When these gene duplications occurred may be assessed by assaying for orthologous genes in various vascular plants. Although systematic searches have not been done in many species, genes that appear to be functionally orthologous to *AG* have been found in evolutionarily diverse angiosperms, including snapdragons (Bradley *et al.*, 1993), petunia (van der Krol and Chua, 1993), and maize (Mena *et al.*, 1996), and apparent functional *AGL11* orthologues have been isolated from petunia (Colombo *et al.*, 1995). Putative *AG*-like genes have also been identified in gymnosperms (Tandre *et al.*, 1995), although in this case, orthology to a particular gene of the *AG* clade is not unequivocal. In this regard, interspecific comparisons of the functions of the genes of the *AG* clade will have to await data gathered from a broader phylogenetic context; however, it is tempting to speculate that diversification of genes within this clade could contribute to the morphological diversity of carpels and fruits of the angiosperms.

H. Genes with Pleiotropic Mutant Phenotypes

Four other genes, mutations in which result in phenotypic defects in the carpel, are worth mentioning, although presently, specific models of their functions are difficult to formulate. Mutations in each of these genes, *TOUSLED* (*TSL*; Roe *et al.*, 1993), *LEUNIG* (*LUG*; Liu and Meyerowitz, 1995; Roe *et al.*, 1997), *AINTEGU-*

MENTA (*ANT*; Elliott *et al.*, 1996; Klucher *et al.*, 1996), and *PERIANTHIA* (*PAN*; Running and Meyerowitz, 1996; Roe *et al.*, 1997), result in defects in fusion at the apex of the carpel. The style is severely affected in both *tsl* and *lug* gynoecia: it is largely absent in *tsl* mutants, whereas in *lug* mutants hornlike projections are present in the lateral style regions, and the medial style regions are reduced. In *ant* gynoecia the defects are variable, in both the frequency and the extent of loss of fusion. In contrast, the defects in *pan* gynoecia are more subtle, with loss of fusion rarely occurring but with the style appearing narrower during its development. Thus, all four genes appear to be involved in the proper fusion of the gynoecium and also in the development of tissues of the style. Three general conclusions can be drawn from the study of these mutants. First, in all double mutant combinations examined to date, either among themselves or with *crc* and *ett* (*crc pan*, *crc lug*, *crc ant*, *crc ett*, *tsl lug*, *tsl ett*, *tsl pan*; Roe *et al.*, 1997; Y. E. and J. L. B., unpublished), the carpels of the gynoecium are essentially unfused to their bases, suggesting that the process of fusion is easily perturbed and that mutations resulting in this phenotype may be affecting the fusion event in an indirect manner, such as resulting in uneven growth of the gynoecial cylinder early in development. Second, in some of the double-mutant combinations, many of the tissues of the carpel fail to develop. An example is the development of placental tissue with ovules independent of the formation of other carpel tissues in *tsl ett* flowers. This suggests that many carpel tissues may be able to develop independent of the formation of other tissues. Third, each of the double mutant combinations analyzed to date exhibits more severe phenotypes than might be predicted from the two single mutants involved, suggesting that single mutant phenotypes will often only partially reveal the nature of that gene's function in development.

IV. Lessons from Analyses of Gene Expression Patterns

Another approach to identify candidate regulators of carpel development is to isolate genes that display distinct temporal or spatial expression patterns in the developing carpel. Primarily, two methods have been employed, differential screening techniques and generation of enhancer trap lines. What follows is by no means a comprehensive review of the field, but rather a discussion focused on two aspects of gene expression: How many genes are involved in carpel development? Do expression patterns delineate compartments within the developing carpel?

Initial differential screening studies focused on identifying genes that are predominantly expressed in tomato carpels, and these studies were limited to some extent by the level of sensitivity inherent in the techniques employed (Gasser, 1991). These studies primarily identified genes expressed at moderate to high levels, with a preponderance of the genes encoding enzymatic proteins and being expressed relatively late in carpel development in the transmitting tract, stigmatic papillae, or the ovules (reviewed in Gasser, 1991; Gasser and Robinson-Beers,

1993; Milligan and Gasser, 1995). Thus, the genes isolated have been interpreted to encode "downstream" genes, rather than regulatory genes. Recently, more sensitive differential display techniques have been employed extending differential screening analyses to the *Arabidopsis* gynoecium (Yung *et al.*, 1999). In this study, potential regulatory proteins were identified along with several genes encoding enzymatic functions. With the advent of the *Arabidopsis* genome project, such studies will soon employ DNA microarray chip technology and, consequently, will be more sensitive and comprehensive. In addition, such technologies could prove useful in the identification of genes downstream of known regulatory genes (e.g., Ito *et al.*, 1997; Sablowski and Meyerowitz, 1998). Despite the promise of being able to identify all genes expressed in the carpels (and determining those that are specific and those that are not), such studies will still require extensive gene expression analyses to determine cellular expression patterns and the application of reverse genetics to elucidate gene function. Nonetheless, such information will be invaluable in identifying candidate regulatory genes and, hence, directing future experiments. A complementary approach is to generate enhancer trap lines utilizing a marker gene that allows detection of gene expression (Topping *et al.*, 1991; Sundaresan *et al.*, 1995). A potential advantage of enhancer traps over differential screening techniques is that enhancer trap lines are not biased to identifying genes that are expressed at a high level but rather should identify genes with approximate equal frequency.

A. How Many Genes?

A general question in development is how many genes are active in any particular organ type. Considering the current estimate of 21,000 genes in the *Arabidopsis* genome (Ecker, 1998), an earlier estimate based on hybridization experiments in tobacco of 25,000 expressed sequences in the carpels (Kamalay and Goldberg, 1980) is a significant overestimate, at least for *Arabidopsis*. More reliable estimates could come from both DNA microarray analyses and studies on enhancer trap lines. Given the sensitivity of the microarray technology, using probes made from the mRNA of a number of genotypes could potentially allow categorization of all 21,000 genes according to their expression patterns in various organ types. The primary limitation (other than the inherent sensitivity of the method) is the extent to which tissues can be dissected, either manually or genetically, to generate potential probes. Using such approaches, it may be feasible to ascertain the number and specificity of genes expressed in each organ type, comparisons between which could prove illuminating. A second question that follows is how many genes exhibit carpel-specific expression? A preliminary figure can be ascertained from analysis of enhancer trap lines; for example, screening of approximately 7000 potential enhancer trap lines yielded 140 lines that exhibit some expression in the carpel, of which 19 lines exhibit expression patterns limited (within the flower) to the carpels (Y. E. and J. L. B., unpublished). It is notable that of these 19 lines, 11 exhibit expression limited to the style or transmitting tract, an observation in accordance with earlier studies utilizing differential screening of mRNAs (Gasser, 1991). Thus, one general conclusion that can be drawn from this study is that there may be few genes whose expression is limited to a specific organ type (exceptions to this might be in the specialized cell types of the pollen, embryo sacs, and transmitting tract; McCormick, 1993; Drews *et al.*, 1998; Grossniklaus and Schneitz, 1998; Howden *et al.*, 1998). Indeed, none of the genes (for which expression patterns are known) described earlier as having critical roles in carpel development (*AG, CRC, ETT, AGL8, AGL1/5*) have their expression limited to the carpel. Such repeated use of genes in different developmental contexts is a common feature in other systems, such as *Drosophila* and *Caenorhabditis elegans*, and has been noted previously in *Arabidopsis* (Sundaresan *et al.*, 1995).

B. Gene Expression Patterns May Delineate Compartments in Developing Carpels

One potential use of surveys of gene expression patterns is to ascertain whether any recurrent patterns emerge that may define developmental compartments or domains. Combining analyses of enhancer trap lines with data on expression patterns of known genes provides a database from which we can begin to address this question. First, the stage 6 carpel primordium is not uniform with respect to gene expression patterns. Already at this stage, the primordium is divided into medial versus lateral domains by the expression of *AGL1* and *AGL5* in the medial domains (Fig. 10B; Savidge *et al.*, 1995; Flanagan *et al.*, 1996) and *CRC* and enhancer trap lines TJ65 and YJ35 in the lateral domains (Fig. 10A; Bowman and Smyth, 1998; Y. E. and J. L. B., unpublished). Likewise, the abaxial and adaxial domains are demarcated by the restriction of *ETT* expression to the abaxial domain of the carpel primordium (Fig. 10C; Sessions *et al.*, 1997). It is unlikely that any of the aforementioned genes (except possibly in the case of *ETT*) are directly involved in establishing these early domains, but rather they are responding to positional information encoded by earlier acting genes. From stages 6 through 8, *AG* mRNA is detected throughout the primordium (Fig. 10D, see color plate; Drews *et al.*, 1991), suggesting that the division of the carpel primordium into regional domains is likely to be independent of *AG*.

Later in carpel development, when tissue and cell type differentiation is occurring, two major domains can also be identified. One domain is composed of the valves, which express *FUL*, as well as enhancer trap lines YJ92 and YJ35 (Fig. 12A, see color plate; Gu *et al.*, 1998; Y. E. and J. L. B., unpublished). A second domain can be considered to be composed of the abaxial replum, the septum with transmitting tract tissue, and the placental tissue, with accompanying ovules, what is traditionally termed the replum. Several genes have their expression limited to regions of this domain: e.g., *AGL1*, *AGL5*, *AGL11*, *AP2*, *KNAT2*, *STM*, and *KNAT1* as well as several enhancer trap lines (Figs. 12B and 12C; Jofuku *et al.*, 1994; Dockx *et al.*, 1995; Rounsley *et al.*, 1995; Savidge *et al.*, 1995; Flanagan *et al.*, 1996; Long *et al.*, 1996; Y. E. and J. L. B., unpublished). A direct connection between the gene expression domains in the stage 6 carpel primordium and the later expression domains has not been established, but considering that many of the genes identified as defining the early expression domains encode transcription factors, it is logical that some of the expression patterns observed later in development are a reflection of the early domains. In addition, many of the genes expressed in the early stage 6 domains become restricted to specific tissue or cell types within these domains later in development. For example, *AGL1* and *AGL5* are expressed in the medial domains during stage 6 and then become localized to parts of abaxial replum, the septum, placenta, and ovules later in development (Savidge *et al.*, 1995; Flanagan *et al.*, 1996). The preceding discussion is by no means meant to imply that there are two mutually exclusive domains (medial vs lateral) with regard to gene expression in the carpels, as there are ample examples of genes whose expression patterns span the two domains (Figs. 10C, 10D, 12D, and 12E; see color plate). Nonetheless, the prevalence of distinct expression patterns is suggestive of defining developmental compartments.

Finally, several enhancer trap lines display tissue- or cell-type-specific expression patterns, providing valuable markers for the analyses of mutant phenotypes in which the locations or prevalence of particular tissues may be altered. In essence, all of the enhancer trap lines (as well as the expression patterns of known genes) provide tools for fine-structure histological analysis of tissues. Classically, tissue and cell types were identified by their unique histology using traditional staining methods. Gene expression patterns merely extend these histological analyses by allowing the examination of another molecular marker, that of gene expression. The major revelation using these types of analyses is the enormous number of discrete cell types, if one defines a cell type as possessing a unique molecular constituency resulting from a distinct pattern of gene expression. Thus, based on gene expression patterns, many of the tissue types defined earlier in this review can be subdivided, as they are complex mosaics of gene expression patterns.

V. Conclusions

A. Pattern Formation in the Carpel

1. Specification of Carpel Identity

The specification of carpel identity appears to be achieved by at least two separate pathways, a pathway mediated by *AG*, and an *AG*-independent pathway(s). Both of these pathways are negatively regulated by A class genes. *AG* likely confers carpel identity in two ways. First, it initiates a cascade of gene expression that leads to cells acquiring a developmental fate of carpel identity. Based on simple loss-of-function and gain-of-function experiments, it is not clear whether *AG* is intimately involved in cell fate decisions during the cellular differentiation of the carpel, but other genetic experiments implicate *AG* in directing some of these decisions (e.g., ovary wall morphology). Second, *AG* directs carpel development by its action in repressing A class function and thus allowing expression of other genes that can act independently of *AG* to specify cell fates within the carpel. At least two other genes, *SPT* and *CRC*, are responsible for promoting differentiation of carpel tissue independent of *AG*. That they can promote carpel differentiation independently of *AG* is demonstrated by comparisons of the phenotypes of *ap2 pi ag*, *crc ap2 pi ag*, and *spt ap2 pi ag* mutant flowers. In the case of *AG*, it has been shown that ectopic *AG* expression is sufficient to impose a fate of carpel identity on organ primordia within the context of the flower. At present, it is not clear if either *CRC* or *SPT* can specify carpel identity in any developmental context. Ectopic expression of *CRC* in the outer whorls of the flower does not usually result in their differentiation as carpels, although some placental tissue with ovules and stigma can develop along the margins of these organs (Bowman and Smyth, 1999). In contrast to *AG*, *CRC* has not been implicated in negatively regulating A class activity, and thus, it may be necessary to examine the effects of ectopic *CRC* expression in other genetic backgrounds, such as *ag* and *spt ap2 pi ag*, to determine whether *CRC* can induce the differentiation of carpel tissues independent of the other known positive and negative regulators of carpel development. The question of whether *SPT* can impart a fate of carpel identity on floral organ primordia awaits the cloning of the gene; however, that the *spt ap2 pi ag* quadruple mutant exhibits such a striking phenotype suggests ectopic expression of *SPT* may promote differentiation of carpel tissues in at least some contexts. It may be worth considering that the primary function of *AG* in specifying carpel identity might be the suppression of A class activity, thus allowing the expression of genes such as *CRC* and *SPT* which are negatively regulated by A class activity.

This brings us back to the question of what makes carpels different from leaves. Based on loss-of-function mutations, *AG*, *CRC*, and *SPT* are three of the factors that distinguish these two organ types. In the case of *AG*, its negative regulation in the leaves is mediated, at least in part, by *CLF* (Goodrich et al., 1997). However, the ectopic expression of *AG* in leaves, either in *clf* mutants or in transgenic plants constitutively expressing *AG*, does not result in the conversion of leaves into carpels but rather results in abnormal curling of the leaves (Mandel et al., 1992; Mizukami and Ma, 1992; Goodrich et al., 1997). This phenotype is unlikely to have any relevance to carpel development since the ectopic expression of many different MADS-box genes results in a similar phenotype (e.g., Mandel and Yanofsky, 1995; Krizek and Meyerowitz, 1996). Likewise, ectopic expression of *CRC* does not result in carpellody of leaves (Bowman and Smyth, 1999), even when combined with *clf* mutants in which *AG* is also ectopically expressed (Y. E. and J. L. B., unpublished). If we consider again that the carpel likely has its evolutionary origin in a leaflike organ, then loss-of-function mutations in genes specifying carpel identity might lead to the conversion of carpels into leaves in the flower. Conversely, gain-of-function mutations in such genes might result in the conversion of leaves into carpels. Since it appears that a combination of loss-of-function mutations is required for the homeotic conversion of carpels into leaflike organs within the flower, it follows that a combination of several gain-of-function mutations might be required to convert leaves into carpels, and thus, the conversion of leaves into carpels requires more than a simple genetic switch.

2. Regional Identity within the Developing Carpel

Gene expression patterns indicate that the stage 6 carpel primordium is already partitioned into several domains. At this stage, the primordium is divided into approximately equally sized medial and lateral domains, whose limits are defined by the expression of genes such as *CRC* and *AGL1/AGL5* (Figs. 10A and 10B; see color plate). Based on morphological analyses, it appears likely that these early gene expression domains demarcate the future replum and valve. In this scenario, the septum, placenta with accompanying ovules, and abaxial replum would be derived from the medial domains, and the valves would be derived from the lateral domains. In support of this, several genes that are expressed early in the medial domains are expressed later in the septum/replum regions, and likewise, genes expressed early in the lateral domains tend to be expressed in the ovary walls later in development. These domains may define developmental compartments, fated to differentiate into the replum and valve tissue types. In animals, a compartment is defined as "an area of the developing or mature organism that is constructed by all the descendants of a founding set of cells" (Lawrence, 1992). To determine whether the domains of the stage 6 carpel primordium represent compartments in this sense, we need to have a better understanding of the cell lineage patterns in the differentiating carpel. The tissues of the replum, the septum, placenta, and abaxial replum can also be considered as the marginal tissues of the carpel. This would imply that the medial and lateral domains observed in the carpel primordium reflect the marginal and central domains, respectively, of the two congenitally fused carpel primordia. One test of this hypothesis would be to examine gene expression patterns in solitary carpels that develop in *ap2* flowers. The subdivision of the stage 6 carpel primordium also occurs along the abaxial–adaxial axis, with the expression of *ETT* possibly defining the abaxial domain (Fig. 10C; see color plate). Thus, one of the initial steps in gynoecium development may be to partition, or compartmentalize, the primordium into regional domains, which then follow distinct developmental programs.

How these domains are established in the first place is a mystery. In some ways the carpel primordium during these stages is reminiscent of a developing flower meristem: how does one partition a group of meristematic cells into distinct domains that will differentiate into different organ or tissue types, and once the cells are divided into developmental fields, how does one specify the identity of the tissues and organs that will develop from these fields? In the case of flower meristems, the second question is beginning to be answered by the analysis of the floral homeotic (ABC) genes; however, the partitioning of the flower meristem into whorls is largely an enigma (Weigel and Meyerowitz, 1994; Meyerowitz, 1997).

Two identified genes, *ETT* and *CRC*, may be involved in interpreting regional information and instructing the differentiation of regional tissue types. The redistribution of tissue types observed in *ett* gynoecia suggests that one role of *ETT* is to restrict the development of several tissue types to the appropriate locations. This could be accomplished by repressing the activity of *SPT*, which appears to promote the growth of adaxial marginal tissues (Alvarez and Smyth, 1998), or by defining apical–basal boundaries that direct the subsequent differentiation of tissue types (Sessions, 1997). Similarly, the development of adaxial tissues in abaxial positions in gynoecia of multiple mutant combinations involving *crc* suggests that *CRC* acts to either specify abaxial fate or, alternatively, prevent adaxial tissues from developing in inappropriate positions. However, none of the genes identified thus far appear to be involved directly in partitioning the primordium into distinct domains. It may be that the regionalization of organ primordia is a general genetic program that operates in all organ primordia, and thus, mutations in such genes would have pleiotropic mutant phenotypes. Pertinent examples are the *phantastica* mutation of *Antirrhinum majus*, which results in loss of adaxial tissues in many organs of the plant (Waites and Hudson, 1995; Waites *et al.*, 1998), and the *narrow sheath* mutation of *Zea mays*, which causes loss of marginal tissues in the leaves and other organs of foliar origin (Scanlon and Freeling, 1998). These general programs would interact with organ-specific genetic programs to generate the diverse tissue types associated with the marginal versus central and abaxial versus adaxial domains of each organ type. Thus, identification of genes that establish these domains in the carpel primordium may be hampered by reasons outlined previously.

3. Differentiation of Tissue Types

Once regional domains have been established, these are likely subdivided as the differentiation of tissue and cell types occurs. Are there distinct genetic factors dedicated to the specification and/or differentiation of particular tissue types? The differentiation of the transmitting tract tissue appears to absolutely require *SPT* function; however, this is not the sole function of *SPT*, since *SPT* promotes the development of all marginal tissues, as revealed by the phenotype of *spt ap2 pi ag* quadruple mutants. In contrast, the MADS-box genes *AGL1/AGL5* and *FUL* may be the clearest examples of genes dedicated to direct the differentiation of specific tissue types, since in these cases, both their mutant phenotypes and their genetic interactions are relatively uncomplicated. The paucity of mutations that simply affect single tissue types suggests that the specification and differentiation of tissue and cell types likely involve multiple regulatory interactions or that these decisions are controlled by redundant factors as indicated by several of the double mutant phenotypes described.

Another question is whether there are any inductive interactions between tissue types in the developing carpel. Presently, clues have come from both the analysis of mutant phenotypes and cell ablation experiments. For example, that the replum and septum of *ful* mutants develop normally despite the inappropriate differentiation of the valves suggests that these tissues are controlled by largely independent developmental programs (Gu *et al.*, 1998). This is corroborated by the phenotype of the carpelloid organs of the *ap2 pi ag* triple mutants in which most of the marginal tissues differentiate normally whereas the central regions have little if any valve identity but instead differentiate with sepal- or leaflike qualities (Alvarez and Smyth, 1999). The independent development of these two regions again supports the idea that the marginal and central domains (or medial and lateral domains, respectively, in the stage 6 carpel primordium) may be delineated early in development and subsequently differentiate independently. Further support for this hypothesis comes from *ett tsl* gynoecia, in which differentiation of placental tissue bearing ovules apparently occurs in the absence of the appropriate development valve and style tissue (Roe *et al.*, 1997). Another approach to uncover inductive interactions is to ablate cells using a cytotoxic agent such as diphtheria toxin, the production of which is controlled by tissue- or region-specific promoters. Currently, the only experiment reported with respect to *Arabidopsis* carpel development is the ablation of the stigmatic papillae (Thorsness *et al.*, 1993), and in this case, the ablation event was likely too late in development to assay for inductive interactions between the stigmatic papillae and other tissues. Using promoters active earlier in development might prove informative; however, such experiments must be interpreted with caution without thorough knowledge of the precise spatial and temporal expression pattern of the promoters used to drive transcription of the toxin gene. In summary, it appears that many of the major carpel tissues might be capable of differentiation independent of the presence of other carpel tissues; however, this does not rule out inductive interactions at a cellular level.

B. Carpel Fusion

The fusion of carpels to create a protected cavity in which to house the ovules is one of the defining features of angiosperms and has been cited as one of the reasons for their evolutionary success (e.g., Doyle and Donoghue, 1986). One striking feature of many of the mutant phenotypes discussed is the tendency for carpel fusion to be disrupted, resulting most often in two free organs comprising the gynoecium. That the loss of fusion usually results in two free carpels supports the hypothesis that the Brassicaceae gynoecium is composed of two carpels (this has been a matter of dispute in the literature; for reviews, see Lawrence (1951) and Meyerowitz *et al.* (1989)), an issue which has been previously addressed by Okada *et al.* (1989). The preponderance of genotypes that result in loss of carpel fusion indicates that the process is easily disrupted, either directly or indirectly. Loss of fusion could result from lack of coordination of growth rates of the two carpels, disruption of marginal tissues of the carpel involved in fusion, or alteration of hypothesized communication pathways connecting the carpels. That there exist specific communication pathways between the two postgenitally fusing carpels of *Catharanthus* has been elegantly demonstrated by Walker (1978) and Siegel and Verbeke (1989). In *Catharanthus*, it has been shown that diffusible water-soluble factors are responsible for the redifferentiation of carpel epidermal cells into parenchymal cells (for a review, see Verbeke, 1992). Only those epidermal surfaces in contact with an opposing carpel undergo redifferentiation, and furthermore, distinct diffusible factors are produced by each of the carpels (Verbeke, 1992). At present, the nature of these factors is not known. It would not be surprising if similar pathways connecting the congenitally fused carpels exist in *Arabidopsis*; however, given the propensity for the fusion process to be disrupted, identification of components of these communication pathways by mutagenesis may prove difficult.

One surprising feature of the solitary carpels developing in the first whorl of *ap2* mutants and all whorls of *ap2 pi ag* triple mutants is the development of stigma and style tissue along the margins of the organs. It appears that the marginal stigma and style tissue develops at the expense of abaxial replum, as this tissue is absent in the solitary carpels found in these genotypes. Thus, two consequences of carpel fusion, at least in these genetic backgrounds, are (1) the localization of stigmatic papillae and style tissue to their appropriate apical positions and (2) the formation of abaxial replum tissue along the basal portions of the margin of fusion. It may be that there are specific communication pathways between the two fusing carpels, as have been proposed for *Catharanthus* (Verbeke, 1992), that result in the induction of marginal abaxial replum tissue and the restriction of stigma and style tissues to the apex of the gynoecial cylinder. In this regard, it is of interest to note that the formation of stigma and style tissue at the apex is localized to those "margins" of the wild-type carpels which are not fused and that the marginal differentiation of these tissue types may be a default pathway in the absence of fusion.

Morphologically the results of congenital and postgenital fusions may be difficult to distinguish in mature organs. The Brassicaceae gynoecium provides an example of congenital fusion as it is composed of two carpels but originates as a single primordium. In contrast, the two carpels of the *Catharanthus* gynoecium provide an example of postgenital fusion, as do the septum and style of the Brassicaceae gynoecium. In comparing the postgenital carpel fusion of *Catharanthus* with congenital carpel fusion in *Arabidopsis*, one similarity that should be noted is that both cases result in a local induction of the differentiation (or redifferentiation) of a tissue type: abaxial replum in *Arabidopsis*, and parenchyma in *Catharanthus*. It may be that in many cases the differences between congenital and postgenital fusions reflect differences in timing rather than fundamental differences in biological processes.

C. Evolutionary Origins of the Carpel

The evolutionary origin of the carpel and its homologies with the reproductive structures of other seed plants are presently enigmatic. A wide variety of living gymnosperms and extinct plants have been postulated to be ancestral to the angiosperms, and thus, a range of evolutionary origins of the carpel have been invoked. Based on cladistic analyses of seed plants (Crane, 1985; Doyle, 1996; Doyle and Donoghue, 1986), two primary hypotheses emerge (reviewed in Friis and Endress (1990) and Doyle (1994, 1996). In one scenario, carpels are derived from Mesozoic seed fern-like megasporophylls (e.g., Caytoniaceae or Glossopteridales). These megasporophylls bore ovules directly on their adaxial surfaces and simple conduplicate folding and fusion of appressed margins of the sporophyll would result in a structure similar to conduplicate carpels found in some fossil and basal angiosperms. Alternatively, angiosperms could be derived from coniferopsids; in this scenario, carpels would be derived from a compound strobilus-like shoot through an intermediate that might resemble the gnetalean female reproductive structure. In this case, ovules would be derived from a reduced shoot system and carpel walls would be derived from bracts of the female gnetalean "flower." Although the cladistic analyses to date suggest that *Gnetales* and the angiosperms are sister groups among extant plants, the positions of the potentially closely related fossil taxa are unresolved (Crane, 1985; Doyle and Donoghue, 1986; Doyle, 1994; Crane *et al.*, 1995). As pointed out by Doyle (1994), three possible sources of data could provide resolution. First, discovery of more complete fossil specimens of known taxa or new taxa could resolve relationships among relevant groups. Second, resolution of relationships of extant angiosperms could suggest an ancestral condition of characters for angiosperms as a whole. Third, orthologs of genes with known roles in angiosperm flower development could be functionally analyzed in other extant seed plants. In this sense, these genes would provide another character state for cladistic analyses. For example, if angiosperms are derived from a gnetalean-like ancestor, then one might expect genes involved in angiosperm carpel morphogenesis to be involved in bract development in the *Gnetales*. Conversely, if both angiosperms and *Gnetales* have a Mesozoic seed fern-like plant as a common ancestor, then genes involved in angiosperm carpel morphogenesis are unlikely to be involved in the development of bracts of the *Gnetales*. Since the origin of the carpels has direct relevance as to the origin of the angiosperms and vice versa, it will be of interest to examine whether *SPT* or *CRC* may be useful in determining organ homologies among extant plants.

AG-independent carpel development is observed in a *lfy ap1* background, in which the floral meristem identity genes are not active (Weigel and Meyerowitz, 1993). In *lfy ap1* plants, the cells that would normally constitute flower meristems instead behave as if they were inflorescence meristems, each of which are subtended by a bractlike organ (Weigel *et al.*, 1992). In the more apical positions of the inflorescence, the bractlike organs are replaced by carpellike organs; however, this carpel tissue is *AG*-independent since the ABC genes do not appear to be active in the *lfy ap1* background (Weigel and Meyerowitz, 1993). One hypothesis is that the *AG*- and *LFY*-independent pathway(s) to carpel differentiation, mediated by genes such as *SPT* and *CRC*, may be controlled directly by factors that mediate floral induction. It is tempting to speculate that these pathways may represent evolutionarily ancient genetic networks directing the development of sporophylls that predate the evolution of flowers. In this scenario, the ABC genetic network would have been recruited to integrate these pre-existing genetic programs involved in the development of the megasporophylls with the development of the flower in angiosperms.

Acknowledgments

The manuscript was greatly improved through discussions with members of our laboratory and David Smyth, Marcus Heisler, Allen Sessions, and Marty Yanofsky. Figure 10C was a generous gift from Allen Sessions, Jennifer Nemhauser, and Patti Zambryski. We thank Tom Jack, Jan Dockx, and Leslie Sieburth for GUS lines and Kelly Bollinger for help in assembling figures. This work was made possible by NSF Grant 96-31458 to J.L.B. and BARD postdoctoral fellowship FI 236-96 to Y.E.

References

Abel, S., and Theologis, A. (1996). Early genes and auxin action. *Plant Physiol.* **111,** 9–17.
Alvarez, J., and Smyth, D. R. (1998). Genetic pathways controlling carpel development in *Arabidopsis thaliana*. *J. Plant Res.* **111,** 295–298.
Alvarez, J., and Smyth, D. R. (1999). *CRABS CLAW* and *SPATULA*, two *Arabidopsis* genes that control carpel development in parallel with *AGAMOUS*. *Development,* in press.
Bailey, I. W., and Swamy, B. G. L. (1951). The conduplicate carpel of dicotyledons and its initial trends of specialization. *Am. J. Bot.* **38,** 373–379.
Bennett, S. R. M., Alvarez, J., Bossinger, G., and Smyth, D. R. (1995). Morphogenesis in *pinoid* mutants of *Arabidopsis thaliana*. *Plant J.* **8,** 505–520.
Boeke, J. H. (1971). Location of the postgenital fusion in the gynoecium of *Capsella bursa-pastoris* (L.). *Acta Bot. Neerl.* **20,** 570–576.
Bossinger, G., and Smyth, D. R. (1996). Initiation patterns of flower and floral organ development in *Arabidopsis thaliana*. *Development* **122,** 1093–1102.
Bowman, J. L. (1994). Genetic interactions among floral homeotic genes: Double mutants. *In* "Arabidopsis: An Atlas of Morphology and Development" (J. Bowman, Ed.), p. 244. Springer-Verlag, New York.

Bowman, J. L., and Smyth, D. R. (1999). *CRABS CLAW*, a gene that regulates carpel and nectary development in *Arabidopsis*, encodes novel protein with zinc finger and helix-loop-helix domains. *Development*, in press.

Bowman, J. L., Smyth, D. R., and Meyerowitz, E. M. (1989). Genes directing flower development in *Arabidopsis*. *Plant Cell* **1,** 37-52.

Bowman, J. L., Smyth, D. R., and Meyerowitz, E. M. (1991a). Genetic interactions among floral homeotic genes of *Arabidopsis*. *Development* **112,** 1-20.

Bowman, J. L., Drews, G. N., and Meyerowitz, E. M. (1991b). Expression of the *Arabidopsis* floral homeotic gene *AGAMOUS* is restricted to specific cell types late in flower development. *Plant Cell* **3,** 749-758.

Bowman, J. L., Alvarez, J., Weigel, D., Meyerowitz, E. M., and Smyth, D. R. (1993). Control of flower development in *Arabidopsis thaliana* by *APETALA1* and interacting genes. *Development* **119,** 721-743.

Bradley, D., Carpenter, R., Sommer, H., Hartley, N., and Coen, E. (1993). Complementary floral homeotic phenotypes result from opposite orientations of a transposon at the *plena* locus of *Antirrhinum*. *Cell* **72,** 85-95.

Carpenter, R., and Coen, E. S. (1990). Floral homeotic mutations produced by transposon-mutagenesis in *Antirrhinum majus*. *Genes Dev.* **4,** 1483-1493.

Chaudhury, A. M., Craig, S., Dennis, E. S., and Peacock, W. J. (1998). Ovule and embryo development, apomixis and fertilization. *Curr. Opin. Plant Biol.* **1,** 26-31.

Clark, S. E., Running, M. P., and Meyerowitz, E. M. (1995). *CLAVATA3* is a specific regulator of shoot and floral meristem development affecting the same processes as *CLAVATA1*. *Development* **121,** 2057-2067.

Coen, E. S., and Meyerowitz, E. M. (1991). The war of the whorls: Genetic interactions controlling flower development. *Nature* **353,** 31-37.

Colombo, L., Franken, J., Koetje, E., van Went, J., Dons, H. J., Angenent, G. C., and van Tunen, A. J. (1995). The petunia MADS box gene *FBP11* determines ovule identity. *Plant Cell* **7,** 1859-1868.

Crane, P. R. (1985). Phylogenetic analysis of seed plants and the origin of angiosperms. *Ann. Missouri Bot. Gard.* **72,** 716-793.

Crane, P. R., Friis, E. M., and Pedersen, K. R. (1995). The origin and early diversification of angiosperms. *Nature* **374,** 27-33.

Cronquist, A. (1988). "The Evolution and Classification of Flowering Plants," 2nd ed. New York Botanical Garden, Bronx.

Dawe, R. K., and Freeling, M. (1991). Cell lineage and its consequences in higher plants. *Plant J.* **1,** 3-8.

Dellaporta, S. L., Moreno, M. A., and Delong, A. (1991). Cell lineage analysis of the gynoecium of maize using the transposable element. *Ac. Development Suppl.* **1,** 141-147.

Derman, H., and Stewart, R. N. (1973). Ontogenetic study of floral organs of peach (*Prunus persica*) utilizing cytochimeral plants. *Am. J. Bot.* **60,** 283-291.

Dockx, J., Quaedvlieg, N., Keultjes, G., Kock, P., Weisbeek, P., and Smeekens, S. (1995). The homeobox gene *ATK1* of *Arabidopsis thaliana* is expressed in the shoot apex of the seedling and in flowers and inflorescence stems of mature plants. *Plant Mol. Biol.* **28,** 723-737.

Doyle, J. A. (1994). Origin of the angiosperm flower: A phylogenetic perspective. *Plant Syst. Evol., Suppl.* **8,** 7-29.

Doyle, J. A. (1996). Seed plant phylogeny and the relationships of Gnetales. *Int. J. Plant Sci.* **157,** S3-S39.

Doyle, J. A., and Donoghue, M. J. (1986). Seed plant phylogeny and the origin of the angiosperms: An experimental cladistic approach. *Bot. Rev.* **52,** 321-431.

Drews, G. N., Bowman, J. L., and Meyerowitz, E. M. (1991). Negative regulation of the *Arabidopsis* homeotic gene *AGAMOUS* by the *APETALA2* product. *Cell* **65,** 991-1002.

Drews, G. N., Lee, D., and Christensen, C. A. (1998). Genetic analysis of female gametophyte development and function. *Plant Cell* **10,** 5-17.

Ecker, J. R. (1998). Genes blossom from a weed. *Nature* **391,** 438–439.
Elleman, C. J., Franklin-Tong, V., and Dickinson, H. G. (1992). Pollination in species with dry stigmas: The nature of the early stigmatic response and the pathway taken by pollen tubes. *New Phytol.* **121,** 413–424.
Elliott, R. E., Betzner, A. S., Huttner, E., Oakes, M. P., Tucker, W. Q. J., Gerentes, D., Perez, P., and Smyth, D. R. (1996). *AINTEGUMENTA*, an *APETALA2*-like gene of *Arabidopsis* with pleiotropic roles in ovule development and floral organ growth. *Plant Cell* **8,** 155–168.
Endress, P. K., and Igersheim, A. (1997). Gynoecium diversity and systematics of the Laurales. *Bot. J. Linnean Soc.* **125,** 93–168.
Fahn, A. (1979). "Secretory Tissues in Plants." Academic Press, London.
Flanagan, C. A., Hu, Y., and Ma, H. (1996). Specific expression of *AGL1* MADS-box gene suggests regulatory functions in *Arabidopsis* gynoecium and ovule development. *Plant J.* **10,** 343–353.
Friis, E. M., and Crepet, W. L. (1987). Time of appearance of floral features. In "The Origins of Angiosperms and Their Biological Consequences" (E. M. Friis, W. G. Chaloner, and P. R. Crane, Eds.), pp. 145–179. Cambridge Univ. Press, Cambridge, UK.
Friis, E. M., and Endress, P. K. (1990). Origin and evolution of angiosperm flowers. *Adv. Bot. Res.* **17,** 99–162.
Furner, I. J., and Pumfrey, J. E. (1992). Cell fate in the shoot apical meristem of *Arabidopsis thaliana*. *Development* **115,** 755–764.
Gasser, C. S. (1991). Molecular studies on the differentiation of floral organs. *Annu. Rev. Plant Physiol. Plant Mol. Biol.* **42,** 621–649.
Gasser, C. S., and Robinson-Beers, K. (1993). Pistil development. *Plant Cell* **5,** 1231–1239.
Gasser, C. S., Broadhvest, J., and Hauser, B. A. (1998). Genetic analysis of ovule development. *Annu. Rev. Plant Physiol. Plant Mol. Biol.* **49,** 1–24.
Goodrich, J., Puangsomlee, P., Martin, M., Long, D., Meyerowitz, E. M., and Coupland, G. (1997). A polycomb-group gene regulates homeotic gene expression in *Arabidopsis*. *Nature* **386,** 44–51.
Goto, K., and Meyerowitz, E. M. (1994). Function and regulation of the *Arabidopsis* floral homeotic gene *PISTILLATA*. *Genes Dev.* **8,** 1548–1560.
Goto, N., Katoh, N., and Kranz, A. R. (1991). Morphogenesis of floral organs in *Arabidopsis*: Predominant carpel formation of the pin-formed mutant. *Jpn. J. Genet.* **66,** 551–567.
Grossniklaus, U., and Schneitz, K. (1998). The molecular and genetic basis of ovule and megagametophyte development. *Sem. Cell Dev. Biol.* **9,** 227–238.
Gu, Q., Ferrándiz, C., Yanofsky, M. F., and Martienssen, R. (1998). The *FRUITFULL* MADS-box gene mediates cell differentiation during *Arabidopsis* fruit development. *Development* **125,** 1509–1517.
Guilfoyle, T. J. (1998). Aux/IAA proteins and auxin signal transduction. *Trends Plant Sci.* **3,** 205–207.
Gustafson-Brown, C., Savidge, B., and Yanofsky, M. F. (1994). Regulation of the *Arabidopsis* floral homeotic gene *APETALA1*. *Cell* **76,** 131–143.
Hardtke, C. S., and Berleth, T. (1998). The *Arabidopsis* gene *MONOPTEROS* encodes a transcription factor mediating embryo axis formation and vascular development. *EMBO J.* **17,** 1405–1411.
Hill, J. P., and Lord, E. M. (1989). Floral development in *Arabidopsis thaliana*: A comparison of the wild type and the homeotic *pistillata* mutant. *Can. J. Bot.* **67,** 2922–2936.
Howden, R., Park, S. K., Moore, J. M., Orme, J., Grossniklaus, U., and Twell, D. (1998). Selection of T-DNA-tagged male and female gametophytic mutants by segregation distortion in *Arabidopsis*. *Genetics* **149,** 621–631.
Hülskamp, M., Schneitz, K., and Pruitt, R. E. (1995). Genetic evidence for a long-range activity that directs pollen tube guidance in *Arabidopsis*. *Plant Cell* **7,** 57–64.
Igersheim, A., and Endress, P. K. (1997). Gynoecium diversity and systematics of the Magnoliales and winteroids. *Bot. J. Linnean Soc.* **124,** 213–271.

Irish, V. F., and Sussex, I. M. (1990). Function of the *apetala1-1* gene during *Arabidopsis* floral development. *Plant Cell* **2**, 741–751.
Irish, V. F., and Sussex, I. M. (1992). A fate map of the *Arabidopsis* embryonic shoot apical meristem. *Development* **115**, 745–753.
Ito, T., Takahashi, N., Simura, Y., and Okada, K. (1997). A serine/threonine protein kinase gene isolated by an *in vivo* binding procedure using the *Arabidopsis* floral homeotic gene product *AGAMOUS*. *Plant Cell Physiol.* **38**, 248–258.
Jack, T., Brockman, L. L., and Meyerowitz, E. M. (1992). The homeotic gene *APETALA3* of *Arabidopsis thaliana* encodes a MADS-box and is expressed in petals and stamens. *Cell* **68**, 683–697.
Jack, T., Fox, G. L., and Meyerowitz, E. M. (1994). *Arabidopsis* homeotic gene *APETALA3* ectopic expression: Transcriptional and post-transcriptional regulation determine floral organ identity. *Cell* **76**, 703–716.
Jofuku, K. D., den Boer, B. G. W., Van Montagu, M., and Okamura, J. K. (1994). Control of *Arabidopsis* flower and seed development by the homeotic gene *APETALA2*. *Plant Cell* **6**, 1211–1225.
Jürgens, G., Mayer, U., Torres Ruiz, R. A., Berleth, T., and Miséra, S. (1991). Genetic analysis of pattern formation in the *Arabidopsis* embryo. *Development Suppl.* **1**, 27–38.
Kamalay, J. C., and Goldberg, R. B. (1980). Regulation of structural gene expression in tobacco. *Cell* **19**, 935–946.
Kandasamy, M. K., Nasrallah, J. B., and Nasrallah, M. E. (1994). Pollen–pistil interactions and developmental regulation of pollen tube growth in *Arabidopsis*. *Development* **120**, 3405–3418.
Kempin, S. A., Mandel, M. A., and Yanofsky, M. F. (1993). Conversion of perianth into reproductive organs by ectopic expression of the tobacco floral homeotic gene *NAG1*. *Plant Physiol.* **103**, 1041–1046.
Kim, J., Harter, K., and Theologis, A. (1997). Protein–protein interactions among Aux/IAA proteins. *Proc. Natl. Acad. Sci. USA* **94**, 11786–11791.
Klucher, K. M., Chow, H., Reiser, L., and Fischer, R. L. (1996). The *AINTEGUMENTA* gene of *Arabidopsis* required for ovule and female gametophyte development is related to the floral homeotic gene *APETALA2*. *Plant Cell* **8**, 137–153.
Komaki, M. K., Okada, K., Nishino, E., and Shimura, Y. (1988). Isolation and characterization of novel mutants of *Arabidopsis thaliana* defective in flower development. *Development* **104**, 195–203.
Krizek, B. A., and Meyerowitz, E. M. (1996). The *Arabidopsis* homeotic genes *APETALA3* and *PISTILLATA* are sufficient to provide the B class organ identity function. *Development* **122**, 11–22.
Kunst, L., Klenz, J. E., Martinez-Zapater, J., and Haughn, G. W. (1989). *AP2* gene determines the identity of perianth organs in flowers of *Arabidopsis thaliana*. *Plant Cell* **1**, 1195–1208.
Landsman, D., and Bustin, M. (1993). A signature for the HMG-1 box DNA-binding proteins. *BioEssays* **15**, 539–546.
Lawrence, G. H. (1951). "Taxonomy of Flowering Plants," pp. 520–524. Macmillan, New York/London.
Lawrence, P. A. (1992). "The Making of a Fly." Blackwell Sci., Oxford.
Lawrence, P. A., and Struhl, G. (1996). Morphogens, compartments, and pattern: Lessons from *Drosophila*? *Cell* **85**, 951–961.
Leitch, I. J., and Bennett, M. D. (1997). Polyploidy in angiosperms. *Trends Plant Sci.* **2**, 470–476.
Lennon, K. A., Roy, S., Hepler, P. K., and Lord, E. M. (1998). The structure of the transmitting tissue of *Arabidopsis thaliana* (L.) and the path of pollen tube growth. *Sex Plant Reprod.* **11**, 49–59.
Leyser, O. (1998). Auxin signalling: Protein stability as a versatile control target. *Curr. Biol.* **8**, R305–R307.
Liljegren, S. J., Ferrándiz, C., Alvarez-Buylla, E. R., Pelaz, S., and Yanofsky, M. F. (1998). *Arabidopsis* MADS-box genes involved in fruit dehiscence. *Flowering Newsl.* **25**, 9–19.

Liu, Z., and Meyerowitz, E. M. (1995). *LEUNIG* regulates *AGAMOUS* expression in *Arabidopsis* flowers. *Development* **121,** 975–991.

Long, J. A., Moan, E. I., Medford, J. I., and Barton, M. K. (1996). A member of the KNOTTED class of homeodomain proteins encoded by the *STM* gene of *Arabidopsis*. *Nature* **379,** 66–69.

Ma, H., Yanofsky, M. F., and Meyerowitz, E. M. (1991). *AGL1-AGL6*, an *Arabidopsis* gene family with similarity to floral homeotic and transcription factor genes. *Genes Dev.* **5,** 484–495.

Mandel, M. A., and Yanofsky, M. F. (1995). A gene triggering flower formation in *Arabidopsis*. *Nature* **377,** 522–524.

Mandel, M. A., Bowman, J. L., Kempin, S. A., Ma, H., Meyerowitz, E. M., and Yanofsky, M. F. (1992). Manipulation of flower structure in transgenic tobacco. *Cell* **71,** 133–143.

McCormick, S. (1993). Male gametophyte development. *Plant Cell* **5,** 1265–1275.

Mena, M., Ambrose, B. A., Meeley, R. B., Briggs, S. P., Yanofsky, M. F., and Schmidt, R. J. (1996). Diversification of C-function activity in maize flower development. *Science* **274,** 1537–1540.

Metcalfe, C. R., and Chalk, L. (1979). "Anatomy of the Dicotyledons," 2nd ed. Clarendon, Oxford.

Mewes, H. W., Albermann, K., Bähr, M., Frishman, D., Gleissner, A., Hani, J., Heumann, K., Kleine, K., Maierl, A., Oliver, S. G., Pfeiffer, F., and Zollner, A. (1997). Overview of the yeast genome. *Nature Suppl.* **387,** 7–65.

Meyerowitz, E. M. (1997). Genetic control of cell division patterns in developing plants. *Cell* **88,** 299–308.

Meyerowitz, E. M., Smyth, D. R., and Bowman, J. L. (1989). Abnormal flowers and pattern formation in floral development. *Development* **106,** 209–217.

Milligan, S. B., and Gasser, C. S. (1995). Nature and regulation of pistil-expressed genes in tomato. *Plant Mol. Biol.* **28,** 691–711.

Mizukami, Y., and Ma, H. (1992). Ectopic expression of the floral homeotic gene *AGAMOUS* in transgenic *Arabidopsis* plants alters floral organ identity. *Cell* **71,** 119–131.

Münster, T., Pahnke, J., Di Rosa, A., Kim, J. T., Martin, W., Saedler, H., and Theissen, G. (1997). Floral homeotic genes were recruited from homologous MADS-box genes preexisting in the common ancestor of ferns and seed plants. *Proc. Natl. Acad. Sci. USA* **94,** 2415–2420.

Nasrallah, J. B., and Nasrallah, M. E. (1993). Pollen–stigma signaling in the sporophytic self-incompatibility response. *Plant Cell* **5,** 1325–1335.

Newbigin, E., Anderson, M. A., and Clarke, A. E. (1993). Gametophytic self-incompatibility systems. *Plant Cell* **5,** 1315–1324.

Okada, K., Komaki, M. K., and Shimura, Y. (1989). Mutational analysis of pistil structure and development of *Arabidopsis thaliana*. *Cell Differ. Dev.* **28,** 27–38.

Okada, K., Ueda, J., Komaki, M. K., Bell, C. J., and Shimura, Y. (1991). Requirement of the auxin polar transport system in early stages of *Arabidopsis* floral bud formation. *Plant Cell* **3,** 677–684.

Parcy, F., Nilsson, O., Busch, M. A., Lee, I., and Weigel, D. (1998). A genetic framework for floral patterning. *Nature* **395,** 561–566.

Purugganan, M. D., Rounsley, S. D., Schmidt, R. J., and Yanovsky, M. F. (1995). Molecular evolution of flower development: Diversification of the plant MADS-box regulatory gene family. *Genetics* **140,** 345–356.

Pyke, K. A., Marrison, J. L., and Leech, R. M. (1991). Temporal and spatial development of the cells of the expanding first leaf of *Arabidopsis thaliana* (L.) Heynh. *J. Exp. Bot.* **42,** 1407–1416.

Riechmann, J. L., and Meyerowitz, E. M. (1997). MADS domain proteins in plant development. *Biol. Chem.* **378,** 1079–1101.

Roe, J. L., Rivin, C. J., Sessions, R. A., Feldmann, K. A., and Zambryski, P. C. (1993). The *tousled* gene of *Arabidopsis thaliana* encodes a protein kinase homolog that is required for leaf and flower development. *Cell* **75,** 939–950.

Roe, J. L., Nemhauser, J. L., and Zambryski, P. C. (1997). *TOUSLED* participates in apical tissue formation during gynoecium development in *Arabidopsis*. *Plant Cell* **9,** 335–353.

Rounsley, S. D., Ditta, G. S., and Yanofsky, M. F. (1995). Diverse roles for MADS box genes in *Arabidopsis* development. *Plant Cell* **7,** 1259–1269.

Rouse, D., Mackay, P., Stirnberg, P., Estelle, M., and Leyser, O. (1998). Changes in auxin response from mutations in an AUX/IAA gene. *Science* **279,** 1371–1373.

Running, M. P., and Meyerowitz, E. M. (1996). Mutations in the *PERIANTHIA* gene of *Arabidopsis* specifically alter floral organ number and initiation pattern. *Development* **122,** 1261–1269.

Sablowski, R. W. M., and Meyerowitz, E. M. (1998). A homolog of *NO APICAL MERISTEM* is an immediate target of the floral homeotic genes *APETALA3/PISTILLATA*. *Cell* **92,** 93–103.

Sakai, H., Medrano, L. J., and Meyerowitz, E. M. (1995). *Arabidopsis* floral boundary maintenance by *SUPERMAN*. *Nature* **378,** 199–203.

Satina, S. (1944). Periclinal chimeras in *Datura* in relation to development and structure (A) of the style and stigma (B) of calyx and corolla. *Am. J. Bot.* **31,** 493–502.

Satina, S., and Blakeslee, A. F. (1943). Periclinal chimeras in *Datura* in relation to the development of the carpel. *Am. J. Bot.* **30,** 453–462.

Satina, S., Blakeslee, A. F., and Avery, A. G. (1940). Demonstrations of the three germ layers in the shoot apex of *Datura* by means of induced polyploidy in periclinal chimeras. *Am. J. Bot.* **27,** 895–905.

Savidge, B., Rounsley, S. D., and Yanofsky, M. F. (1995). Temporal relationship between the transcription of two *Arabidopsis* MADS box genes and the floral organ identity genes. *Plant Cell* **7,** 721–733.

Scanlon, M. J., and Freeling, M. (1998). The narrow sheath leaf domain deletion: A genetic tool used to reveal developmental homologies among modified maize organs. *Plant J.* **13,** 547–561.

Schmidt, A. (1924). Histologische studien an phanerogamen vegetationspunkten. *Bot. Arch.* **8,** 345–404.

Schultz, E. A., Pickett, F. B., and Haughn, G. W. (1991). The *FLO10* gene product regulates the expression domain of homeotic gene *AP3* and *PI* in *Arabidopsis* flowers. *Plant Cell* **3,** 1221–1238.

Schwarz-Sommer, Z., Huijser, P., Nacken, W., Saedler, H., and Sommer, H. (1990). Genetic control of flower development: Homeotic genes of *Antirrhinum majus*. *Science* **250,** 931–936.

Sessions, A., Nemhauser, J. L., McColl, A., Roe, J. L., Feldmann, K. A., and Zambryski, P. C. (1997). *ETTIN* patterns the *Arabidopsis* floral meristem and reproductive organs. *Development* **124,** 4481–4491.

Sessions, R. A. (1997). *Arabidopsis* (Brassicaceae) flower development and gynoecium patterning in wild type and *ettin* mutants. *Am. J. Bot.* **84,** 1179–1191.

Sessions, R. A., and Zambryski, P. C. (1995). *Arabidopsis* gynoecium structure in the wild type and in *ettin* mutants. *Development* **121,** 1519–1532.

Shevchenko, V. V., Grinikh, L. I., Grigor'eva, G. A., and Draginskaya, L. Y. (1976). Chimerism of reproductive tissue in *Arabidopsis thaliana* following irradiation of seedlings. *Genetika* **12,** 180–182.

Sieburth, L. E., Running, M. P., and Meyerowitz, E. M. (1995). Genetic separation of third and fourth whorl functions of *AGAMOUS*. *Plant Cell* **7,** 1249–1258.

Siegel, B. A., and Verbeke, J. A. (1989). Diffusible factors essential for epidermal cell redifferentiation in *Catharanthus roseus*. *Science* **244,** 580–582.

Smyth, D. R., Bowman, J. L., and Meyerowitz, E. M. (1990). Early flower development in *Arabidopsis*. *Plant Cell* **2,** 755–767.

Sundaresan, V., Springer, P., Volpe, T., Haward, S., Jones, J. D. G., Dean, C., Ma, H., and Martienssen, R. (1995). Patterns of gene action in plant development revealed by enhancer trap and gene trap transposable elements. *Genes Dev.* **9,** 1797–1810.

Tandre, K., Albert, V. A., Sundas, A., and Engstrom, P. (1995). Conifer homologues to genes that control floral development in angiosperms. *Plant Mol. Biol.* **27,** 69–78.

Taylor, D. W. (1991). Angiosperm ovules and carpels: Their characters and polarities, distribution in basal clades, and structural evolution. *Postilla* **208,** 1–23.

Thorsness, M. K., Kandasamy, M. K., Nasrallah, M. E., and Nasrallah, J. B. (1993). Genetic ablation of floral cells in *Arabidopsis*. *Plant Cell* **5,** 253–261.

Topping, J. F., Wei, W., and Lindsey, K. (1991). Functional tagging of regulatory elements in the plant genome. *Development* **112,** 1009–1019.

Ulmasov, T., Hagen, G., and Guilfoyle, T. (1997). ARF1, a transcription factor that binds to auxin response elements. *Science* **276,** 1865–1868.

van der Krol, A. R., and Chua, N.-H. (1993). Flower development in petunia. *Plant Cell* **5,** 1195–1203.

van Heel, W. A. (1981). A SEM investigation on the development of free carpels. *Blumea* **27,** 499–522.

Verbeke, J. A. (1992). Fusion events during floral morphogenesis. *Annu. Rev. Plant Physiol. Plant Mol. Biol.* **43,** 583–598.

Waites, R., and Hudson, A. (1995). *phantastica*: a gene required for dorsiventrality of leaves in *Antirrhinum majus*. *Development* **121,** 2143–2154.

Waites, R., Selvadurai, H. R. N., Oliver, I. R., and Hudson, A. (1998). The *PHANTASTICA* gene encodes a MYB transcription factor involved in growth and dorsiventrality of lateral organs in *Antirrhinum*. *Cell* **93,** 779–789.

Walker, D. B. (1978). Morphogenetic factors controlling differentiation and dedifferentiation of epidermal cells in the gynoecium of *Catharanthus roseus*. *Planta* **142,** 181–186.

Webb, M. (1994). Morphology of the stigma and pollen tube growth in the pistil. *In* "*Arabidopsis*: An Atlas of Morphology and Development" (J. Bowman, Ed.) pp. 338–339, 342–345. Springer-Verlag, New York.

Weberling, F. (1989). "Morphology of Flowers and Inflorescences," pp. 139–192. Cambridge Univ. Press, Cambridge, UK.

Weigel, D., and Meyerowitz, E. M. (1993). Activation of floral homeotic genes in *Arabidopsis*. *Science* **261,** 1723–1726.

Weigel, D., and Meyerowitz, E. M. (1994). The ABCs of floral homeotic genes. *Cell* **78,** 203–209.

Weigel, D., Alvarez, J., Smyth, D. R., Yanofsky, M. F., and Meyerowitz, E. M. (1992). *LEAFY* controls floral meristem identity in *Arabidopsis*. *Cell* **69,** 843–859.

Yanofsky, M. F., Ma, H., Bowman, J. L., Drews, G. N., Feldmann, K. A., and Meyerowitz, E. M. (1990). The protein encoded by the *Arabidopsis* homeotic gene *agamous* resembles transcription factors. *Nature* **346,** 35–40.

Yung, M.-H., Schaffer, R., and Putterill, J. (1999). Identification of genes expressed during early *Arabidopsis* carpel development by mRNA differential display: Characterization of ATCEL2, a novel endo-1,4-β-D-glucanase gene. *Plant J.* **17,** 203–208.

5

Digging out Roots: Pattern Formation, Cell Division, and Morphogenesis in Plants

Ben Scheres and Renze Heidstra
Department of Molecular Cell Biology
Utrecht University
3584 CH Utrecht, The Netherlands

I. Introduction
II. Embryonic Pattern Formation
   A. Formation of the Apical–Basal Axis
   B. Apical–Basal Pattern Formation
   C. Formation of the Radial Axis
   D. Radial Pattern Formation
III. Postembryonic Perpetuation of Cellular Pattern
   A. Prepatterning of Meristem Cells?
   B. What Are the Relevant Prepatterning Cues?
IV. Control of Cell Division during Development
   A. Pattern Formation Can Be Uncoupled from Cell Division
   B. Cell Fate Specification Can Regulate Cell Division
   C. Activation of Meristems
   D. Maintaining Meristem Activity
V. Growth and Organ Morphogenesis
   A. Growth Can Be Uncoupled from Cell Division
   B. Organ Morphogenesis Can Be Uncoupled from Cell Division
   C. What Are the Molecules Involved in Determining Organ Size and Shape?
VI. Concluding Remarks
   References

The analysis of plant development by genetic, molecular, and surgical approaches has accumulated a large body of data, and yet it remains a challenge to uncover the basic mechanisms that are operating. Early steps of development, when the zygote and its daughter cells organize the embryonic plant, are poorly understood despite considerable efforts toward the identification of relevant genes. Reported cases of genetic redundancy suggest that the difficulty in uncovering patterning genes may reflect overlapping gene activities. Our current knowledge on plant embryo development still leaves open whether mechanisms for axis formation and subsequent pattern formation are fundamentally different in animals and plants. Axis formation may follow the general principle of establishing a peripheral asymmetric cue and mobilizing the cytoskeleton toward this cue—in the case of plants possibly located in the cell wall—but the molecules involved may be entirely different. Embryonic pattern formation involves the establishment of different domains, but although there are candidates, it is not clear whether genes that define these domains are identified yet. Pattern formation continues postembryonically in the meristem, and the flexibility of this process may be explained by a feed-forward system of patterning cues originating from more mature cells. Control of cell division and differentiation, which is important in the meristems—regions of continuous development—has been studied intensively and appears to involve short-range signaling and transmembrane receptor kinase activation. Finally, although high importance of control of cell division rates and planes for plant morphogenesis have been often inferred, recent genetic studies as well as comparative morphological data point to a less decisive role of cell division and to global controls of as yet unknown nature. © 1999 Academic Press.

I. Introduction

Plants and animals alike develop from a single fertilized egg cell into a multicellular organism with a three-dimensional pattern of specialized cell types. In animals, the route from egg cell to organism can be formally subdivided into distinct steps (Gurdon, 1992), most of which pertain to the gradual emergence of specified regions and cells at appropriate places: (i) Axis formation establishes localized cues. (ii) Cues are interpreted to pattern body regions and tissue and cell types. (iii) Cell fate is progressively determined. (iv) Overt differentiation takes place. Control of cell division operates in parallel with these patterning and specification steps to ensure the appropriate production of new cells.

To identify analogous developmental problems, plant development can be schematized in the same way. Such an exercise does not presuppose similarities at the mechanistic level: several plant-specific features may require novel developmental strategies. For example, individual cells are encased in walls and do not rearrange significantly. The presence of the cell wall may influence the biochemical mechanisms that are available for cell–cell signaling, and the lack of cell movement implies a more significant role for oriented cell division and expansion in development. Furthermore, plant development continues after embryogenesis through the activity of small groups of continuously dividing cells, the meristems. Control of the developmental capacities of these meristems is likely to require special mechanisms.

In the plant sciences, development has been studied for decades by comparative anatomy and surgical experiments, and the plant hormones that modulate development have been a prominent topic of concurrent physiological studies (Steeves and Sussex, 1989). It is, however, only within the past decade that the application of molecular genetic approaches is yielding insights into plant development. For example, the analysis of flower formation has provided valuable information on the establishment of regional identity in plants. A network of transcription factors involved in floral organ identity has been identified, and new details on their action and regulation continuously emerge (Weigel and Meyerowitz, 1994; Weigel,

1998). Nevertheless, flower formation is a late event in the development of the multicellular plant, and connections with early patterning events are just beginning to be established (Parcy *et al.*, 1998).

This review aims to provide a perspective on plant development in terms of the formal steps outlined above. To illustrate these steps, we mainly take examples from the development of one anatomically uncomplicated plant organ, the *Arabidopsis* root. Previous reviews have covered several aspects of root formation, including pattern formation, cell specification, control of cell division, and the execution of cell differentiation and morphogenesis (Dolan and Roberts, 1995a, 1995b; Scheres *et al.*, 1996; Malamy and Benfey, 1997a; Schiefelbein *et al.*, 1997). As root development originates in the embryo, we will include information on embryogenesis where appropriate. Comprehensive reviews on plant embryogenesis have appeared recently (Laux and Jürgens, 1997; Mordhorst *et al.*, 1997), including alternative viewpoints (Kaplan and Cooke, 1997).

II. Embryonic Pattern Formation

A short overview of axis formation in plants serves to provide a framework for the discussion on embryo development. Plant embryos contain a primary body axis in the apical–basal dimension and they establish a secondary radial axis (Fig. 1). The embryonic organs, including the root, and two oppositely located groups of stem cells, the meristems, are patterned along the apical–basal axis. The three major tissue types are patterned in the radial dimension. Further (bilateral, dorsoventral) axes emerge later and these will not be discussed here.

A. Formation of the Apical–Basal Axis

The mechanisms by which the apical–basal axis is specified are hitherto unknown. The multicellular haploid ("gametophytic") phase of the life cycle during which the oocyte and associated cells are formed (Drews *et al.*, 1998) may be relevant for axis specification: the apical–basal axis of the embryo is invariably aligned with the major female gametophyte axis. However, plants can make embryos in the absence of maternal tissue under special circumstances such as tissue culture, implying that axial information can be generated without maternal input. Nevertheless, it is reasonable to propose a role for maternal information in normal axis specification (for further discussion, see Jürgens (1995) and Jürgens *et al.* (1997)).

There are some clues on how the main axis may become fixed. In the brown alga *Fucus*, with early embryo development reminiscent of that of higher plants, the primary axis is established by external cues, and axis fixation is dependent on the cell wall (Kropf *et al.*, 1988). Targeted secretion is required for cell wall polariza-

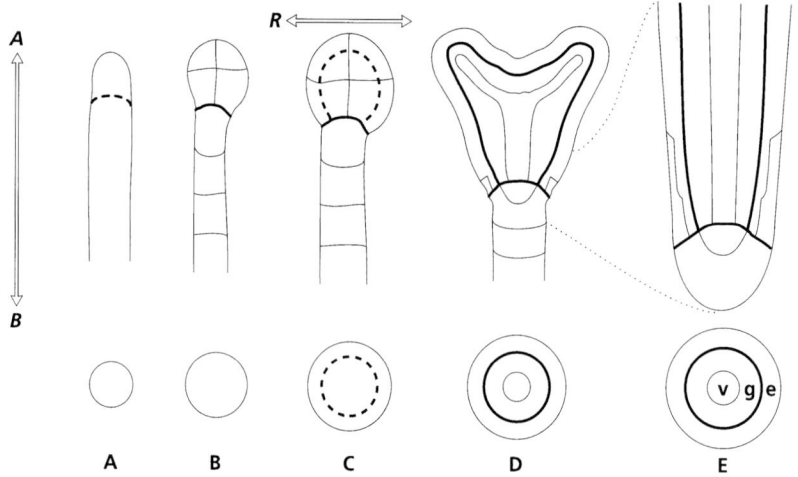

Fig. 1 Axis formation and the formation of the main tissues in plant embryogenesis. The first asymmetric cell divisions that manifest apical–basal and radial polarity are shown in *Arabidopsis* embryos of successive stages and in the postembryonic root. A, first division of zygote; B, octant stage; C, 16-cell stage; D, heart stage; E, seedling root. Upper row, longitudinal view; lower row, transverse view. *A*, apical; *B*, basal; *R*, radial. v, vascular tissue; g, ground tissue e, epidermal tissue.

tion, and the polarized wall orients the first division plane of the zygote (Shaw and Quatrano, 1996). In analogy, the *GNOM* gene is required to position the zygotic division plane in *Arabidopsis* (Mayer *et al.*, 1993) and, using expression of an early apical marker gene as a criterion, also for fixation of the apical–basal axis (Vroemen *et al.*, 1996). The GNOM protein shares homology with yeast guanine nucleotide exchange proteins involved in vesicle transport (Shevell *et al.*, 1994; Bush *et al.*, 1996; Peyroche *et al.*, 1996; Mossessova *et al.*, 1998). This may suggest that targeted secretion to the cell wall plays a role in axis fixation in plants as it does in *Fucus*, although the large evolutionary distance and the differences in habitat between vascular plants and algae argue against an overly optimistic view on shared mechanisms.

Which cues are established during axis specification to initiate pattern formation in higher plants? Laser ablation studies in *Fucus* again implicate the cell wall. Zygotic daughter cells can switch fate if they are manipulated to contact a different cell wall (Berger *et al.*, 1994). Therefore the wall has a capacity to instruct cell fate, in addition to its role in orienting the zygotic division plane. It remains to be investigated whether localized patterning cues similarly reside in the cell wall of higher plant zygotes. Notably, the asymmetric segregation of cell fate determinants and concomitant control of cell division plane also take place in yeast, nematodes, and flies (Yan and Yan, 1998).

B. Apical–Basal Pattern Formation

Whatever the cues may be that initiate apical–basal pattern formation, in many plant species asymmetry is immediately evident when the first embryonic cell division generates a large basal cell and a small apical cell with different destinies (Fig. 2A). In *Arabidopsis*, embryogenesis has been described extensively (Mansfield and Briarty, 1991; Jürgens and Mayer, 1994). The basal cell will form the extraembryonic suspensor and the hypophysis that will become part of the embryo proper (Figs. 2B and 2C). The apical cell will give rise to the remainder of the embryo. Although *twin* (*twn*) mutants can give rise to basal cell-derived embryos (Vernon and Meinke, 1994; Zhang and Sommerville, 1997), demonstrating that the basal cell retains the capacity to adopt apical fates, the corresponding genes do not specify cell fates. For example, the molecular lesion in *twn2* causes altered expression of a valyl-tRNA-synthase gene and apical cell progeny arrests, followed by the development of embryos from the basal cell. It is postulated that the apical cell suppresses the alternative fate of the basal cell in wild-type embryos and that apical cell defects in the *twn* mutants obliterate this control (Vernon and Meinke, 1994).

The embryo proper divides into apical, central, and basal domains with stereotyped cell division patterns (Fig. 2C; Mayer *et al.*, 1991). Subsequently, the embryo is partitioned in regions that will give rise to the seedling shoot apical meristem, embryonic leaves (cotyledons), hypocotyl, root, and root apical meristem (Figs. 2D and 2E). The prospective organ primordia do not correlate with the three early domains, showing that the final definition of organs and the positioning of organ boundaries are later events (Scheres *et al.*, 1994). The root, for example, derives from cell groups that have been separated from the first zygotic division onward (Fig. 2). The first region, encompassing most of the root, derives from the central domain and consists of concentric layers of the main tissue types (Fig. 1E). A small second region, located distally, derives from the basal domain—the hypophysis—and contains two unique cell types. A second, even more conspicuous, example to illustrate the lack of correspondence between early domains and seedling regions is the formation of the cotyledons. These originate from the apical domain but also incorporate cells from the central domain (Fig. 2D). The imperfect correspondence of early embryo domains and seedling regions that emerges from the fate map suggests that non-cell-autonomous mechanisms operate to refine the apical–basal pattern.

Despite the absence of a strict correlation between the three early domains and the different seedling regions, genetic analysis in *Arabidopsis* suggests that the early domains are relevant for pattern formation. Mutants defective in the formation of the main seedling regions along the apical–basal axis display early aberrations in cell division patterns within the corresponding domains (Mayer *et al.*, 1991; Berleth and Jürgens, 1993; Torres-Ruiz *et al.*, 1996; Willemsen *et al.*, 1998). *gurke*

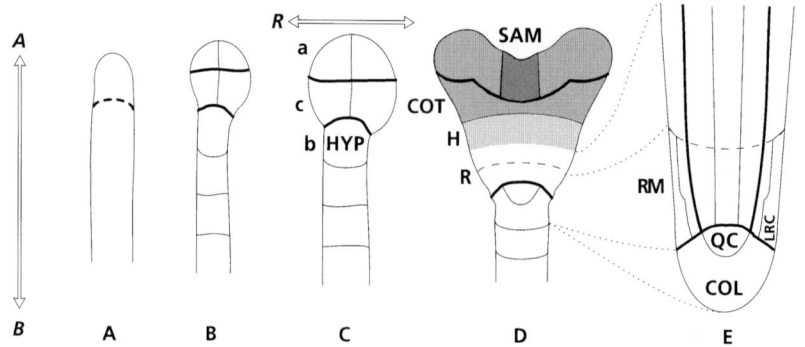

Fig. 2 Fate map of the *Arabidopsis* embryo. Stages and axes as in Fig. 1. a, apical region; c, central region; b, basal region; HYP, hypophyseal cell; SAM, shoot apical meristem; COT, cotyledons; H, hypocotyl; R, root; RM, root meristem; QC, quiescent center; COL, columella; LRC, lateral root cap.

(*gk*) mutants are affected in the apical domain and are defective in the formation of cotyledons and shoot apical meristem (Torres-Ruiz *et al.*, 1996). *fackel* (*fk*) mutants affect the central domain and hypocotyl formation (Mayer *et al.*, 1991). *monopteros* (*mp*) mutants are defective in the formation of the central domain and the basal domain, and strong alleles lack root and hypocotyl (Berleth and Jürgens, 1993). *hobbit* (*hbt*) mutants are disturbed in basal domain formation and they lack the root meristem (Willemsen *et al.*, 1998). All four genes do not necessarily control region identity directly, but may be involved in region-specific aspects of cellular differentiation. Phenotypic defects at early stages are described for the *mp* and *hbt* mutants, both required for root development. In contrast, the *gk* and *fk* defects become apparent at a later stage of embryogenesis (Torres-Ruiz *et al.*, 1996; Laux and Jürgens, 1997). These differences in manifestation of first detectable phenotypes may originate from a sequential patterning mechanism, with early specification of the basal region and later specification of more apical regions. Alternatively, they may reflect the occurrence of characteristic cell divisions in the basal domain that allow the detection of perturbations at an earlier stage. In any case, root-defective embryo mutants display the earliest reported domain-specific embryo defects correlated with a loss of regional identity, and we will discuss them in some detail.

1. The *HOBBIT* Gene and the Basal Domain

The basal domain encompasses the hypophyseal cell, and anatomical as well as clonal analysis shows that this cell divides stereotypically to give rise to quiescent center and columella root cap cells in the seedling root tip (Figs. 2C–E; Dolan *et al.*, 1994; Scheres *et al.*, 1994). Mutations in the *HBT* gene result in aberrant development of the basal domain (Willemsen *et al.*, 1998). In strong *hbt* mutants, the characteristic cell division pattern of the hypophyseal cell is not observed, a quiescent center is not present structurally and functionally, and markers for columella root cap are not expressed (van den Berg *et al.*, 1997; Willemsen *et al.*, 1998). The *hbt* mutant phenotype suggests that the corresponding gene is required in the basal domain to establish the appropriate cell types. Early molecular markers of regional and cellular identity will be required to assess whether aberrant cell division patterns in the basal region of *hbt* embryos represent lost or changed regional identity. Molecular cloning and expression analysis of the *HBT* gene should contribute further to this issue.

The *hbt* embryo phenotype is consistent with the view that cell–cell interactions refine the apical–basal pattern. The defects in *hbt* mutants are restricted to the progeny of a single cell in the basal region at early stages but spread to an immediately adjacent cell tier at the heart stage of embryogenesis (Fig. 2D). At that stage, the stem cells ("initials") of the root meristem should be defined in this cell tier. In wild-type plants, the initials perform ordered cell divisions that extend cell files to give rise to the root meristem. The characteristic cell division pattern of initial cells is not observed in the corresponding cell tier of *hbt* mutants. Furthermore, a new cell type, the lateral root cap that should originate from this tier, is not specified in *hbt* mutants by both anatomical and marker gene expression criteria (Willemsen *et al.*, 1998).

An attractive view is that *hbt*-dependent cell signaling across the domain boundaries defines the adjacent cell layer to become the root meristem with lateral root cap, although the nonautonomy of *HBT* requirement in these cells remains to be proven. A patterning mechanism with early establishment of three domains, followed by interactions at the boundaries of the domains to specify seedling regions, is reminiscent of signaling events at boundaries during animal development. For example, *glp-1* transmembrane signaling acts during the specification of cells at the boundary of anterior and posterior cell progeny in *C. elegans* (Priess *et al.*, 1987; Evans *et al.*, 1994) and *hedgehog* signaling instructs cells at the boundary of anterior and posterior compartments in *Drosophila* (Lee *et al.*, 1992).

2. The *MONOPTEROS* Gene and the Central Domain

monopteros mutants are defective in root and hypocotyl formation and display aberrant cell division patterns in the basal and central domains of the early embryo (Figs. 2C and 2D; Berleth and Jürgens, 1993). In the basal domain, the normal division pattern of the hypophyseal cell does not occur. In the central domain, cells do not elongate properly to establish the embryo axis that should be formed by this region. *mp* mutants display defects in vascular tissue formation at all stages of development. Vascular strands do not show the linear continuity that characterizes wild-type vascular strand formation, which is attributed to the failure of *mp* vascular cells to properly establish a cellular axis ("axialize") (Przemeck *et al.*, 1996).

The formation of continuous vascular strands can be promoted by the plant hormone auxin and has been postulated to require auxin transport (Sachs, 1991). Interestingly, the recent cloning of the *MP* gene provides a link to auxin, as it encodes a transcription factor with a motif that has been shown to bind to promoter elements of auxin-inducible genes (Ulmasov *et al.*, 1997; Hardtke and Berleth, 1998). Therefore, the primary role of the *MP* gene may be to respond to auxin transport to mediate the expression of downstream genes involved in appropriately oriented cell axialization. In early embryo development, the *MP* gene would in this view promote axialization of cells in the central domain, consistent with the gradual restriction of its embryonic expression to the most axialized—vascular—cells (Hardtke and Berleth, 1998). How does a role of the *MP* gene in cell axialization relate to its requirement for the development of the basal domain? One possibility is the *MP*-mediated axialization in the central domain promotes a signal that is required to activate genes like *HBT* that in turn specify the basal domain. The question whether *MP* requirement in the basal domain is indeed non-cell-autonomous should be addressed by mosaic analysis, but as a preliminary step, it will be interesting to verify whether *MP* expression can be detected in the hypophyseal cell region.

In summary, two genes are now identified that are required for the development of the basal region from the stage that this region is recognizable in the embryo. A number of genes which mutate to similar phenotypes have been reported, but their mutation frequency is low, suggesting that they are not simple loss-of-function mutants (Berleth *et al.*, 1996; Scheres *et al.*, 1996). Future goals will be to establish the relationship between the *HBT* and *MP* gene activities and to identify upstream regulators. Given the early appearance of the *hbt* and *mp* phenotypes, such regulators may uncover primary steps in apical–basal patterning.

C. Formation of the Radial Axis

While at least some ideas concerning the formation of the apical–basal axis have emerged from the studies mentioned in Section IIA, it remains mysterious how the radial axis is specified during embryogenesis. The first manifestation of the radial axis is the formation of the outer protoderm layer, the incipient epidermis (Fig. 1C). It may be relevant that plant zygotes have been shown to be coated with a cuticle layer, a feature that is retained in epidermal cells only (Bruck and Walker, 1985). The *ATML1* gene, encoding a homeodomain protein, is also expressed prior to the separation of an outer layer, and downregulated in inner daughter cells thereafter (Lu *et al.*, 1996). Hence, epidermal cell fate could be positively instructed by their position at the periphery. Therefore it has been speculated that radial axis formation results in an "outer" cue to specify future epidermis (Laux and Jürgens, 1997). Such a cue could be localized to the outer cell wall (or membrane) or it could be the result of signals from surrounding endosperm cells.

D. Radial Pattern Formation

Radial axis specification may mark the outer embryo surface or result in some other initial asymmetry, and the next question is how such cues lead to pattern formation. One possibility is that the withdrawal of an outer cue leads to a second cell type. Analysis of expression of the *ATLTP1* gene in mutant backgrounds has provided support for this notion. The *ATLTP1* gene is normally expressed in the outer layer of the globular stage embryo but expression becomes ubiquitous in *knolle* mutants that do not complete cytokinesis due to a defective syntaxin (Lukowitz *et al.*, 1996; Vroemen *et al.*, 1996; Lauber *et al.*, 1997). Cytokinesis may thus be required to separate inner cells from an outer cue to allow alternative fates. The inner cells continue to form a new layer, which could be specified in turn by signal exchange with the outer cell layer, leading to the three main tissues, epidermis, ground tissue, and vascular tissue (Figs. 1C–E). In theory, genetic dissection should provide genes involved in such a patterning scenario, for example genes whose disruption causes the absence or duplication of major tissue layers. However, no genes that are involved in specification of the major tissue layers in response to early radial patterning have been identified by phenotype. Candidate tissue identity genes may be identified by alternative means, e.g., based on their expression pattern. An example is the homeodomain-encoding *ATML1* gene that is expressed in the prospective embryonic epidermis (Lu *et al.*, 1996). Reverse genetics approaches now provide the means to analyze such genes at the functional level.

Genes may mutate to a phenotype that at first sight suggests involvement in radial pattern formation, but they may have different functions upon closer inspection. For example, the defective vascular development in *mp* mutants warrants some discussion on the relation between *MP* gene function and radial pattern formation. *mp* mutant organs contain all classes of differentiated vascular cell types, but the cells do not axialize properly (Przemeck *et al.*, 1996). Therefore, the *MP* gene seems not involved in specification per se of vascular cells during radial patterning. However, recessive *mp* alleles with a similar strong phenotype may not be nulls (Hardtke and Berleth, 1998), which does not entirely rule out that complete loss of *MP* function may interfere with vascular cell specification. Nevertheless, the initially broad *MP* gene expression domain in the embryo that becomes confined to vascular tissue only at later stages (Hardtke and Berleth, 1998) is consistent with the notion that the *MP* gene has no primary role in cell type specification.

The inability to identify radial patterning mutants involved in early steps may reflect early lethality of such mutants. A class of early lethal mutants has been described in *Arabidopsis* (Meinke, 1986) but the lack of early markers hampers a detailed analysis of patterning defects within this class. Recent advances in finding genes that mark the major tissues by their expression should alleviate this problem. The lack of radial patterning mutants may also reflect extensive redundancy in embryo patterning mechanisms. One striking example of such redundancy has been reported recently in a case where two genes involved in embryonic organ separation and meristem establishment were identified only because a double mutant was fortuitously identified (Aida et al., 1997). Again, gene isolation methods such as gene or enhancer trap strategies that are independent of mutant phenotype provide an alternative approach to cope with redundant genes.

1. Subspecification of Radial Pattern Elements

While the patterning of major tissue layers remains almost uncharted territory, more information is becoming available about later steps. Genes involved in epidermal subspecification were identified in both shoot- and root-mutant screens, but many of the ones involved in shoot epidermal patterning function in roots also (Hülskamp et al., 1994; Dolan and Roberts, 1995a). Other genetic loci involved in the subspecification of the embryonic tissue layers have in several instances emerged from screens for root mutants. The root contains the most simple and basic extension of the radial organization of the plant embryo axis in three main tissues (Fig. 1E). Underneath the outer epidermal layer there are two layers of ground tissue, the cortical parenchyma (cortex hereafter) and endodermis. The inner vascular bundle, with a small number of specialized conducting cells and associated cell types, is enclosed by the pericycle layer. Because of the simple tissue pattern in roots, defects within tissue layers can be readily recognized and traced back to corresponding defects in the embryo (Benfey et al., 1993; Scheres et al., 1995). In the following sections, we will discuss genes that are relevant for subspecification events, with emphasis on the uncomplicated root system.

a. Epidermis. In *Arabidopsis* roots, the epidermis consists of alternating files with two different cell fates: trichoblast and atrichoblast (hair-bearing and hairless, Fig. 3). Alterations in this pattern of epidermal subspecification can be readily detected and several genes have been shown to be involved. Recessive mutations in the *TRANSPARENT TESTA GLABRA* (*TTG*) gene and in the homeobox gene *GLABRA2* (*GL2*) result in roots with hair cells only (Galway et al., 1994; Rerie et al., 1994; Masucci et al., 1996). These genes therefore act as negative regulators of the hair cell fate. *ttg* mutants can be rescued by overexpression of the maize *R* gene (Lloyd et al., 1994), which has sequence similarity to *myc*-related transcription factors, indicating that an *Arabidopsis* homologue of the *R* gene operates downstream of *TTG*.

Two observations implicate the *TTG* gene as an upstream regulator of GL2. First, in *gl2* mutants but not in *ttg* mutants some cellular characteristics of hairless cells remain. Notably, the remaining hairless cell-specific traits in *gl2* mutants appear at the correct position (Masucci et al., 1996), suggesting that GL2 is not involved in setting up the alternating pattern but responds to it and mediates a subset of hairless cell differentiation processes. In addition, *GL2* transcription is

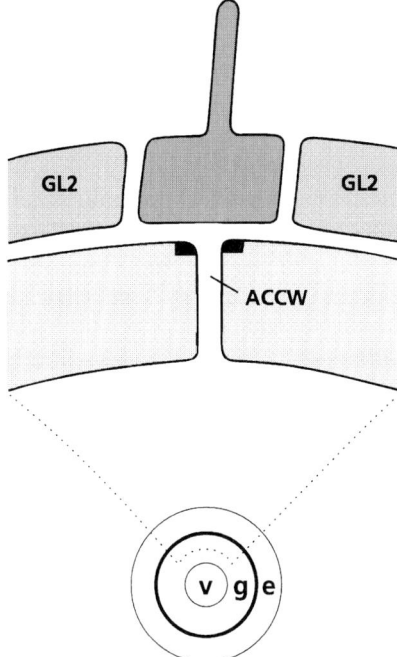

Fig. 3 Hair cell patterning in the root epidermis. In *Arabidopsis* roots, the epidermis consists of alternating files of hair-bearing and hairless cells. Hair cells are invariantly located over anticlinal cortical cell walls (ACCW) between two cells of the underlying cortex layer. Hair cell fate is dependent on the distance to the ACCW, and extracellular cues (black) may reside in a narrow domain centering on the anticlinal wall. The *GL2* gene acts to repress hair fate and is expressed in hairless cells. Transverse section as in Fig. 1E.

lowered in *ttg* mutants (Di Christina *et al.*, 1996). The *GL2* gene is expressed predominantly in hairless cell files, implying that it acts cell-autonomously to repress hair fate, and this expression in alternating cell files is set up during embryogenesis (Fig. 3; Berger *et al.*, 1998a).

Recessive mutations in the *CAPRICE* (*CPC*) gene show that it is required for the hair cell fate (Wada *et al.*, 1997). *CPC* encodes a *myb*-like transcription factor, and one awaits data on its localization and interaction with *GL2* as well as localization of the site of action of TTG, to understand how these transcription factors regulate epidermal subspecification.

In summary, subspecification of the root epidermis involves a suite of transcription factors and one upstream regulator. Notably, these genes are also involved in the specification of hairs of the shoot epidermis (trichomes) but with opposite roles: *TTG* is required to specify the *trichome* fate and to regulate trichome distribution, and *GL2* promotes trichome outgrowth (Hülskamp *et al.*, 1994). *CPC* overexpression has a negative effect on trichome formation (Wada *et al.*, 1997), but it is unclear whether this reflects a negative input of *CPC in vivo* or is an effect of promiscuous interactions. Two genes without apparent function in the root epidermis are also involved in trichome patterning. *GLABROUS1* (*GL1*), with sequence similarity to *myb*-like transcription factors, is required together with *TTG* to specify trichome fate (Oppenheimer *et al.*, 1991; Hülskamp *et al.*, 1994). Mutations in these genes result in leaves without trichomes. *TRIPTYCHON* (*TRY*), with as yet unknown molecular identity, acts as a negative regulator of trichome distribution (Hülskamp *et al.*, 1994; Schnittger *et al.*, 1998): in *try* mutants trichomes often appear in clusters. An interesting finding is that ubiquitous expression of *GL1* in *try* mutant background induced ectopic trichomes, not only on the epidermis of organs where trichomes are normally not formed but also in subepidermal ground tissue cells (Schnittger *et al.*, 1998). The subepidermal trichomes are distributed without guidance of the epidermal trichomes (although lateral inhibition between layers seems to occur), which indicates that also the trichome patterning mechanism is induced in the ground tissue layer. This important finding shows that developmental constraints in plant tissue layers are not rigid and that activation of a cell-type specific pathway can mediate both cell specification and patterning.

b. Ground Tissue. The embryonic ground tissue gives rise to cortical parenchyma, of which two layers are present in the hypocotyl and one in the root, and to the inner endodermis. To achieve this, the single layer of ground tissue in the globular embryo performs two periclinal divisions in the hypocotyl region and a single periclinal division in the root region. These divisions are asymmetric in that the daughter cells have different fates. In the root meristem of many *Arabidopsis* ecotypes, a single stem or initial cell layer continues after embryogenesis to produce daughter cells that give rise to both endodermis and cortex by asymmetric cell division (Fig. 4).

Genes first identified on the basis of a root phenotype have enabled an analysis of the subdivision of ground tissue in two different cell types. Mutations in the *SHORTROOT* (*SHR*) and *SCARECROW* (*SCR*) genes interfere with both embryonic and postembryonic asymmetric divisions (Benfey *et al.*, 1993; Scheres *et al.*, 1995). *shr* and *scr* mutants were identified by virtue of their secondary phenotype: a reduction in root growth. In *shr* mutants, the remaining cell layer lacks endodermal attributes whereas the layer that remains in *scr* mutants expresses both cortical and endodermal markers (Di Laurenzio *et al.*, 1996). Apparently, *shr* is required for both the asymmetric cell division and the acquisition of endodermal cell fate, whereas *scr*, based on these observations, would be required exclusively for the execution of the cell division. The *SCR* gene encodes a putative transcription factor, which provides no direct relation to asymmetric cell division. Interestingly, the gene is not only expressed in embryonic ground tissue cells and root meri-

5. Morphogenesis in Plants

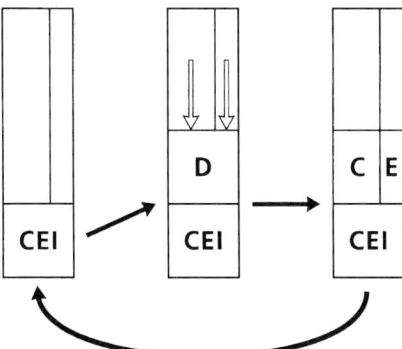

Fig. 4 Patterning and signaling in the ground tissue. In the *Arabidopsis* root meristem, a cortex/endodermal initial (CEI) divides to produce daughter cells (D) that give rise to both cortex (C) and inner endodermis (E) by asymmetric cell division. Continuous positional information in the root apex reaches the position of the initial cells, indicating that more mature cells act as a patterning template for newly generated cells (open arrows).

stem initials at the time of asymmetric cell division, but transcription remains active in the endodermal layer (Di Laurenzio *et al.*, 1996). The significance of the endodermal expression in roots is not clear from the root phenotype, but newly isolated *scr* and *shr* alleles have hinted to a *scr* function in the inner ground tissue layer. These alleles were recovered in a screen for shoot gravitropic mutants. It was shown that the mutants lack a normal endodermis based on the absence of sedimented (gravity-sensing) amyloplasts in the hypocotyl and in the shoot meristem-derived stem (Fukaki *et al.*, 1998). This observation makes two important points. First, the *SCR* and *SHR* genes are involved in the radial organization of not only the embryo axis but also the shoot meristem-derived organs. Second, an abnormal marker in the endodermis implies that the *SCR* gene, just like *SHR*, is involved in asymmetric cell division but could play a role in aspects of inner cell fate that are not readily detectable in the root as well. A continued expression in the inner cell layer is consistent with a dual function for SCR.

c. Vascular Tissue. Within the vascular tissue, several cell types arise in an ordered spacing pattern. In the *Arabidopsis* root, two metabolite-conducting phloem elements are located on opposite poles of the vascular bundle, and between them there are single water-conducting xylem elements. In other regions of the plant, these elements also coexist within the vascular bundle in a different arrangement. What regulates the spacing pattern of phloem and xylem cells within the vascular tissue? The *wooden leg* (*wol*) mutant provides a genetic entrance into this issue. In this mutant, only xylem forms within the vascular bundle of the root (Scheres *et al.*, 1995). Phloem cells are, however, present in the upper part of the hypocotyl and in leaves, demonstrating the ability of the mutant to differentiate this cell type. During embryogenesis, *wol* mutants form too few cells in the vascular tissue. The number of cells in this region of *wol* mutants can be increased by constructing double mutants with *fass* (*fs*), a mutant that results in supernumerary cell divisions (Torres-Ruiz and Jürgens, 1994). In the *wol fs* double mutant the root now produces phloem cells again, indicating that the amount of available cells in the vascular region is critical for correct pattern formation in *wol* mutants and that xylem prevails over other cell fates in that case (Scheres et al., 1995). The spacing pattern of phloem and xylem during normal vascular development could signify that such first-specified xylem becomes a source of lateral inhibitory signals for phloem development.

In summary, a handful of genes have been identified with a role in subspecification of the primary tissue layers, and the patterning mechanisms involved are now open for investigation.

III. Postembryonic Perpetuation of Cellular Pattern

While insights into axis formation and early patterning will have to emerge from studies on embryogenesis, during which a mature embryo is produced in the protective seed coat, the story of plant development goes well beyond this phase. When a seed germinates and the mature embryo gives rise to a juvenile plant (the seedling), most of its cells fully differentiate. However, the descendants of the stem cell populations within the shoot and root meristems are exceptional. These are mitotically reactivated and they extend and modify the basic embryonic tissue pattern to give rise to the adult plant (Steeves and Sussex, 1989). This process continues throughout the entire life span of the plant. To understand the postembryonic chapter of plant development, it is necessary to address the question how newly produced cells in the meristems differentiate into the appropriate cell type. Clonal analysis in shoot meristems demonstrates that the fate of newly generated cells is position- and not lineage-dependent (for *Arabidopsis*, see Furner and Pumfrey (1992), Irish and Sussex (1992), and Schnittger et al. (1996)). In the root meristem, cell lineages are more regular but clonal analysis demonstrates that also in this meristem position determines cell identity (Scheres et al., 1994; Grierson, 1997). Moreover, the transparent *Arabidopsis* root has facilitated the manipulation of cells with a laser beam to analyze positional components in cell specification. Laser ablation results in replacement and position-dependent differentiation of daughter cells from neighboring tissues in all instances tested (van den Berg et al., 1995).

Several mechanisms can account for the ability of meristematic cells to continuously assess position and differentiate appropriately. The perfect continuity of pattern in tissue formed before and after embryogenesis implies that the challenge in understanding meristems seems to be not whether but how they elaborate on embryonic mechanisms or prepatterns.

A. Prepatterning of Meristem Cells?

In animal systems, prepatterning genes define regions prior to visible differences, either by dictating fate cell-autonomously or by conferring competence to respond to further patterning cues. In *Drosophila*, for example, transcriptional regulators of the Iroquois complex allow expression of the proneural genes that select bristle-forming cells in specific domains (Gomez-Skarmeta *et al.*, 1996; Kehl *et al.*, 1998). In the context of plant meristems, prepatterning genes should define tissues before the meristems become active. What is the evidence for such genes and how would they dictate pattern?

In the root meristem, the stem cells of the different tissues—epidermis, ground tissue, vascular tissue, and root cap—are anatomically distinct from the moment that they are born, indicating that they are specified differently (Dolan *et al.*, 1993). Gene expression patterns also reveal extensive differences between stem cell groups (van den Berg *et al.*, 1995; Berger *et al.*, 1998a; Scheres and Wolkenfelt, 1998). The *SCR* gene, encoding a putative transcription factor essential for the formation of specific cell types (Section IID1b), marks the cortical stem cells and their endodermal daughters, and it displays layer-specific expression prior to the formation of the root meristem (Di Laurenzio *et al.*, 1996). The *GL2* gene, encoding a homeodomain transcription factor, is expressed in epidermal nonhair files within the meristem and it is required to maintain the nonhair fate of these cells (Section IID1a; Masucci *et al.*, 1996). *GL2* expression is restricted to the non-hair-forming cell files in the mature embryo, but only when the root meristem initials already display their characteristic stem cell-like division pattern. In conclusion, anatomical and gene expression data show that cells within the postembryonic root meristem are not naïve and that some but not all genes with a patterning phenotype are expressed prior to the emergence of the meristem.

In the shoot meristem, separate cell layers, in the case of the epidermal L1 layer fated to give rise to a single tissue, can be detected by anatomical criteria (Fig. 5). The *ATML1* gene, encoding a putative transcription factor, is an example of a gene expressed specifically in all cells of the L1 layer, prior to the emergence of the shoot meristem (Lu *et al.*, 1996). As in the root meristem, the emerging picture is that many, if not all, cells within meristems express tissue-specific genes, most of them already during embryogenesis.

The clonal analyses and laser ablations discussed in Section III, however, show that cells within meristems have flexible fate and might constitute single equivalence groups, indicating that the previously established expression patterns do not dictate the fate of cells in a stable, cell-autonomous way.

B. What Are the Relevant Prepatterning Cues?

Cells that switch position postembryonically can start to express genes whose expression is normally initiated during embryogenesis. A well-documented example

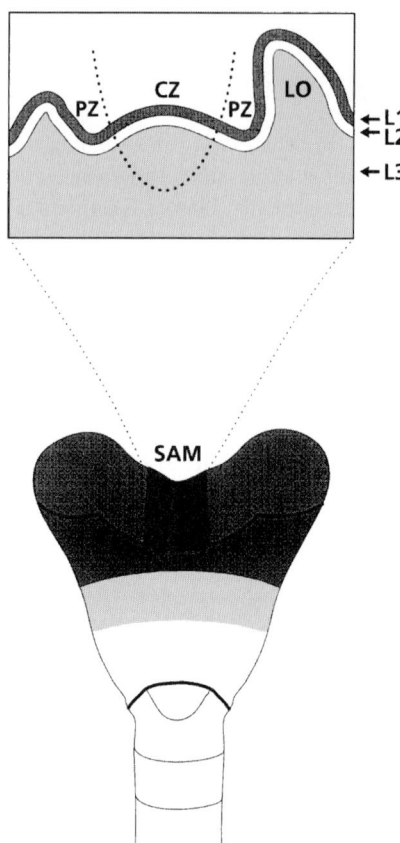

Fig. 5 Shoot meristem organization. The shoot meristem can be divided into layers and zones. Separate cell layers (L1, L2, and L3) can be detected by anatomical criteria. The region corresponding to the central zone (CZ) consists of stem cells with reduced division rates and is surrounded by a peripheral zone (PZ) of more rapidly dividing cells where lateral organs (LO) or meristems initiate.

involves hair cell patterning in the root epidermis. Hair cells are invariantly located over anticlinal cell walls between two cells of the underlying cell layer (Fig. 3). The *GL2* gene, required to execute the nonhair fate in the root epidermis, is expressed accordingly in the remaining cells already in the mature embryo (Berger *et al.*, 1998a). With low frequency, cells overlying the anticlinal cell wall perform a longitudinal division that creates two cells, which subsequently form two clones. One of these clones is displaced from the underlying anticlinal cell wall. Analysis of *GL2* transcription in such displaced cells reveals that gene is now expressed (Berger *et al.*, 1998b). Hence *GL2* gene expression does not form a positional reference in the meristem by itself, but it rather remains responsive to cues of unknown nature. We should like to know the nature of these and other cues that have

5. Morphogenesis in Plants 223

been demonstrated to direct cell fate in meristems. The first clues have been obtained about the direction of such positional cues involved in the specification of cell types in all three major root tissues, the hair/nonhair cells in the epidermal cell layer, the cortex/endodermis derived from the ground tissue, and the vascular tissue. These will be discussed in the following sections.

1. Epidermis

The cues that govern the hair/nonhair cell fate decision are present throughout the root meristem, as revealed by the inspection of the early and late hair/hairless clones discussed in Section IIIB (Berger *et al.*, 1998a, 1998b). Cell fate is dependent in a precise way on the distance to the anticlinal wall, and a variety of ablations of neighboring cells are not able to disturb position-dependent differentiation. Although rigorous proof of the lack of signals emanating from neighbors requires complete isolation of epidermal cells—an experiment that is obstructed by technical difficulties—it is tempting to conclude that extracellular cues may reside in a narrow domain centering on the anticlinal cortical wall (Fig. 3; Berger *et al.*, 1998a).

The plant hormone ethylene governs a number of developmental processes in plants, and its synthesis and signal transduction are beginning to be understood in some detail (Ecker, 1995; Fluhr, 1998). Root hair formation is affected by modulation of ethylene synthesis and perception, either genetically or with precursors/ inhibitors (Dolan *et al.*, 1994; Masucci and Schiefelbein, 1994; Tanimoto *et al.*, 1995). These studies indicate that ethylene signaling is necessary and sufficient for root hair formation. Ethylene can still affect cells at relatively advanced stages of cell differentiation (Masucci and Schiefelbein, 1996). Is this plant hormone involved in maintaining a—potentially extracellular—cue that determines epidermal cell fate, is it involved in the initial establishment of this cue, or is ethylene acting in a pathway independent from cortical domains and transcription factors such as *GL2*? More experiments will be necessary to elucidate the relation between transcription factors, cortical wall-dependent differentiation, and ethylene.

2. Cortex and Endodermis

Continuous positional information in the root apex reaches the position of the stem cells for all the different tissues, as cell invasions in this region result in position-dependent fate switching (van den Berg *et al.*, 1995). What is the origin of these cues? In the ground tissue, it has been investigated how the meristematic stem cells obtain the information to give rise to two different cell types (Fig. 4). When ground tissue stem cells that normally give rise to cortex and endodermis are isolated from their own tissue by laser ablation of more mature daughters, they can still proliferate but no longer perform the characteristic asymmetric cell division to create both layers (van den Berg et al., 1995). This indicates that more mature cells act as a patterning template for newly generated cells (Fig. 4). A guiding effect of more mature cells on newly added cells seems a reasonable strategy to ensure that a pattern that is set up in the protective environment within the seed coat is perpetuated throughout development. More evidence and studies on the molecular mechanisms involved are needed to substantiate this "patterning template" idea.

It may be of interest that cytoplasmic connections between cells in the root meristem reminiscent of gap junctions—plasmodesmata—have been shown in several instances to establish layer-specific connections (Duckett et al., 1994; Oparka et al., 1994; McLean et al., 1997). In the shoot meristem, it has been proposed that transcription factors may be routed through plasmodesmata (Lucas et al., 1995). Additional experiments will be required to test whether positional cues utilize plasmodesmata, thereby providing a direction to positional signaling. Relevant questions are whether prepatterned genes like *SCR* are also required to maintain the postembryonic pattern and whether involvement in maintenance of pattern by transcription factors like *SCR* could involve directional routing via plasmodesmata. Continuous deployment of patterning genes is not unprecedented. In *Drosophila*, the Hedgehog (HH) gene product seems to signal continuously across A/P boundaries to mediate patterning. HH is first active in the cellularized embryo and continues to be active in the developing imaginal discs (Lee et al., 1992; Basler and Struhl, 1994; Strigini and Cohen, 1997). A critical test whether patterning genes act continuously in plants has to await accurate assessment of the temporal requirement of embryonic patterning genes.

An alternative mechanism could underlie the patterning template scenario that is outlined above. Embryonic pattern formation can yield a prepattern of cells that subsequently mediate position-dependent cell specification by postembryonic mechanisms unrelated to those that are used to set up the prepattern. An example of a new mechanism operating after prepatterning may be the "community effect" that is proposed to enable *Xenopus* mesoderm cells to progress to muscle or notochord differentiation. It has been shown that a minimal number of mesoderm cells is required for muscle cell differentiation (Gurdon et al., 1993). A mesoderm-derived "community factor" is thought to be in part responsible for activating mesoderm differentiation (Carnac and Gurdon, 1997). One may describe this effect in more general terms: as soon as embryonic events have established a sufficiently large pool of a particular cell type, these cells mutually reinforce their fate. In plant tissues, reinforcement mechanisms would have to be directional—as indicated by laser ablation experiments (van den Berg et al., 1995)—to ensure the maintenance of patterns of single-cell width.

3. Vascular Tissue

A feed-forward mechanism to recruit newly generated cells into existent tissue layers may also operate in the vascular tissue. As discussed in Section IIB2, the *MP* gene is required to promote vascularization in the root/hypocotyl region of the embryo (Berleth and Jürgens, 1993). When MP levels are reduced, basal cells are not recruited to enter the vascular differentiation pathway, although the gene is not required for vascular differentiation per se but rather for the connection of vascular cells to give continuous strands (Przemeck *et al.*, 1996). Apparently, vascular cell continuity is required for ongoing specification. Hence, also in this case, an existing pattern may be maintained by directional recruitment, perhaps by directional auxin transport (Hardtke and Berleth, 1998).

The overall conclusion is that as yet no evidence has been found for prepatterning genes as cell-autonomous organizers, such as at the A/P boundaries in fly development. There is, however, evidence for the localization of positional cues or for the direction in which cues are delivered. In the case of epidermal subspecification, positional information may be laid down in localized extracellular domains. For continuous vascular development and for ground tissue subspecification, maturing pattern elements may be involved in the directional transport of signals to recruit new cells. Are reinforcement mechanisms community effects, with the additional feature that tissue layers can promote differentiation of their own kind in a directional fashion? In that case, the mature tissues would be the "organizers" of cellular pattern in the meristems. This exciting issue will have to be addressed by analysis at the molecular level.

IV. Control of Cell Division during Development

Throughout the development of all multicellular organisms, cell divisions in time and space accompany the formation of a specific pattern of different cell types, tissues, and organs. Unlike the situation in animals, plant development is predominantly postembryonic and relies on the activity of apical meristems, pattern elements that are first established during embryogenesis. Meristems are defined as localized clusters of dividing cells. Two types of cell divisions contribute to the development of the mature plant. Proliferative cell divisions lead to the production of more cells of the same type within a cell file or tissue. Formative cell divisions include asymmetric cell divisions that give rise to different cell types and oriented cell divisions whereby additional cell layers are created (reviewed in Jürgens, 1996).

Cell division is involved in three aspects of postembryonic plant development occurring in overlapping stages. (i) Different cell fates continue to be generated within the meristems involving positional information and asymmetric cell divisions. Cell fate specification during pattern formation may regulate cell division. In plants, it is an important issue whether pattern formation in turn depends on precise number and sequence of cell division. (ii) A second feature concerns the activation and maintenance of the meristems. Postembryonic development begins when the seed germinates and cell division is reactivated in the shoot and root apical meristems. To ensure continued cell production, it is essential that an active meristem is maintained. (iii) The third aspect involves the growth and morphogenesis that give rise to the size and distinct shape of cells, organs, and the organism itself. Growth is a combination of increasing numbers of cells and cell expansion and involves mostly proliferative cell divisions. Morphogenesis shapes the organism through the combined activities of proliferative and formative cell division, cell expansion, cell death, and, in animals, cell movement. Since cell walls constrain cell movements, the relevant issue here is whether precise control of cell division is essential for plant morphogenesis.

We will focus mainly on the *Arabidopsis* root meristem to illustrate the role of cell division during the development stages outlined here, and we will also take into account important advances in identifying genes involved in regulation of cell division in the shoot meristem.

A. Pattern Formation Can Be Uncoupled from Cell Division

During *Arabidopsis* embryogenesis the sequence of cell division is very regular, enabling seedling structures to be traced back to groups of cells in the early embryo. Recessive mutations in the *FASS* (*FS*) and *TONNEAU* (*TON-1*, *TON-2*) genes alter cell division sequences and orientation from the zygote stage onward, and by the octant stage the mutant embryos look abnormal (Torres-Ruiz and Jürgens, 1994; Traas et al., 1995). Whereas no primordia of seedling structures can be recognized by morphological criteria at the heart stage, all elements of the body pattern are differentiated in *fs* and *ton* seedlings, although their morphology is abnormal. At the cellular level, *fs* and *ton* mutations affect cell elongation and orientation of cell division. For example, the root lacks the regular arrangement of cell files found in the wild type and shows irregularly enlarged cells. Nevertheless, all cell types are present. These results show that *fs* and *ton* mutants do not affect pattern formation, suggesting that the regularity of cell division is not instrumental in pattern formation. Similarly, irregularity in cell division coupled to precise patterning appears to exist in nature; in embryos of plant species such as cotton the pattern of cell division is not at all regular and yet the seedling has the same body pattern as *Arabidopsis* (Pollock and Jensen, 1964).

The interdependence of patterning from cell division is supported by analysis of tobacco plants transformed with a dominant negative (DN) form of the heterologous *Arabidopsis CDC2A* gene (Hemerly et al., 1995). Cell division is slowed down but DN plants appear only slightly smaller and morphologically normal due to increased cell sizes. Cross sections through a leaf, cotyledon, and root show that all cell layers and cell types are present in a normal arrangement, indicating pattern formation does not depend on precise cell numbers. In addition, the normally very regular cell division sequences in the root meristem of tobacco is disturbed in DN plants, particularly in the apical region affecting the quiescent center and initials, apparently without interfering in the patterning of the different root tissues.

Pattern formation and cell division have also been uncoupled in animals. In *Drosophila*, a temperature-sensitive (TS) *CDC2* mutation has been used to inhibit cell division locally and during different developmental stages (Weigmann *et al.*, 1997). Wing size and shape are normal when few cell divisions are blocked because cell size increases. Notably, the presence of fewer larger cells has no significant effect on the pattern of veins in the wing and expression of the patterning gene *DPP* and its targets *OMB* and *SPALT* in the wing disc or *ODD* in the leg disk. Thus the processes that determine the pattern and sizes of expression domains of these genes can continue in the absence of cell division.

Taken together, the results suggest that in plants and animals patterning cues are relatively independent of the number of cells and, in plants, plane of cell division. For animals this translates to long-range cues that are assessed in a distance- or volume-dependent manner. Such long-range cues have not been identified in plants. However, a patterning mechanism based on cell–cell communication and reinforcement signals (Section III) would still allow patterning when cell numbers and sizes are deregulated.

B. Cell Fate Specification Can Regulate Cell Division

Mechanisms whereby cell fate specification leads to patterned regulation of cell division are common in animal development (Cohen, 1996; Kim *et al.*, 1996; Schnabel, 1996; Folette and O'Farrell, 1997). The perhaps simplest form of this regulation is exemplified by $CDC25^{string}$ expression during fly embryogenesis (Edgar *et al.*, 1994). $CDC25^{string}$ is a phosphatase that removes inhibitory phosphorylation from CDC2. Normal embryonic expression of *string* requires extensive *cis*-acting regulatory sequences. *string* expression is controlled at the transcriptional level and patterned by positional information supplied by a large set of genes that determine many aspects of cell fate. For example, in embryos that are mutant in *twist*, a gene encoding a transcription factor that is specifically expressed in the prospective mesoderm, there is a specific absence of *string* expression and hence cell division in the mesodermal cells. Apparently, combinations of transcription factors bind to the *string* promoter and drive its expression in unique patterns resulting in local cell division (Edgar *et al.*, 1994).

Such a relationship between fate specification and cell division may also be characteristic of plant meristems. Recall, for example, that the *SCR* and *SHR* genes are involved in asymmetric divisions in the root that generate cortex and endodermis as well as in aspects of endodermal cell fate (Di Laurenzio *et al.*, 1996; Fukaki *et al.*, 1998). We will discuss in some detail recent experiments on the *Arabidopsis* root epidermis that provide further evidence for control of cell division by cell fate specification.

1. Cell Fate Regulates Cell Division in the Root Epidermis

The alternating files of hair and hairless cells in the root epidermis consist of cells of different sizes. Hairless cells are longer than hair cells at maturity (Dolan *et al.*, 1993), which indicates that hairless cell files produce fewer cells. Clones originating from single hair cells, which give rise to both hair and hairless cells (Section IIIB), also display differential cell length together with differential *GL2* expression (Berger *et al.*, 1998b). As the clones increase in size, the number of hairless cells gradually trails the number of hair cells. This is reflected by the ratio of the number of cells in either clone, which increases until it reaches an average maximum. Thus, the cell division rate initially slows down in the hairless clone, allowing these cells to reach their normal larger size. These results show that a change in cell fate can influence cell division rate. Hairless cells were reported to be longer throughout the meristem except around the initials (Beemster and Baskin, 1998; Berger *et al.*, 1998b), showing that differential cell division rate and cell size is normally regulated close to the initial cells of the meristem. Do genes controlling hair cell fate, such as *TTG* (Section IID1a; Galway *et al.*, 1994), regulate cell division? In *ttg* mutants, which display ectopic root hair formation, epidermal cell sizes in all cell files are similar (Berger *et al.*, 1998b). These results show that *TTG* is involved in the control of cell division rate in the epidermis. Furthermore, the normally markedly higher occurrence of longitudinal anticlinal divisions observed in hair cells, resulting in a higher number of clones being derived from these cells, is significantly altered in *ttg* mutants (Berger *et al.*, 1998b).

In conclusion, in plants cell fate specification can regulate cell division rate and orientation and this mechanism may contribute to generating different cell sizes.

C. Activation of Meristems

Following fertilization, the zygote undergoes cell divisions to produce the embryo with the primary meristems of the shoot and root set up at opposite poles of the apical–basal axis (Fig. 2D). The embryo ceases cell proliferation upon maturation and then enters dormancy until seed germination, when postembryonic development starts and cell division is reactivated in the meristems. As the plant grows, shoots set aside axillary meristems that may initiate secondary branches. The root system branches by forming lateral roots via new meristems generated from nonmeristematic tissue (Steeves and Sussex, 1989; Malamy and Benfey, 1997b).

It is not yet clear whether specific positive input is required for embryonic cell divisions as opposed to postembryonic ones. The *PROLIFERA* (*PRL*) gene is zygotically required for embryo development in *Arabidopsis* (Springer *et al.*, 1995). *PRL* is likely to represent a general cell division factor since it is related to the *MCM2-3-5* family of yeast genes that are required for the initiation of DNA replication. Accordingly, reporter gene expression data reveal that *PRL* is expressed in dividing cells throughout the embryo and mature plant. *medea* (*mea*) mutant embryos grow excessively and this growth regulation phenotype is strictly dependent on maternal contribution of the mutant gene (Grossniklaus *et al.*, 1998). The *MEA* gene, therefore, provides a negative maternal input for the controlled cell proliferation in the embryo. *MEA* encodes a *Polycomb*-related protein and it has been suggested that its expression is regulated by genomic imprinting.

1. Activating the Shoot Meristem

Activation of the shoot meristem in maize may involve the *KNOTTED1* (*KN1*) gene. *KN1* encodes a homeobox protein and is expressed in the cells of the shoot meristem but disappears rapidly from portions of the meristem where leaf primordia or floral organs initiate (Fig. 5; Vollbrecht *et al.*, 1991; Jackson *et al.*, 1994). Transgenic tobacco plants ectopically expressing *KN1* in leaves can have adventitious shoot meristems and shoots forming on the leaves (Sinha *et al.*, 1993), showing that *KN1* is sufficient for induction of cell division when ectopically expressed. The *Arabidopsis SHOOT MERISTEMLESS* (*STM*) gene is a homolog of *KN1* and shares a closely related RNA expression pattern with *KN1* (Long *et al.*, 1996). *stm* mutants are already defective in the initial formation of the meristem during embryogenesis, indicating a function in establishing the meristem (Long and Barton, 1998). Although the role of *STM* in activation of the meristem is unclear, analysis of loss-of-function mutants has revealed a role for both *STM* and *KN1* in meristem maintenance (Section IVD1; Clark *et al.*, 1996; Kerstetter *et al.*, 1997).

2. Activating the Root Meristem

Several mutants have been obtained that display arrested root growth following germination. The *root meristemless* (*rml*) mutants have a root meristem with a normal cellular architecture that undergoes no (*rml1*) or limited (*rml2*) postembryonic cell division following germination whereas shoot development is reported to be unaffected (Cheng *et al.*, 1995). Interestingly, *rml1* lateral roots are able to initiate normally but cell division ceases when the same number of cells as the embryonic root and the *rml1* primary root is reached. Histological data on wild-type roots show the number of cells constituting the lateral root prior to meristem activation corresponds to the number in the *rml1* mutant (Malamy and Benfey, 1997b). The *aberrant lateral root formation 3* (*alf3*) mutants form densely spaced lateral root primordia that are arrested at a similar stage (Celenza *et al.*, 1995). Unlike the *rml* mutants, *alf3* mutants can be rescued by applying the plant hormone auxin or its precursor indole, suggesting that the absence of auxin/indole is responsible for the premature arrest in cell division.

These studies indicate that both primary and lateral root formation is a two-stage process: (1) formation of the meristem involving mostly formative cell divisions:

(2) activation of cell division in the meristem involving mostly proliferative cell divisions. Primary root development and secondary root development seem to share a common end point at which time the *RML1* and *ALF3* gene products are required for the activation of the meristem and the continuation of root growth.

a. Environmental Factors Control Cell Division. A characteristic feature of plants is their plastic development in response to external factors, e.g., nutrition, light, and stress. It is now well established that environmental adversity results, at least partly, in the enhanced formation of active oxygen species. Adaptation of plants to stressful conditions involves the recruitment of ascorbic acid (AA), glutathione (GSH), and NAD(P)H for efficient removal of the oxidizing agents (Bowler *et al.*, 1992; May and Leaver, 1993; Babiychuk *et al.*, 1995; Conklin *et al.*, 1996). These molecules that undergo reversible alterations are ideal candidates to act as sensors or even effectors in the control of cell division.

Three lines of evidence indicate that GSH participates in the regulation of cell division in the *Arabidopsis* root meristem (Sanchez-Fernandez *et al.*, 1997). First, artificially increased endogeneous GSH levels stimulate cell divisions. More mitotic figures are observed within the same zone, indicating an increased cell division rate, but also the length of the meristem is extended compared to controls, resulting in more dividing cells. Both processes contribute to an increased cell production. Under these conditions root growth is slightly enhanced even though cell size is decreased. Second, artificially decreasing GSH levels through treatment with L-buthionine (S, R)-sulfoximine (BSO), a specific inhibitor of GSH synthesis, reduces cell divisions. Root growth does not significantly decrease because cell size increases dramatically. Importantly, the inhibition of cell division through BSO can be completely reversed by supplementing GSH, indicating that GSH levels are specifically lowered. Third, *in vivo* fluorescence labeling of GSH shows that high levels of endogenous GSH are associated with actively dividing cells but not the mitotically inactive quiescent center. This distribution closely parallels that described for the pattern of AA in the maize root meristem (Kerk and Feldman, 1995). The low levels of AA in the quiescent center appear to be actively maintained by the local high level of ascorbate oxidase, suggesting a function for AA in the regulation of cell division. This is consistent with the observation that cells in the quiescent center have extended G1 phases (Clowes, 1975) and that AA may be necessary for G1/S transition in the cell cycle (Liso *et al.*, 1984, 1988; Citterio *et al.*, 1994). Taken together, these results suggest a mechanism that operates to lower the GSH and AA pools or shift their redox balance not only in the quiescent center but also in proximally located cells of the meristem to restrict the extent of the meristematic zone.

These observations imply a direct link between the levels of GSH and/or AA and the plant cell cycle, which parallels results in animal systems. For example, specifically depleting the GSH level in human cells induces oxidative stress. This results in the induction of WAF1 expression, encoding a potent inhibitor of cyclin-dependent kinases required for progression from the G1 into the S phase of the cell cycle, and cell cycle progression is arrested (Russo et al., 1995). Induction of WAF1 is prevented by pretreating cells with a precursor of GSH. Similarly, other studies indicate the requirement for *de novo* synthesis of GSH for both entry and progression through the cell cycle (Suthanthiran et al., 1990; Poot et al., 1995). GSH may be a conserved factor in modulating the activity of transcription factors responsible for cell division. Whether this means that, in plants, control of environmental factors has evolved as a mechanism for developmental control is unclear.

D. Maintaining Meristem Activity

Maintaining a pool of stem cells is important to ensure a continued existence of the meristem. Substantial progress has been made in identifying genes involved in the maintenance of the stem cell population in the shoot meristem (reviewed in Clark (1997), Meyerowitz (1997a), and Laux and Mayer (1998)). Also in the root, stem cell maintenance by central quiescent cells may be one of the ways to ensure continuous meristem activity.

1. Shoot Meristem Maintenance

The shoot meristem contains a central region of stem cells with reduced division rates which is surrounded by a peripheral zone of more rapidly dividing cells (Fig. 5; Steeves and Sussex, 1989; Meyerowitz 1997a). The stem cells are relatively undifferentiated in the sense that they lack morphological features associated with cells in mature tissues. The *CLAVATA* (*CLV1* and *CLV3*), *WIGGUM* (*WIG*), and *SHOOT MERISTEMLESS* (*STM*) genes are implicated in regulation of the balance between maintaining the stem cell population and initiation of organ formation (Clark et al., 1996; Running et al., 1998). *clv* and *wig* mutants display overproliferation of the central and peripheral zones, respectively, thereby creating an enlarged meristem. These phenotypes implicate the *CLV* and *WIG* genes in the control of cell division and/or promoting differentiation leading to organ formation (Clark et al., 1993, 1995, 1997; Running et al., 1998). Double-mutant analysis shows that *WIG* acts in a separate pathway from the *CLV* genes (Running et al., 1998). In *stm* mutants, adventitious meristems form postembryonically but these always terminate after producing fewer organs than usual (Barton and Poethig 1993; Clark et al., 1996; Endrizzi et al., 1996), suggesting a role for the *STM* gene in maintenance of the meristem by stimulating cell proliferation and/or suppressing differentiation in the central zone. *stm clv1* double mutants display intermediate phenotypes (Clark et al., 1996), indicating that *STM* and *CLV1* competitively regulate the balance between undifferentiated cells and organ formation; i.e., they do not act in the same pathway. The *STM* and *CLV1* genes are first expressed in the developing meristem of the globular and heart stage embryo, respectively, and continue to be expressed in the central meristem (Long et al., 1996; Clark et al., 1997; Long and Barton, 1998). *STM* activity is required for maintenance but not onset of *CLV1* expression. These results show that the mechanism of shoot meristem maintenance is set up early in development and is essential for proper meristem activity (Long and Barton, 1998).

The *CLV1* gene encodes a putative leucine-rich repeat containing receptor kinase, suggesting a function in a signaling pathway (Clark et al., 1997). The *CLV3* gene that genetically interacts with *CLV1* is a candidate for encoding the CLV1 ligand (Clark et al., 1995).

2. Root Meristem Maintenance

In the root, new cells are added to files of different cell types by dividing cells in the root meristem. The initial cells at the end points of these files function as stem cells: with each division they set off new cell tiers while one daughter cell retains its position within the meristem (Dolan et al., 1993). In analogy with the shoot meristem, root meristems also contain a group of slowly dividing cells, the quiescent center (QC; Fig. 6). In *pickle* (*pkl*) mutants the primary root meristem develops into an abnormally thickened and green structure (Ogas et al., 1997). Root tissue carrying the *pkl* mutation spontaneously regenerates new embryos and plants because it has retained characteristics of embryonic tissue. Expression of this aberrant differentiation state was suppressed by the plant hormone gibberellin. Although genes like *PKL* but also *RML1*, *ALF3*, and *HBT* are required for maintaining root meristem activity, it is not clear whether this reflects their primary function or is a consequence of more general or earlier effects.

In the *Arabidopsis* root, all stem or initial cells surround and contact the QC (Fig. 6), which consists of only four mitotically inactive cells. Laser ablation of one QC cell results in the cessation of division and the progression of cell differentiation in the underlying columella initial in a contact-dependent manner. The columella initials contacting the neighboring intact QC cell still perform their normal asymmetric division (van den Berg et al., 1997). Cortex initials contacting the ablated QC cell also progress in differentiation status: they behave as daughters and divide asymmetrically. Support for direct suppression of differentiation comes from ablation experiments in mutants that lack postembryonic cell divisions. QC ablation in these mutants also results in differentiation, which is therefore independent on the initial cell's ability to divide (van den Berg et al., 1997). Thus the QC may directly inhibit differentiation of all surrounding initials in a contact-dependent manner, thereby maintaining their stem cell status (Fig. 6). A possibly contact-dependent signaling mechanism is reminiscent of control of differentiation in animal systems by the NOTCH–DELTA receptor–ligand interaction that represses the differentiation of neuroblasts in *Drosophila* (Muskavitch, 1994; Rooke and Xu, 1998).

When is a differentiation control mechanism set up in the root meristem? As

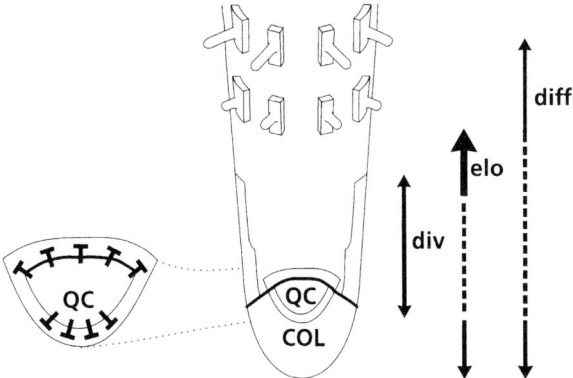

Fig. 6 Cellular activities in root development. The seedling root encompasses three partly overlapping developmental zones (arrows): the division (div), elongation (elo), and differentiation (diff) zones. Arrow thickness indicate elongation rate and terminal differentiation. QC, quiescent center; COL, columella. Detachment: the initial cells of the different cell types surround and contact the QC, which inhibits differentiation, thereby ensuring their stem cell status.

mentioned in Section IIB1, *hbt* mutants are affected in the development of the basal domain, resulting in the absence of the QC by anatomical and functional criteria. The characteristic cell division pattern for initial cells is not observed in *hbt* embryos; e.g., cells at the position of the cortex initials have already divided asymmetrically, and meristem activity is absent (Willemsen *et al.*, 1998). These observations indicate that the mechanism for maintaining undifferentiated initials in the root meristem is set up during embryogenesis and is essential for meristem activity.

When signals from the QC inhibit differentiation, there may also be signals inducing differentiation. Root initials are not undifferentiated as they can be marked with cell type specific promoter/enhancer traps (van den Berg *et al.*, 1995; Malamy and Benfey, 1997b). Therefore, progression of differentiation is regulated and not the maintenance of a group of undifferentiated cells. Does differentiation require positive input or do cells differentiate by default in a cell-autonomous manner? In the shoot meristem, where *STM* and *CLV1* are implicated in regulating the differentiation status of meristem cells, one interpretation is that STM inhibits and CLV1 activity promotes cell differentiation (Laux and Schoof, 1997). In analogy, recently identified root meristem specific *CLV1*-like genes (our unpublished data) may be involved in promoting differentiation of root meristem cells once they no longer contact the QC or once the influence of a QC-derived inhibitory signal is eliminated.

In conclusion, maintaining root and shoot meristem activity may require a balance of signals regulating the differentiation status of the stem cells and their daughters. The *CLV1* gene and its root-specific homologues provide tools to test this model.

V. Growth and Organ Morphogenesis

As cells become committed to a developmental fate, they undergo differentiation, which is accompanied by changes in cell size and shape. A recent review extensively discusses the current knowledge on cell morphogenesis—i.e., the acquisition of a particular shape—of different genetically well-characterized cell types in *Arabidopsis* (Hülskamp et al., 1997). Instead we will concentrate here on organ morphogenesis.

The collective changes in cell number and shape equal the size and shape of the entire organism. Therefore, one can easily imagine cell division and expansion being the driving force behind these processes, especially in plants, where cell movement does not occur.

Traditionally, the regulation of growth and morphogenesis in plants is viewed from two perspectives (Fig. 7; reviewed in Kaplan (1992) and Jacobs (1997)). (i) The organismal theory holds that growth and morphogenesis are driven by the expansion of organs whose constituent cells merely partition the resultant volumes (Fig. 7A). Instead of cell division and expansion being the cause of growth and morphogenesis, they represent the markers of it. This represents a spatial view stating that cell division and expansion are specified by global positional signals. However, we emphasize that these signals may be the result of communication between cells. "Global signaling" may be an emerent property of a short-range cellular communication network. (ii) The cellular theory holds that autonomous division and expansion of cells drive growth and morphogenesis (Fig. 7B). Cells are considered individual entities that together make up the organism. This implies that the time and number of cell divisions and the time and extent of expansion are specified cell-autonomously when a cell is formed. Whichever theory is close to reality, both are intimately coupled with patterning. Pattern formation may determine the amount of signal made or its threshold in a cell, thereby indirectly influencing division, or, alternatively, determine the autonomy of division in a cell.

Whether growth and morphogenesis drive or depend on cell division may be determined by directly interfering with the cell cycle through disruption or manipulation of the function of key regulators, e.g., CDC2 protein kinase. In addition, the influence of developmental and, particularly in plants, environmental factors allows the analysis of the relation between growth and cell division. Alternatively, mutants that uncouple growth or morphogenesis from cell division or expansion are useful for investigating whether these processes directly depend on the precise control of cell division and expansion.

A. Growth Can Be Uncoupled from Cell Division

Growth is the combined activity of cell expansion and cell production. The rate of cell production has two distinct components: the number of dividing cells and their

5. Morphogenesis in Plants

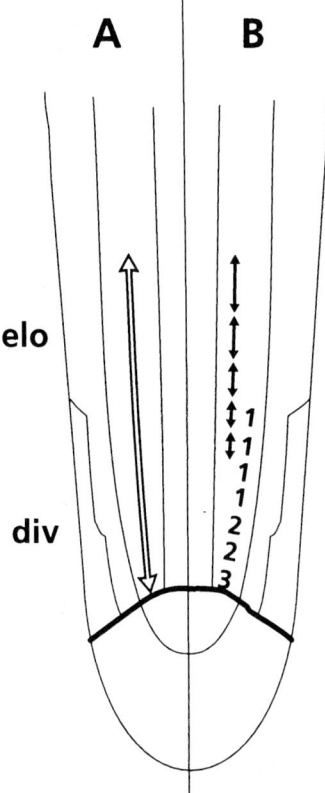

Fig. 7 Mechanisms of growth. (A) Organismal theory. Cells communicate and respond to the total of signaling that determines the position where cells divide and elongate. The arrow represents the outcome of signaling that determines the size of the division and elongation zone. (B) Cellular theory. Division and elongation are cell-autonomous (e.g., counting). Number and arrows represent single cells.

rate of division. An increase in cell production can be caused by an increase in the number of dividing cells or by enhancing the passage through the cell cycle, increasing the division rate, or both. Experiments performed in *Drosophila* and tobacco in which the function of CDC2 has been altered allow the analysis of cell division defects on growth.

In the fly, when the last two cell divisions in entire imaginal discs, in anterior compartments, or in clones of cells homozygous for *TS-CDC2* are specifically inhibited, wings of normal size are formed (Section IVA; Weigmann *et al.*, 1997). However, the induced mutant wings, compartments, and clones are composed of cells considerably larger than those seen in wild-type controls. Recent experiments show that besides lowered cell division, also increased cell division rates has little effect on clone or compartment size of *Drosophila* wings (Neufeld *et al.*, 1998). These are examples in a series of experiments in different animal systems that show that organ growth control is independent of the exact cell number (Frankhauser, 1941, 1945; Santamaria, 1983; Fero *et al.*, 1996; Kipreos *et al.*, 1996). However, growth is limited in the absence of cell division. In *Drosophila*, early inactivation of CDC2 function in the anterior compartment results in a distorted disc shape with few large cells (Weigmann *et al.*, 1997).

In plants, the relation between growth and cell division was examined for the organs of the DN tobacco plants (Section IVA; Hemerly *et al.*, 1995). The introduction of the heterologous *DN-ATCDC2A* slows down but does not completely abolish cell division in tobacco. Developmental timing is not affected, new leaves initiate and grow at the same rate, and mutants flower at the same time. They exhibit slightly smaller organs, and microscopic analysis reveals that cells in all tissues are larger than those from wild type, except in flowers and embryos (Hemerly *et al.*, 1995). The severity of the phenotype in DN plants is directly correlated with the levels of the mutant protein, indicating that titration of regulatory factors reduces the frequency of cell division. In agreement with the experiments in flies, the observation that DN plants are smaller supports the notion that cells cannot grow indefinitely. Nevertheless, experiments in both systems show that when cell division rate is altered, cells compensate for this defect by growing in size so that the overall size of the organs is relatively normal. Taken together, the picture emerges that normal organ size can be reached independent of the exact cell number, consistent with an organismal theory on growth.

1. Root Growth and Meristem Size Can Be Uncoupled from Cell Division Control

How do plants modify their growth rate during their postembryogenic development? The observation that the primary roots of plants grow more rapidly with time (Baskin *et al.*, 1995) has been exploited to achieve insight into this question. The seedling root can be divided into three partly overlapping developmental regions: the division zone (meristem), the elongation zone, and the differentiation zone (Fig. 6). The meristem encompasses part of the elongation zone as cell division continues well beyond the location where cell elongation rates start to increase (Dolan and Roberts, 1995b; Jacobs, 1997; Beemster and Baskin, 1998). Differentiation occurs gradually throughout the different developmental zones.

Both cell expansion and cell production have been quantified for cortical cells in growing *Arabidopsis* roots with a kinematic method, using time-lapse video imaging (Beemster and Baskin, 1998). The acceleration of root growth with time correlates with an increased cell number in the meristem with little change in cellular expansion rate or cell length. This is reflected by an increase in the length of the division and elongation zones and a steady increase of cell numbers in these zones. However, the average cell division rates are approximately constant

5. Morphogenesis in Plants 237

throughout the meristem and remain constant with time (Beemster and Baskin, 1998). The developmentally enhanced growth is compatible with the cellular theory on growth. Namely, a longer competence of a meristematic cell to divide will enhance the number of dividing cells and lead to an enlarging cell division zone. Concomitantly, more cells will be expanding at a given time, increasing also the size of the expansion zone. However, the data can also be explained using the organismal theory: global positional signals allowing cell division may cover an increasing distance over the course of development. As elongation rates and final cell sizes are only slightly altered, the elongation zone has to be similarly extended to allow all cells to reach their normal mature size. Therefore, the sizes of the division and elongation zones should be coupled.

The redox experiments, whereby environmental conditions are simulated by alteration of GSH levels, show that root growth is not or only slightly affected by the decreased or increased cell divisions. Obviously, an increase in cell division results in a decrease of mature cell size and vice versa (Section IVC2a; Sanchez-Fernandez *et al.*, 1997). These results indicate that the size of the division and elongation zones is determined. This way, alterations in rates of cell division will lead to more smaller or fewer larger cells in the division zone but these cells will not reach their normal size because they elongate either too little or too much due to the unaltered elongation zone. Interestingly, regulation of root growth in this way is used by the plant to its own advantage. Addition of GSH produces short hair cells with long hairs, whereas depletion of GSH leads to long hair cells with short or no hairs (Sanchez-Fernandez *et al.*, 1997). In nature, this response allows the root to maintain normal growth and explore the soil under adverse conditions with minimal cell divisions and root hair production or capitalize on a favorable local environment by production of numerous long root hairs.

Together, the foregoing results support the hypothesis that the division and elongation zones are coupled and may actually constitute one developmental zone. This suggests that the signals regulating the sizes of the division and elongation zones are coupled. For example, they may constitute a graded version of one signal or two opposing signals, acting over a distance or by means of coupled short-range interactions.

B. Organ Morphogenesis Can Be Uncoupled from Cell Division

Although animal cells can migrate and in this way aid in creating organ shape, plant cells cannot and an invariable sequence of cell division and expansion has often been considered crucial for morphogenesis. Nevertheless, early studies in which cell division was arrested chemically or by irradiation already demonstrated that leafs can initiate and that existing leaves undergo properly oriented, albeit limited, growth in the absence of cell division (Haber, 1962; Haber and Foard, 1963). What is striking in the examples manipulating the cell cycle in tobacco and *Drosophila* is that a normal shape of the organs is maintained when cell division is inhibited, supporting the notion that within a certain range, cell number is not crucial for morphogenesis.

Moreover, mutants obtained in maize in which cell division orientation or cell expansion is affected do not affect organ shape. In mutant *tangled-1* (*tan-1*) leaves of corn, the majority of longitudinal divisions, normally correlated with leaf widening, are substituted by aberrantly oriented divisions (Smith et al., 1996). Other organs are similarly affected, showing that TAN-1 functions throughout the plant in controlling the orientation of cell division. Although mutant leaves grow more slowly and their final size is smaller than normal, their shapes at all stages of leaf growth are similar to those of normal leaves of the same length. In the *warty-1* (*wty-1*) mutant, mature leaves have raised warts that appear in patchy distribution, consisting of excessively enlarged cells, but cell sizes are also abnormal in internal leaf tissues (Reynolds et al., 1998). The primary mutant defect is overexpansion of cells. When the cells are enlarged more than 133% relative to nonmutant cells, they perform abnormally oriented and incomplete cell divisions. Cells adjoining these abnormally enlarged cells in *wty-1* mutants divide more rapidly and expand less than comparably positioned cells in wild-type plants. Despite these defects, *wty-1* mutant leaves are normal in size and shape, showing that the cell expansion defects are compensated by neighboring cells.

In conclusion, these results demonstrate that organ morphogenesis is relatively independent of the precise control of cell division rate and, at least in plants, cell division orientation and cell expansion. Therefore, it follows that morphogenesis is regulated at the supracellular level, either by a separate category of signals or by a cellular communication network.

C. What Are the Molecules Involved in Determining Organ Size and Shape?

To identify growth determinants requires compounds to be able to alter organ sizes due to more or less incorporated cells, without affecting cell division or elongation characteristics. They would have to act in a concentration-dependent manner and may originate from and affect communicating cells. A candidate for such a growth agent in animals is nitric oxide (NO), which may be involved in controlling the size of body structures during *Drosophila* development (Kuzin et al., 1996). NO is a diffusible multifunctional second messenger that has been implicated in numerous physiological functions in animals (Schmidt and Walter, 1994). In addition, NO can act in cells as an efficient antiproliferative agent (Garg and Hassid, 1989; Chen et al., 1998). In flies, NO synthase (NOS) is expressed at high levels in developing imaginal discs. Inhibition of NOS in larvae causes an increase in size of some organs in adult flies, corresponding to an increase in cell number, whereas ectopic expression of NOS has the opposite effect (Kuzin et al., 1996). Interestingly, in the developing leg, the segments that are most affected by NOS

5. Morphogenesis in Plants

inhibition and ectopic induction corresponded to those that normally have high and low expression of NOS, respectively. This suggests that NO may play a causative role in growth arrest in normal development. In addition, the pattern of NOS expression correlates with the growth capacity of the particular segment, indicating that pattern formation and growth are normally coordinated.

In plants, the embryos, leafs, and flowers grow to a particular size. In shoots the size of the meristem is important for the formation of correct organ numbers, and in roots the size of the meristematic and elongation zone needs to be determined to regulate the growth rate. Compounds that would qualify for a growth determinant, analogous to NO, remain to be identified in plants. However, this growth determinant may be a composition of more integrated intercellular signals directing rate and orientation of division and elongation. Mutations in one signaling pathway can be compensated for by adjusting the other processes, but only up to a certain level. Accordingly, in animals and plants, growth and cell division can be uncoupled but when cell division is severely inhibited, growth is affected.

VI. Concluding Remarks

After the first decade of intense investigation of plant development by molecular genetic approaches, the mechanisms of axis formation and pattern formation are still uncertain and it cannot yet be decided whether they will differ significantly from mechanisms used outside the plant kingdom. The identification of only a low number of genes involved in these processes, despite large-scale screening efforts, can be explained by embryo lethality and by extensive genetic redundancy. For the latter case, some evidence has emerged. The generation of a large collection of gene/enhancer traps in combination with the rapid progress of the *Arabidopsis* genome project should change this situation in the future.

The scarce knowledge on axis formation in the plant embryo, assembled with the help of complementary studies in the brown alga *Fucus*, hint at the importance of localized cues in the cell wall. The nature of the cues is unknown, and the mechanisms involved may turn out to bear resemblance to asymmetry-generating mechanisms in yeast (Chant, 1994, 1996), nematodes (Guo and Kemphues, 1996), or flies (Yan and Yan, 1998) as these all involve localized cues in the cell cortex. Although all these organisms utilize mobilization of the cytoskeleton to a localized cue, the mechanisms seem to differ considerably in the genes that are deployed, so a novel variant in plants may be expected. Molecular genetic dissection of embryo pattern formation has not yet established rigorous concepts, in part because the mutants that have been analyzed do not allow the straightforward deduction of underlying mechanisms. The genes that have, however, been identified, some of which are required at an early stage, will undoubtedly serve as an aid to bootstrap researches into the regulatory mechanisms of early patterning.

The ongoing process of patterning in meristems, one of the most fascinating aspects of plant development, is receiving increasing attention. The *Arabidopsis* root meristem is suited to investigate the continuous signaling events that facilitate the flexible acquisition of cell fate and their relation to embryo patterning. Molecular genetic analysis of the shoot meristem is providing important clues on how meristems maintain themselves by regulating the balance between cell differentiation and cell division, and the emerging information can be utilized to study similar questions in the root meristem. The availability of genes involved in cell division and the potential to study their function in transgenic plants are a second avenue that is actively pursued by plant scientists to understand meristems, and major progress can be expected in the near future.

The role of cell division in plant development, previously almost unanimously considered important for patterning, growth, and morphogenesis, is beginning to be critically questioned, as plants and animals turn out be astonishingly alike in their capacity to regulate perturbations in cell division rate and orientation.

The limited complexity of the plant body bears the inherent promise that, despite a slow start compared to the animal model systems, valuable knowledge can be gained in plants on the interaction of different aspects of multicellular development (Meyerowitz, 1997a). When we have some insight into this interplay, we can start to understand at a deeper level the perhaps most remarkable aspect of plant development: its flexibility and its continuous ability to respond to the environment.

References

Aida, M., Ishida, T., Fukaki, H., Fujisawa, H., and Tasaka, M. (1997). Genes involved in organ separation in *Arabidopsis*: An analysis of the *cup-shaped cotyledon* mutant. *Plant Cell* **9**, 841–857.

Babiychuk, E., Kushnir, S., Belles-Boix, E., Van Montagu, M., and Inzé, D. (1995). *Arabidopsis thaliana* NADPH oxidoreductase homologs confer tolerance of yeasts toward the thiol-oxidizing drug diamide. *J. Biol. Chem.* **270**, 26224–26231.

Barton, M. K., and Poethig, R. S. (1993). Formation of the shoot apical meristem in *Arabidopsis thaliana*: An analysis of development of the wild type in the *shoot meristemless* mutant. *Development* **119**, 823–831.

Baskin, T. I., Cork, A., Williamson, R. E., and Gorst, R. (1995). *STUNTED PLANT1*, a gene required for expansion in rapidly elongating but not in dividing cells and mediating root growth responses to applied cytokinin. *Plant Physiol.* **107**, 233–243.

Basler, K., and Struhl, G. (1994). Compartment boundaries and the control of *Drosophila* limb pattern by *hedgehog* protein. *Nature* **368**, 208–214.

Beemster, G. T., and Baskin, T. I. (1998). Analysis of cell division and elongation underlying the developmental acceleration of root growth in *Arabidopsis thaliana*. *Plant Physiol.* **116**, 1515–1526.

Benfey, P., Linstead, P., Roberts, K., Schiefelbein, J., Hauser, M.-T., and Aeschbacher, R. (1993). Root development in *Arabidopsis*: Four mutants with dramatically altered root morphogenesis. *Development* **119**, 57–70.

Berger, F., Taylor, A., and Brownlee, C. (1994). Cell fate determination by the cell wall in early *Fucus* development. *Science* **263**, 1421–1423.

Berger, F., Haselhoff, J., Schiefelbein, J., and Dolan, L. (1998a). Positional information in root epi-

5. Morphogenesis in Plants   241

dermis is defined during embryogenesis and acts in domains with strict boundaries. *Curr. Biol.* **17,** 421–430.

Berger, F., Hung, C. Y., Dolan, L., and Schiefelbein, J. (1998b). Control of cell division in the root epidermis of *Arabidopsis thaliana*. *Dev. Biol.* **194,** 235–245.

Berleth, T., and Jürgens, G. (1993). The role of the *monopteros* gene in organising the basal body region of the *Arabidopsis* embryo. *Development* **118,** 575–587.

Berleth, T., Hardtke, C. S., Przemeck, G. K. H., and Müller, J. (1996). Mutational analysis of root initiation in the *Arabidopsis* embryo. *Plant Soil* **187,** 1–9.

Bowler, C., Van Montagu, M., and Inzé, D. (1992). Superoxide dismutase and stress tolerance. *Annu. Rev. Plant Physiol. Plant Mol. Biol.* **43,** 83–116.

Bruck, D. K., and Walker, D. B. (1985). Cell determination during embryogenesis in *Citrus jambhiri*. I. Ontogeny of the epidermis. *Bot. Gaz.* **146,** 188–195.

Bush, M., Mayer, U., and Jürgens, G. (1996). Molecular analysis of the *Arabidopsis* pattern formation gene *GNOM*: Gene structure and intragenic complementation. *Mol. Gen. Genet.* **250,** 681–691.

Carnac, G., and Gurdon, J. B. (1997). The community effect in *Xenopus* myogenesis is promoted by dorsalizing factors. *Int. J. Dev. Biol.* **41,** 521–524.

Celenza, J. L., Jr., Grisafi, P. L., and Fink, G. R. (1995). A pathway for lateral root formation in *Arabidopsis thaliana*. *Genes Dev.* **9,** 2131–2142.

Chant, J. (1994). Cell polarity in yeast. *Trends Genet.* **10,** 328–333.

Chant, J. (1996). Septin scaffolds and cleavage planes in *Saccharomyces*. *Cell* **84,** 187–190.

Chen, L., Daum, G., Forough, R., Clowes, M., Walter, U., and Clowes, A. W. (1998). Overexpression of human endothelial nitric oxide synthase in rat vascular smooth muscle cells and in balloon-injured carotid artery. *Circ. Res.* **82,** 862–870.

Cheng, J. C., Seeley, K. A., and Sung, Z. R. (1995). *RML1* and *RML2*, *Arabidopsis* genes required for cell proliferation at the root tip. *Plant Physiol.* **107,** 365–376.

Citterio, S., Sgorbati, S., Scrippa, S., and Sparvoli, E. (1994). Ascorbic acid effect on the onset of cell proliferation in pea root. *Physiol. Plant.* **92,** 601–607.

Clark, S. E. (1997). Organ formation at the vegetative shoot meristem. *Plant Cell* **9** (Special Issue), 1067–1076.

Clark, S. E., Running, M. P., and Meyerowitz, E. M. (1993). *CLAVATA1*, a regulator of meristem and development in *Arabidopsis*. *Development* **119,** 397–418.

Clark, S. E., Running, M. P., and Meyerowitz, E. M. (1995). *CLAVATA3* is a specific regulator of shoot and floral meristem development affecting the same processes as *CLAVATA1*. *Development* **121,** 2057–2067.

Clark, S. E., Jacobsen, S. E., Levin, J. Z., and Meyerowitz, E. M. (1996). The *CLAVATA* and *SHOOT MERISTEMLESS* loci competitively regulate meristem activity in *Arabidopsis*. *Development* **122,** 1567–1575.

Clark, S. E., Williams, R. W., and Meyerowitz, E. M. (1997). The *CLAVATA1* gene encodes a putative receptor kinase that controls shoot and floral meristem size in *Arabidopsis*. *Cell* **89,** 575–585.

Clowes, F. A. L. (1975). The quiescent centre. *In* "The Development and Function of Roots" (J. G. Torrey and D. T. Clarkson, Eds.), pp. 3–19. Academic Press, New York.

Cohen, S. M. (1996). Controlling growth of the wing: *Vestigial* integrates signals from the compartment boundaries. *BioEssays* **18,** 855–858.

Conklin, P. L., Williams, E. H., and Last, R. L. (1996). Environmental stress sensitivity of an ascorbic acid-deficient *Arabidopsis* mutant. *Proc. Natl. Acad. Sci. USA* **93,** 9970–9974.

Di Christina, M. D., Sessa, G., Dolan, L., Linstead, P., Baima, S., Ruberti, I., and Morelli, G. (1996). The *Arabidopsis* Athb-10 (GLABRA2) is an HD-Zip protein required for regulation of root hair development. *Plant J.* **10,** 393–402.

Di Laurenzio, L., Wysockadiller, J., Malamy, J. E., Pysh, L., Helariutta, Y., Freshour, G., Hahn,

M. G., Feldmann, K. A., and Benfey, P. N. (1996). The *SCARECROW* gene regulates an asymmetric cell division that is essential for generating the radial organization of the *Arabidopsis* root. *Cell* **86,** 423–433.

Dolan, L., and Roberts, K. (1995a). Two ways to skin a plant: The analysis of root and shoot epidermal development in *Arabidopsis*. *BioEssays* **17,** 865–872.

Dolan, L., and Roberts, K. (1995b). Plant development: Pulled up by the roots. *Curr. Opin. Genet. Dev.* **5,** 432–438.

Dolan, L., Janmaat, K., Willemsen, V., Linstead, P., Poethig, S., Roberts, K., and Scheres, B. (1993). Cellular organisation of the *Arabidopsis* root. *Development* **119,** 71–84.

Dolan, L., Duckett, C. M., Grierson, C., Linstead, P., Schneider, K., Lawson, E., Dean, C., Poethig, S., and Roberts, K. (1994). Clonal relationships and cell patterning in the root epidermis of *Arabidopsis*. *Development* **120,** 2465–2474.

Drews, G. N., Lee, D., and Christensen, C. A. (1998). Genetic analysis of female gametophyte development and function. *Plant Cell* **10,** 5–17.

Duckett, C., Oparka, K., Prior, D., Dolan, L., and Roberts, K. (1994). Dye-coupling in the root epidermis of *Arabidopsis* is progressively reduced during development. *Development* **120,** 3247–3255.

Ecker, J. (1995). The ethylene signal transduction pathway in plants. *Science* **268,** 667–675.

Edgar, B. A., Lehman, D. A., and O'Farrell, P. H. (1994). Transcriptional regulation of *string* (*cdc25*): A link between developmental programming and the cell cycle. *Development* **120,** 3131–3143.

Endrizzi, K., Moussian, B., Haecker, A., Levin, J. Z., and Laux, T. (1996). The *SHOOT MERISTEMLESS* gene is required for maintenance of undifferentiated cells in *Arabidopsis* shoot and floral meristems and acts at a different regulatory level than the meristem genes *WUSCHEL* and *ZWILLE*. *Plant J.* **10,** 967–979.

Evans, T., Crittenden, S., Kodoyianni, V., and Kimble, J. (1994). Translational control of *glp-1* mRNA establishes an asymmetry in the *C. elegans* embryo. *Cell* **77,** 183–194.

Fero, M. L., Rivkin, M., Tasch, M., Porter, P., Carow, C. E., Firpo, E., Polyak, K., Tsai, L. H., Broudy, V., Perlmutter, R. M., Kaushansky, K., and Roberts, J. M. (1996). A syndrome of multiorgan hyperplasia with features of gigantism, tumorigenesis, and female sterility in *p27(Kip1)*-deficient mice. *Cell* **85,** 733–744.

Fluhr, R. (1998). Ethylene perception: From two-component signal transducers to gene induction. *Trends Plant Sci.* **3,** 141–146.

Folette, P. J., and O'Farrell, P. H. (1997). Connecting cell behaviour to patterning: Lessons from the cell cycle. *Cell* **88,** 309–314.

Frankhauser, G. (1941). Cell size, organ and body size in triploid newts (*Triturus viridescens*). *J. Morphol.* **68,** 161–177.

Frankhauser, G. (1945). Maintenance of normal structure in heteroploid salamander larvae, through compensation of changes in cell size by adjustment of cell number and cell shape. *J. Exp. Zool.* **100,** 445–455.

Fukaki, H., Wysocka-Diller, J., Kato, T., Fujisawa, H., Benfey, P. N., and Tasaka, M. (1998). The endodermis is essential for shoot gravitropism in *Arabidopsis thaliana*. *Plant J.* **14,** 425–430.

Furner, I. J., and Pumfrey, J. E. (1992). Cell fate in the shoot apical meristem of *Arabidopsis thaliana*. *Development* **115,** 755–764.

Galway, M. E., Masucci, J. D., Lloyd, A. M., Walbot, V., Davis, R. W., and Schiefelbein, J. W. (1994). The *TTG* gene is required to specify epidermal cell fate and cell patterning in the *Arabidopsis* root. *Dev. Biol.* **166,** 740–754.

Garg, U. C., and Hassid, A. (1989). Inhibition of rat mesangial cell mitogenesis by nitric oxide-generating vasodilators. *Am. J. Physiol.* **257,** F60–F66.

Gomez-Skarmeta, J. L., del Corral, R. D., de la Calle-Mustienes, E., Ferres-Marco, D., and Modolell, J. (1996). *araucan* and *caupolican*, two members of the novel iroquois complex, encode homeoproteins that control proneural and vein-forming genes. *Cell* **85,** 95–105.

Grierson, C. S., Roberts, K., Feldmann, K. A., and Dolan, L. (1997). The *COW1* locus of *Arabidopsis* acts after *RHD2*, and in parallel with *RHD3* and *TIP1*, to determine the shape, rate of elongation, and number of root hairs produced from each site of hair formation. *Plant Physiol.* **115**, 981–990.

Grossniklaus, U., Vielle-Calzada, J. P., Hoeppner, M. A., and Gagliano, W. B. (1998). Maternal control of embryogenesis by *MEDEA*, a *Polycomb* group gene in *Arabidopsis*. *Science* **280**, 446–450.

Guo, S., and Kemphues, K. J. (1996). Molecular genetics of asymmetric cleavage in the early *Caenorhabditis elegans* embryo. *Curr. Opin. Genet. Dev.* **6**, 408–415.

Gurdon, J. B. (1992). The generation of diversity and pattern in animal development. *Cell* **68**, 185–199.

Gurdon, J. B., Tiller, E., Roberts, J., and Kato, K. (1993). A community effect in muscle development. *Curr. Biol.* **3**, 1–11.

Haber, A. H. (1962). Nonessentiality of concurrent cell divisions for degree of polarization of leaf growth. I. Studies with radiation-induced mitotic inhibition. *Am. J. Bot.* **49**, 583–589.

Haber, A. H., and Foard, D. E. (1963). Nonessentiality of concurrent cell divisions for degree of polarization of leaf growth. II. Evidence from untreated plants and from chemically induced changes of the degree of polarization. *Am. J. Bot.* **50**, 937–944.

Hardtke, C. S., and Berleth, T. (1998). The *Arabidopsis* gene *MONOPTEROS* encodes a transcription factor mediating embryo axis formation and vascular development. *EMBO J.* **17**, 1405–1411.

Hemerly, A., de Almeida Engler, J., Bergounioux, C., Van Montagu, M., Engler, G., Inzé, D., and Ferreira, P. (1995). Dominant negative mutants of the *Cdc2* kinase uncouple cell division from iterative plant development. *EMBO J.* **14**, 3925–3936.

Hülskamp, M., Misera, S., and Jürgens, G. (1994). Genetic dissection of trichome cell development in *Arabidopsis*. *Cell* **76**, 555–566.

Hülskamp, M., Folkers, U., and Grini, P. E. (1997). Cell morphogenesis in *Arabidopsis*. *BioEssays* **20**, 20–29.

Irish, V. E., and Sussex, I. M. (1992). A fate map of the *Arabidopsis* embryonic shoot apical meristem. *Development* **115**, 745–753.

Jackson, D., Veit, B., and Hake, S. (1994). Expression of maize *KNOTTED1* related homeobox genes in the shoot apical meristem predicts patterns of morphogenesis in the vegetative shoot. *Development* **120**, 405–413.

Jacobs, T. (1997). Why do plant cells divide? *Plant Cell* **9** (Special Issue), 1021–1029.

Jürgens, G. (1995). Axis formation in plant embryogenesis: Cues and clues. *Cell* **81**, 467–470.

Jürgens, G. (1996). Cell division and morphogenesis in angiosperm embryogenesis. *Semin. Cell Dev. Biol.* **7**, 867–872.

Jürgens, G., and Mayer, U. (1994). *Arabidopsis*. In "EMBRYOS. Colour Atlas of Development" (J. Bard, Ed.), pp. 7–21. Wolfe Publications, London.

Jürgens, G., Grebe, M., and Steinmann, T. (1997). Establishment of cell polarity during early plant development. *Curr. Opin. Cell Biol.* **9**, 849–852.

Kaplan, D. R. (1992). The relationship of cells to organisms in plants: Problem and implications of an organismal perspective. *Int. J. Plant Sci.* **153**, S28–S37.

Kaplan, D. R., and Cooke, T. J. (1997). Fundamental concepts in the embryogenesis of dicotyledons: A morphological interpretation of embryo mutants. *Plant Cell* **9**, 1903–1919.

Kehl, B. T., Cho, K. O., and Choi, K. W. (1998). *mirror*, a *Drosophila* homeobox gene in the Iroquois complex, is required for sensory organ and alula formation. *Development* **125**, 1217–1227.

Kerk, N. M., and Feldman, L. J. (1995). A biochemical model for the initiation and maintenance of the quiescent centre; implications for organization of root meristems. *Development* **121**, 2825–2833.

Kerstetter, R. A., Laudencia-Chingcuanco, D., Smith, L. G., and Hake, S. (1997). Loss-of-function mutations in the maize homeobox gene, *knotted1*, are defective in shoot meristem maintenance. *Development* **124**, 3045–3054.

Kim, J., Sebring, A., Esch, J. J., Kraus, M. E., Vorwerk, K., Magee, J., and Carroll, S. B. (1996). Integration of positional signals and regulation of wing formation and identity by the *Drosophila vestigial* gene. *Nature* **382**, 133–138.
Kipreos, E. T., Lander, L. E., Wing, J. P., He, W. W., and Hedgecock, E. M. (1996). *cul-1* is required for cell cycle exit in *C. elegans* and identifies a novel gene family. *Cell* **85**, 829–839.
Kropf, D. L., Kloareg, B., and Quatrano, R. S. (1988). Cell wall is required for fixation of the embryonic axis in *Fucus* zygotes. *Science* **239**, 187–190.
Kuzin, B., Roberts, I., Peunova, N., and Enikolopov, G. (1996). Nitric oxide regulates cell proliferation during *Drosophila* development. *Cell* **87**, 639–649.
Lauber, M. H., Waizenegger, I., Steinmann, T., Schwarz, H., Mayer, U., Hwang, I., Lukowitz, W., and Jürgens, G. (1997). The *Arabidopsis* KNOLLE protein is a cytokinesis-specific syntaxin. *J. Cell Biol.* **139**, 1485–1493.
Laux, T., and Jürgens, G. (1997). Embryogenesis: A new start in life. *Plant Cell* **9**, 989–1000.
Laux, T., and Mayer, K. F. (1998). Cell fate regulation in the shoot meristem. *Semin. Cell. Dev. Biol.* **9**, 195–200.
Laux, T., and Schoof, H. (1997). Maintaining the root meristem—The role of *CLAVATA1*. *Trends Plant Sci.* **2**, 325–327.
Lee, J. J., von Kessler, D. P., Parks, S., and Beachy, P. A. (1992). Secretion and localized transport suggests a role in positional signaling for products of the segmentation gene *hedgehog*. *Cell* **71**, 33–50.
Liso, B. R., Calabrese, G., Bitoni, M. B., and Arrigoni, O. (1984). Relationship between ascorbic acid and cell division. *Exp. Cell Res.* **150**, 314–320.
Liso, B. R., Innocenti, A. M., Bitoni, M. B., and Arrigoni, O. (1988). Ascorbic acid induced progression of quiescent centre cells from G1 to S phase. *New Phytol.* **110**, 469–471.
Lloyd, A. M., Schena, M., Walbot, V., and Davis, R. W. (1994). Epidermal cell fate determination in *Arabidopsis*: Patterns defined by steroid-inducible regulator. *Science* **266**, 436–439.
Long, J. A., and Barton, K. (1998). The development of embryonic pattern in *Arabidopsis*. *Development* **126**, 3027–3035.
Long, J. A., Moan, E. I., Medford, J. I., and Barton, M. K. (1996). A member of the KNOTTED class of homeodomain proteins encoded by the *STM* gene of *Arabidopsis*. *Nature* **379**, 66–69.
Lu, P., Porat, R., Nadeau, J., and O'Neill, S. (1996). Identification of a meristem L1 layer specific gene in *Arabidopsis* that is expressed during embryonic pattern formation and defines a new class of homeobox genes. *Plant Cell* **8**, 2155–2168.
Lucas, W. J., Bouche-Pillon, S., Jackson, D. P., Nguyen, L., Baker, L., Ding, B., and Hake, S. (1995). Selective trafficking of KNOTTED1 homeodomain protein and its mRNA through plasmodesmata. *Science* **270**, 1980–1983.
Lukowitz, W., Mayer, U., and Jürgens, G. (1996). Cytokinesis in the *Arabidopsis* embryo involves the syntaxin-related KNOLLE gene product. *Cell* **84**, 61–71.
Malamy, J. E., and Benfey, P. N. (1997a). Down and out in *Arabidopsis*: The formation of lateral roots. *Trends Plant Sci.* **2**, 390–396.
Malamy, J., and Benfey, P. N. (1997b). Organization and cell differentiation in lateral roots of *Arabidopsis thaliana*. *Development* **124**, 33–44.
Mansfield, S., and Briarty, L. (1991). Early embryogenesis in *Arabidopsis thaliana*. II. The developing embryo. *Can. J. Bot.* **69**, 461–467.
Masucci, J. D., and Schiefelbein, J. W. (1994). The *rhd6* mutation of *Arabidopsis thaliana* alters root-hair initiation through an auxin- and ethylene-associated process. *Plant Physiol.* **106**, 1335–1346.
Masucci, J. D., and Schiefelbein, J. W. (1996). Hormones act downstream of *TTG* and *GL2* to promote root hair outgrowth during epidermis development in the *Arabidopsis* root. *Plant Cell* **5**, 1505–1517.
Masucci, J. D., Rerie, W. G., Foreman, D. R., Zhang, M., Galway, M. E., Marks, M. D., and Schiefelbein, J. W. (1996). The homeobox gene *GLABRA 2* is required for position-dependent cell differentiation in the root epidermis of *Arabidopsis thaliana*. *Development* **122**, 1253–1260.

5. Morphogenesis in Plants

May, M. J., and Leaver, C. J. (1993). Oxidative stimulation of glutathione synthesis in *Arabidopsis thaliana* suspension cultures. *Plant Physiol.* **103,** 621–627.

Mayer, U., Torres-Ruiz, R., Berleth, T., Misera, S., and Jürgens, G. (1991). Mutations affecting body organization in the *Arabidopsis* embryo. *Nature* **353,** 402–407.

Mayer, U., Büttner, G., and Jürgens, G. (1993). Apical–basal pattern formation in the *Arabidopsis* embryo: Studies on the role of the *gnom* gene. *Development* **117,** 149–162.

McLean, B. G., Hempel, F. D., and Zambryski, C. (1997). Plant intercellular communication via plasmodesmata. *Plant Cell* **9,** 1043–1054.

Meinke, D. W. (1986). Embryo-lethal mutants and the study of plant embryo development. *Oxf. Surv. Plant Mol. Cell Biol.* **3,** 122–165.

Meyerowitz, E. M. (1997a). Genetic control of cell division patterns in developing plants. *Cell* **88,** 299–308.

Meyerowitz, E. M. (1997b). Plants and the logic of development. *Genetics* **145,** 5–9.

Mordhorst, A. P., Toonen, M. A. J., and de Vries, S. C. (1997). Plant embryogenesis. *Crit. Rev. Plant Sci.* **16,** 535–576.

Mossessova, E., Gulbis, J. M., and Goldberg, J. (1998). Structure of the guanine nucleotide exchange factor Sec7 domain of human Arno and analysis of the interaction with ARF GTPase. *Cell* **92,** 415–423.

Muskavitch, M. A. T. (1994). Delta-Notch signaling and *Drosophila* cell fate choice. *Dev. Biol.* **166,** 415–430.

Neufeld, T. P., de la Cruz, A. F., Johnston, L. A., and Edgar, B. A. (1998). Coordination of growth and cell division in the *Drosophila* wing. *Cell* **93,** 1183–1193.

Ogas, J., Cheng, J. C., Sung, Z. R., and Somerville, C. (1997). Cellular differentiation regulated by gibberellin in the *Arabidopsis thaliana pickle* mutant. *Science* **277,** 91–94.

Oparka, K. J., Duckett, C. M., Prior, D. A. M., and Fischer, D. B. (1994). Real-time imaging of phloem unloading in the root tip of *Arabidopsis. Plant J.* **6,** 759–766.

Oppenheimer, D. G., Herman, P. L., Sivakumaran, S., Esch, J., and Marks, M. D. (1991). A *myb* gene required for leaf trichome differentiation in *Arabidopsis* is expressed in stipules. *Cell* **67,** 483–493.

Parcy, F., Nilsson, O., Busch, M. A., Lee, I., and Weigel, D. (1998). A genetic framework for floral patterning. *Nature* **395,** 561–566.

Peyroche, A., Paris, S., and Jackson, C. L. (1996). Nucleotide exchange on ARF mediated by yeast Gea1 protein. *Nature* **384,** 479–484.

Pollock, E. G., and Jensen, W. A. (1964). Cell development during early embryogenesis in *Capsella* and *Gossypium. Am. J. Bot.* **51,** 915–921.

Poot, M., Teubert, H., Rabinovitch, P. S., and Kavanagh, T. J. (1995). De novo synthesis of glutathione is required for both entry into and progression through the cell cycle. *J. Cell Physiol.* **163,** 555–560.

Priess, J., Schnabel, H., and Schnabel, R. (1987). The *glp-1* locus and cellular interactions in early *C. elegans* embryos. *Cell* **51,** 601–611.

Przemeck, G. K. H., Mattsson, J., Hardtke, C. S., Sung, Z. R., and Berleth, T. (1996). Studies on the role of the *Arabidopsis* gene *MONOPTEROS* in vascular development and plant cell axialization. *Planta* **200,** 229–237.

Rerie, W. G., Feldmann, K. A., and Marks, M. D. (1994). The *GLABRA2* gene encodes a homeodomain protein required for normal trichome development in *Arabidopsis. Genes Dev.* **8,** 1388–1399.

Reynolds, J. O., Eisses, J. F., and Sylvester, A. W. (1998). Balancing division and expansion during maize leaf morphogenesis: Analysis of the mutant, *warty-1. Development* **125,** 259–268.

Rooke, J. E., and Xu, T. (1998). Positive and negative signals between interacting cells for establishing neural fate. *BioEssays* **20,** 209–214.

Running, M. P., Fletcher, J. C., and Meyerowitz, E. M. (1998). The *WIGGUM* gene is required for proper regulation of floral meristem size in *Arabidopsis. Development* **125,** 2545–2553.

Russo, T., Zambrano, N., Esposito, F., Ammendola, R., Cimino, F., Fiscella, M., Jackman, J., O'Connor, P. M., Anderson, C. W., and Appella, E. (1995). A p53-independent pathway for activation of *WAF1/CIP1* expression following oxidative stress. *J. Biol. Chem.* **270**, 29386–29391.

Sachs, T. (1991). Cell polarity and tissue patterning in plants. *Development. Suppl.* **1**, 83–93.

Sánchez-Fernández, R., Fricker, M., Corben, L. B., White, N. S., Sheard, N., Leaver, C. J., Van Montagu, M., Inzé, D., and May, M. J. (1997). Cell proliferation and root hair tip growth in the *Arabidopsis* root are under mechanistically different forms of redox control. *Proc. Natl. Acad. Sci. USA* **94**, 2745–2750.

Santamaria, P. (1983). Analysis of haploid mosaics in *Drosophila. Dev. Biol.* **9**, 285–295.

Scheres, B., and Wolkenfelt, H. (1998). The *Arabidopsis* root as a model to study plant development. *Plant Physiol. Biochem.* **36**, 21–32.

Scheres, B., Wolkenfelt, H., Willemsen, V., Terlouw, M., Lawson, E., Dean, C., and Weisbeek, P. (1994). Embryonic origin of the *Arabidopsis* primary root and root meristem initials. *Development* **120**, 2475–2487.

Scheres, B., Di Laurenzio, L., Willemsen, V., Hauser, M. T., Janmaat, K., Weisbeek, P., and Benfey, P. N. (1995). Mutations affecting the radial organisation of the *Arabidopsis* root display specific defects throughout the embryonic axis. *Development* **121**, 53–62.

Scheres, B., McKhan, H. I., and van den Berg, C. (1996). Roots redefined: Anatomical and genetic analysis of root development. *Plant Physiol.* **111**, 959–964.

Schiefelbein, J., Masucci, J. D., and Wang, H. (1997). Building a root: The control of patterning and morphogenesis during root development. *Plant Cell* **9** (Special Issue), 1089–1098.

Schmidt, H. H. W., and Walter, U. (1994). NO at work. *Cell* **78**, 919–925.

Schnabel, R. (1996). Pattern formation: Regional specification in the early *C. elegans* embryo. *BioEssays* **18**, 591–594.

Schnittger, A., Grini, P. E., Folkers, U., and Hülskamp, M. (1996). Epidermal fate map of the *Arabidopsis* shoot meristem. *Dev. Biol.* **175**, 248–255.

Schnittger, A., Jürgens, G., and Hülskamp, M. (1998). Tissue layer and organ specificity of trichome formation are regulated by *GLABRA1* and *TRIPTYCHON* in *Arabidopsis. Development* **125**, 2283–2289.

Shaw, S. L., and Quatrano, R. S. (1996). The role of targeted secretion in the establishment of cell polarity and orientation of the division plane in *Fucus* zygotes. *Development* **122**, 2623–2630.

Shevell, D., Leu, W.-M., Stewart Gillmor, S., Xia, G., Feldmann, K., and Chua, N.-H. (1994). *EMB030* is essential for normal cell division, cell expansion, and cell adhesion in *Arabidopsis* and encodes a protein that has similarity to Sec7. *Cell* **77**, 1051–1062.

Sinha, N. R., Williams, R. E., and Hake, S. (1993). Overexpression of the maize homeobox gene, *KNOTTED-1*, causes a switch from determinate to indeterminate cell fates. *Genes Dev.* **7**, 787–795.

Smith, L. G., Hake, S., and Sylvester, A. W. (1996). The *tangled-1* mutation alters cell division orientations throughout maize lead development without altering leaf shape. *Development* **122**, 481–489.

Springer, P. S., McCombie, W. R., Sundaresan, V., and Martienssen, R. A. (1995). Gene trap tagging of *PROLIFERA*, an essential *MCM2-3-5*-like gene in *Arabidopsis. Science* **268**, 877–880.

Steeves, T. A., and Sussex, I. M. (1989). "Patterns in Plant Development." Cambridge Univ. Press, Cambridge, UK.

Strigini, M., and Cohen, S. M. (1997). A Hedgehog activity gradient contributes to AP axial patterning of the *Drosophila* wing. *Development* **124**, 4697–4705.

Suthanthiran, M., Anderson, M. E., Sharma, V. K., and Meister, A. (1990). Glutathione regulates activation-dependent DNA synthesis in highly purified normal human T lymphocytes stimulated via the CD2 and CD3 antigens. *Proc. Natl. Acad. Sci. USA* **87**, 3343–3347.

Tanimoto, M., Roberts, K., and Dolan, L. (1995). Ethylene is a positive regulator of root hair development in *Arabidopsis thaliana. Plant J.* **8**, 943–948.

Torres-Ruiz, R. A., and Jürgens, G. (1994). Mutations in the *FASS* gene uncouple pattern formation and morphogenesis in *Arabidopsis* development. *Development* **120**, 2967–2978.
Torres-Ruiz, R. A., Lohner, A., and Jürgens, G. (1996). The *GURKE* gene is required for normal organisation of the apical region in the *Arabidopsis* embryo. *Plant J.* **10**, 1005–1016.
Traas, J., Bellini, C., Nacry, P., Kronenberger, J., Bouchez, D., and Caboche, M. (1995). Normal differentiation patterns in plants lacking microtubular preprophase bands. *Nature* **375**, 676–677.
Ulmasov, T., Hagen, G., and Guilfoyle, T. J. (1997). *ARF1*, a transcription factor that binds to auxin response elements. *Science* **276**, 1865–1868.
van den Berg, C., Willemsen, V., Hage, W., Weisbeek, P., and Scheres, B. (1995). Cell fate in the *Arabidopsis* root meristem determined by directional signalling. *Nature* **378**, 62–65.
van den Berg, C., Willemsen, V., Hendriks, G., Weisbeek, P., and Scheres, B. (1997). Short-range control of cell differentiation in the *Arabidopsis* root meristem. *Nature* **390**, 287–289.
Vernon, D., and Meinke, D. (1994). Embryogenic transformation of the suspensor in *twin*, a polyembryonic mutant of *Arabidopsis*. *Dev. Biol.* **165**, 566–573.
Vollbrecht, E., Veit, B., Sinha, N., and Hake, S. (1991). The developmental gene *Knotted-1* is a member of a maize homeobox gene family. *Nature* **350**, 241–243.
Vroemen, C., Langeveld, S., Mayer, U., Ripper, G., Jürgens, G., Van Kammen, A., and de Vries, S. (1996). Pattern formation in the *Arabidopsis* embryo revealed by position-specific lipid transfer protein gene expression. *Plant Cell* **8**, 783–791.
Wada, T., Tachibana, T., Shimura, Y., and Okada, K. (1997). Epidermal cell differentiation in *Arabidopsis* determined by a *Myb* homolog, *CPC*. *Science* **277**, 1113–1116.
Weigel, D. (1998). From floral induction to floral shape. *Curr. Opin. Plant Biol.* **1**, 55–59.
Weigel, D., and Meyerowitz, E. M. (1994). The ABCs of floral homeotic genes. *Cell* **78**, 203–209.
Weigmann, K., Cohen, S. M., and Lehner, C. F. (1997). Cell cycle progression, growth and patterning in imaginal discs despite inhibition of cell division after inactivation of *Drosophila* Cdc2 kinase. *Development* **124**, 3555–3563.
Willemsen, V., Wolkenfelt, H., de Vrieze, G., Weisbeek, P., and Scheres, B. (1998). The *HOBBIT* gene is required for formation of the root meristem in the *Arabidopsis* embryo. *Development* **125**, 521–531.
Yan, Y. N., and Yan, L. Y. (1998). Asymmetric cell division. *Nature* **392**, 775–778.
Zhang, J., and Sommerville, C. R. (1997). Suspensor-derived polyembryony caused by altered expression of valyl-tRNA-synthase in the *twn2* mutant of *Arabidopsis*. *Proc. Natl. Acad. Sci. USA* **94**, 7349–7355.

Index

A

Actin, gene expression in sea urchin development, 47–55
  CyI gene, 53
  CyIIa gene, 54–55
  CyIIIa gene, 47–53
  CyIIIb gene, 53–54
  S9 gene, 54–55
Adhesion proteins, expression in sea urchin development, 64–65
AG gene, gynoecium development role in *Arabidopsis*, 173–177, 192–194
AGL genes, gynoecium development role in *Arabidopsis*, 187–188, 192
Animal–vegetal axis, gene expression in sea urchin development, 78–81
Antirrhinum, phan gene, 10–11
Apical–basal axis, *see* Pattern formation
Arabidopsis
  apical–basal pattern formation, 211
  gynoecium development genetics, 155–199
    carpel
      evolution, 158, 170–171, 198–199
      functions, 156–158
      fusion, 196–198
      identity specification, 173–185, 192–194
      pattern formation, 192–196
      tissue type origins, 168–171, 195–196
      tissue types, 159–165
    cell lineage, 168–169
    developmental compartments, 169–170, 191–192
    mechanisms, 165–168
    molecular genetics, 171–192
      AGAMOUS gene, 173–177, 192–194
      AGAMOUS gene-independent development, 176–177
      AGL genes, 187–188
      carpel identity specification, 173–177
      A and C class homeotic gene interactions, 175
      CRABS CLAW gene, 180–182
      ETTIN gene, 182–185
      expression pattern analysis, 189–192
      floral homeotic mutants, 172–173
      FRUITFULL gene, 185–187
      pleiotropic mutant phenotypes, 188–189
      regional identity response, 182–185
      SPATULA gene, 177–180
      value maturation regulation, 185–187
      value–replum boundary tissue definition, 187–188
    overview, 156–159, 192–199
    pattern elements, 159–165
      gynophore, 165
      ovary wall, 161–163
      ovules, 164
      placenta, 164
      septum, 163–164
      stigma, 161
      style, 161
      vasculature, 164–165
  leaf epidermis development
    cuticle, 29–32
    cytoplasmic connections, 17–18
    determined state, potency, 27–28
    growth, 12, 14–15
    identity, 9–10
    morphology, 3–4
    ontogeny
      cell lineages, 9
      embryogenesis period, 5–7
    overview, 32
    pattern formation, 19, 21–24
    radial pattern formation, 216–220
Arylsulfatase, gene expression in sea urchin development, 81
ATLTP1 gene, radial pattern formation in plants, 215
ATML1 gene, radial axis formation in plants, 214, 221

249

B

Bindin, gene expression in sea urchin development, 85
Bone morphogenetic proteins, gastrulation regulation in mouse, 129–131, 136, 139, 143

C

Calmodulin, gene expression in sea urchin development, 84
Carpels, *see* Gynoecium development
CAT gene, expression in sea urchin development, 56
Cell differentiation
  carpel development in *Arabidopsis*, 168–171, 195–196
  leaf epidermis development, 18–24
    stomates, 18–22
    trichomes, 22–24
Cell division
  gene expression in sea urchin development, 74–75
  plants development regulation, 225–233
    cell fate specification, 227–228
    environmental factors, 230–231
    epidermis growth in plants, 11–13
    meristem activation, 228–231
    meristem activity maintenance, 231–233
    uncoupling
      organ morphogenesis, 234–238
      pattern formation, 226–227
Cell fate, *see* Fate mapping
Cell lineages
  gastrulation organizer in mouse embryo, 117–118
  gynoecium in *Arabidopsis*, 168–169
  leaf epidermis, 8–9, 24–25
Ciliogenesis, gene expression in sea urchin development, 73–74
Coelomocytes, gene expression, 83–84
Col2a1 gene, gastrulation regulation in mouse, 133
Collagen, gene expression in sea urchin development, 61–63
Cortical granule content, gene expression in sea urchin development, 65–66
CPC gene, radial pattern formation in plant hair cells, 217
CRABS CLAW gene, gynoecium development role in *Arabidopsis*, 180–182
Cuticle, development in plants, 29–32
Cyclin, gene expression in sea urchin development, 74–75
Cytoplasm, leaf epidermis connections in development, 17–18
Cytoskeleton, actin gene expression in sea urchin development, 47–55
  CyI gene, 53
  CyIIa gene, 54–55
  CyIIIa gene, 47–53
  CyIIIb gene, 53–54
  S9 gene, 54–55

D

Development, *see specific aspects*
Distrophin, gene expression in sea urchin development, 85

E

Early gastrula organizer, *see* Gastrulation
Echinonectin protein, expression in sea urchin development, 63–64
Ectoderm, gene expression in sea urchin development, 56–59
eed gene, gastrulation regulation in mouse, 133
EGF domain, gene expression in sea urchin development, 82
Egg receptor, gene expression in sea urchin development, 83
Embryogenesis
  gastrulation, *see* Gastrulation
  leaf epidermis development, 5–7
  pattern formation, *see* Pattern formation
Endoderm
  gastrulation organizer in mouse embryo, 136–138
  gene expression in sea urchin development, 55–56
    CAT gene, 56
    Endo16 gene, 55–56
    SM50 gene, 56
  morphogenesis in plants, 136, 138, 223–224
Endo16 gene, expression in sea urchin development, 55–56
Epidermis, development in leaves
  cuticle, 29–32
  determined state, 24–29
  fate determination, 24–25
  potency, 26–29
  growth, 11–15

cell division, 11–13
cell expansion, 13–14
hormone targets, 14–15
morphogenesis relations, 15
identity, 9–11
heteroblasty, 9–10
regional identity within leaves, 10–11
internal tissue interactions, 16–18
cytoplasmic connections, 17–18
genetic mosaics, 16–17
structure correlations, 16
morphology, 2–5
dicots, 3–4
monocots, 4–5
ontogeny, 5–9
cell lineages, 8–9
embryogenesis period, 5–7
leaf development period, 7–8
overview, 1–2, 32
pattern formation, 18–24
stomates, 18–22
trichomes, 22–24
root-hair fate, 223
ETTIN gene, gynoecium development role in *Arabidopsis*, 182–185
Evolution, carpels in *Arabidopsis*, 158, 170–171, 198–199
Extracellular matrix, gene expression in sea urchin development, 59–65
adhesion proteins, 64–65
collagen genes, 61–63
echinonectin protein, 63–64
hyaline protein, 63–64
PM27 gene, 60
primary mesenchyme cell proteins, 59–65
SM30 gene, 60
SM50 gene, 60–62

F

Fascin, gene expression in sea urchin development, 84
Fate mapping
gastrulation organizer in mouse embryo, 125–126
leaf epidermis, 8–9, 24–25
plant cell specification, 227–228
Fibroblast growth factor, gene expression in sea urchin development, 76
Flower development, *see* Gynoecium development

Follistatin gene, gastrulation regulation in mouse, 130
FRUITFULL gene, gynoecium development role in *Arabidopsis*, 185–187

G

Gastrulation
gene expression in sea urchin development, 67–70
organizer in mouse embryo, 117–145
alternative sources of organizing activity, 136–138
anterior patterning, 136–137
molecular regionalization, 137–138
primitive endoderm, 136–138
body pattern diversity, 117–118
cell fate, 125–126
head and trunk organizing activity, 131–138
mutant embryo evidence, 133–136
regionalization, 131–133
lineage diversity, 117–118
molecular properties, 128–129
morphogenetic movement, 126–128
overview, 117–121, 144–145
patterning activity, 119, 121–125
regulatory signals, 129–131
tissue patterning, 139–144
asymmetry, 143–144
extraembryonic tissues role, 142–143
mutational studies, 140–141
Nieuwkoop signal, 139–140
organizer formation, 143–144
transgenic studies, 140–141
GL2 gene, radial pattern formation in plant epidermis, 216–218, 221–223
Growth
leaf epidermis development, 11–15
cell division, 11–13
cell expansion, 13–14
hormone targets, 14–15
morphogenesis relations, 15
organ morphogenesis in plants, 234–239
cell division uncoupling, 234–238
growth determinants, 238–239
Gynoecium development, genetics of *Arabidopsis*, 155–199
carpel
evolution, 158, 170–171, 198–199
functions, 156–158

Gynoecium development (*cont.*)
  fusion, 196–198
  identity specification, 173–185, 192–194
  pattern formation, 192–196
  tissue type origins, 168–171, 195–196
  tissue types, 159–165
  cell lineage, 168–169
  developmental compartments, 169–170, 191–192
  mechanisms, 165–168
  molecular genetics, 171–192
    AGAMOUS gene, 173–177, 192–194
    AGAMOUS gene-independent development, 176–177
    AGL genes, 187–188
    carpel identity specification, 173–177
    A and C class homeotic gene interactions, 175
    CRABS CLAW gene, 180–182
    ETTIN gene, 182–185
    expression pattern analysis, 189–192
    floral homeotic mutants, 172–173
    FRUITFULL gene, 185–187
    pleiotropic mutant phenotypes, 188–189
    regional identity response, 182–185
    SPATULA gene, 177–180
    value maturation regulation, 185–187
    value–replum boundary tissue definition, 187–188
  overview, 156–159, 192–199
  pattern elements, 159–165
    gynophore, 165
    ovary wall, 161–163
    ovules, 164
    placenta, 164
    septum, 163–164
    stigma, 161
    style, 161
    vasculature, 164–165

H

Heat shock proteins, gene expression in sea urchin development, 77–78
Heteroblasty, leaf epidermis, 9–10
Histones, gene expression in sea urchin development, 43–47
  CS histones, 44
  H2A and H2B histones, 44–45
  H1 histone genes, 45–46
  H3 histone genes, 46

H4 histone genes, 46
Hnf3β genes, gastrulation regulation in mouse, 133, 136–137, 140
HOBBIT gene, apical–basal pattern formation in plants, 212–213
Homeobox genes, expression in sea urchin development, 70–72
Hormones, *see also specific types*
  leaf epidermis growth targets, 14–15
Hyaline protein, expression in sea urchin development, 63–64

J

Jelly protein, gene expression in sea urchin development, 83

K

Kinase, gene expression in sea urchin development, 75–76

L

Lamin, gene expression in sea urchin development, 86
Leaves, epidermis development, 1–32
  cuticle, 29–32
  determined state, 24–29
    fate determination, 24–25
    potency, 26–29
  growth, 11–15
    cell division, 11–13
    cell expansion, 13–14
    hormone targets, 14–15
    morphogenesis relations, 15
  identity, 9–11
    heteroblasty, 9–10
    regional identity within leaves, 10–11
  internal tissue interactions, 16–18
    cytoplasmic connections, 17–18
    genetic mosaics, 16–17
    structure correlations, 16
  morphology, 2–5
    dicots, 3–4
    monocots, 4–5
  ontogeny, 5–9
    cell lineages, 8–9
    embryogenesis period, 5–7
    leaf development period, 7–8
    overview, 1–2, 32

pattern formation, 18–24
  stomates, 18–22
  trichomes, 22–24
LFY gene, gynoecium development role in *Arabidopsis*, 175, 177
Lim1 gene, gastrulation regulation in mouse, 131, 133, 137, 140
Lineage differentiation, *see* Gastrulation
Lineages, *see* Cell lineages

M

Meristem cells
  cellular pattern postembryonic perpetuation, 221
  division regulation
    activity maintenance, 231–233
    division activation, 228–231
Mesenchyme cell proteins, gene expression in sea urchin development, 59–65
Metallothionein, gene expression in sea urchin development, 76–77
Mitochondria, gene expression in sea urchin development, 85
Mitotic apparatus, gene expression in sea urchin development, 73–74
MONOPTEROS gene, apical–basal pattern formation in plants, 213–214
Morphogenesis
  gastrulation organizer in mouse embryo, 126–128
  plants
    cellular pattern postembryonic perpetuation, 220–225
      cortex, 223–224
      cues, 221–225
      endodermis, 136, 138, 223–224
      epidermis, 223
      meristem cell prepatterning, 221
      vascular tissue, 224–225
    development
      cell division regulation, 225–233
      cell fate specification, 227–228
      environmental factors, 230–231
      leaf epidermis relations, 15
      meristem activation, 228–231
      meristem activity maintenance, 231–233
      pattern formation uncoupling, 226–227
    embryonic pattern formation
      apical–basal axis formation, 209–210
      apical–basal pattern formation, 211–214
        basal domain, 212–213
        central domain, 213–214
        epidermis subspecification, 216–218
        ground tissue subspecification, 218–219
        HOBBIT gene, 212–213
        MONOPTEROS gene, 213–214
        overview, 207–209
        radial axis formation, 214
        radial pattern formation, 215–220
        vascular tissue subspecification, 219–220
      growth, 234–239
        cell division uncoupling, 234–238
        determinants, 238–239
Mosaics, leaf epidermal tissue interactions, 16–17
Mouse, gastrulation organizer, 117–145
  alternative sources of organizing activity, 136–138
    anterior patterning, 136–137
    molecular regionalization, 137–138
    primitive endoderm, 136–138
  body pattern diversity, 117–118
  cell fate, 125–126
  head and trunk organizing activity, 131–138
    mutant embryo evidence, 133–136
    regionalization, 131–133
  lineage diversity, 117–118
  molecular properties, 128–129
  morphogenetic movement, 126–128
  overview, 117–121, 144–145
  patterning activity, 119, 121–125
  regulatory signals, 129–131
  tissue patterning, 139–144
    asymmetry, 143–144
    extraembryonic tissues role, 142–143
    mutational studies, 140–141
    Nieuwkoop signal, 139–140
    organizer formation, 143–144
    transgenic studies, 140–141
Myogenic factor, gene expression in sea urchin development, 81–82

N

Nieuwkoop signal, gastrulation organizer in mouse embryo, tissue patterning, 139–140
nodal gene, gastrulation regulation in mouse, 130, 133, 136, 140
noggin gene, gastrulation regulation in mouse, 130

O

Organizer, *see* Gastrulation
Oxt2 gene, gastrulation regulation in mouse, 133, 137, 140

P

Paracentrotus lividus, see Sea urchin
Pattern formation
  embryonic plant development
    apical–basal axis formation, 209–210
      basal domain, 212–213
      central domain, 213–214
      HOBBIT gene, 212–213
      MONOPTEROS gene, 213–214
    apical–basal pattern formation, 211–214
    cell division uncoupling, 226–227
    cellular pattern postembryonic perpetuation, 220–225
      cortex, 223–224
      cues, 221–225
      endodermis, 223–224
      epidermis, 223
      meristem cell prepatterning, 221
      vascular tissue, 224–225
    overview, 207–209
    radial axis formation, 214
    radial pattern formation, 215–220
      epidermis subspecification, 216–218
      ground tissue subspecification, 218–219
      vascular tissue subspecification, 219–220
  gastrulation organizer in mouse embryo
    body pattern diversity, 117–118
    head and trunk organizing activity, 131–138
    mutant embryo evidence, 133–136
    regionalization, 131–133
    organizing activity, 119, 121–125, 136–137
    tissue patterning, 139–144
      asymmetry, 143–144
      extraembryonic tissues role, 142–143
      mutational studies, 140–141
      Nieuwkoop signal, 139–140
      organizer formation, 143–144
      transgenic studies, 140–141
  gynoecium development genetics in *Arabidopsis*
    carpels
      identity specification, 173–185, 192–194
      regional identity, 194–195
      tissue type differentiation, 195–196
      tissue types, 159–165
        gynophore, 165
        ovary wall, 161–163
        ovules, 164
        placenta, 164
        septum, 163–164
        stigma, 161
        style, 161
        vasculature, 164–165
  leaf epidermis development, 18–24
    stomates, 18–22
    trichomes, 22–24
Phan gene, *Antirrhinum* leaf development, 10–11
Plants
  cellular pattern postembryonic perpetuation, 220–225
    cortex, 223–224
    cues, 221–225
    endodermis, 223–224
    epidermis, 223
    meristem cell prepatterning, 221
    vascular tissue, 224–225
  development, cell division regulation, 225–233
    cell fate specification, 227–228
    environmental factors, 230–231
    meristem activation, 228–231
    meristem activity maintenance, 231–233
    pattern formation uncoupling, 226–227
  embryonic pattern formation
    apical–basal axis formation, 209–210
    apical–basal pattern formation, 211–214
      basal domain, 212–213
      central domain, 213–214
      HOBBIT gene, 212–213
      MONOPTEROS gene, 213–214
    overview, 207–209
    radial axis formation, 214
    radial pattern formation, 215–220
      epidermis subspecification, 216–218
      ground tissue subspecification, 218–219
      vascular tissue subspecification, 219–220
  gynoecium development genetics in *Arabidopsis*, 155–199
    carpel
      evolution, 158, 170–171, 198–199
      functions, 156–158

Index 255

fusion, 196–198
identity specification, 173–185, 192–194
pattern formation, 192–196
tissue type origins, 168–171, 195–196
tissue types, 159–165
cell lineage, 168–169
developmental compartments, 169–170, 191–192
mechanisms, 165–168
molecular genetics, 171–192
  AGAMOUS gene, 173–177, 192–194
  AGAMOUS gene-independent development, 176–177
  AGL genes, 187–188
  carpel identity specification, 173–177
  A and C class homeotic gene interactions, 175
  CRABS CLAW gene, 180–182
  ETTIN gene, 182–185
  expression pattern analysis, 189–192
  floral homeotic mutants, 172–173
  FRUITFULL gene, 185–187
  pleiotropic mutant phenotypes, 188–189
  regional identity response, 182–185
  SPATULA gene, 177–180
  value maturation regulation, 185–187
  value–replum boundary tissue definition, 187–188
overview, 156–159, 192–199
pattern elements, 159–165
  gynophore, 165
  ovary wall, 161–163
  ovules, 164
  placenta, 164
  septum, 163–164
  stigma, 161
  style, 161
  vasculature, 164–165
leaf epidermis development, 1–32
cuticle, 29–32
determined state, 24–29
  fate determination, 24–25
  potency, 26–29
growth, 11–15
  cell division, 11–13
  cell expansion, 13–14
  hormone targets, 14–15
  morphogenesis relations, 15
identity, 9–11
  heteroblasty, 9–10

regional identity within leaves, 10–11
internal tissue interactions, 16–18
  cytoplasmic connections, 17–18
  genetic mosaics, 16–17
  structure correlations, 16
morphology, 2–5
  dicots, 3–4
  monocots, 4–5
ontogeny, 5–9
  cell lineages, 8–9
  embryogenesis period, 5–7
  leaf development period, 7–8
overview, 1–2, 32
pattern formation, 18–24
  stomates, 18–22
  trichomes, 22–24
organ morphogenesis and growth, 234–239
  cell division uncoupling, 234–238
  growth determinants, 238–239
PM27 gene, expression in sea urchin development, 60
Poly(A)-binding protein, gene expression in sea urchin development, 82
Profilin, gene expression in sea urchin development, 84
Proto-oncogenes, expression in sea urchin development, 87
Psammechinus miliaris, see Sea urchin

R

Radial patterning, *see* Pattern formation
Retroposons, expression in sea urchin development, 87
Ribosomal protein genes, expression in sea urchin development, 82

S

SCR gene, radial pattern formation in plant ground tissue, 218–219
Sea urchin, gene expression in development, 41–87
  animal–vegetal axis-expressed genes, 78–81
  arylsulfatase genes, 81
  bindin-related genes, 85
  calmodulin gene, 84
  cell division-related genes, 74–75
  ciliogenesis-related genes, 73–74
  coelomocytes-expressed genes, 83–84
  cortical granule content genes, 65–66

Sea urchin (*cont.*)
  cyclin gene, 74–75
  cytoskeletal *actin* genes, 47–55
    CyI gene, 53
    CyIIa gene, 54–55
    CyIIIa gene, 47–53
    CyIIIb gene, 53–54
    S9 gene, 54–55
  distrophin-related genes, 85
  ectoderm-specific genes, 56–59
  EGF domain genes, 82
  egg receptor genes, 83
  endoderm-specific genes, 55–56
    CAT gene, 56
    Endo16 gene, 55–56
    SM50 gene, 56
  extracellular matrix-specific genes, 59–65
    adhesion proteins, 64–65
    collagen genes, 61–63
    echinonectin protein, 63–64
    hyaline protein, 63–64
    PM27 gene, 60
    primary mesenchyme cell proteins, 59–65
    SM30 gene, 60
    SM50 gene, 60–62
  fascin gene, 84
  fibroblast growth factor gene, 76
  gastrulation-related genes, 67–70
  heat shock protein genes, 77–78
  histones, 43–47
    CS histones, 44
    H2A and H2B histones, 44–45
    H1 histone genes, 45–46
    H3 histone genes, 46
    H4 histone genes, 46
  homeobox-containing genes, 70–72
  jelly protein genes, 83
  kinase genes, 75–76
  lamin genes, 86
  metallothionein genes, 76–77
  mitochondria-related genes, 85
  mitotic apparatus-related genes, 73–74
  myogenic factor genes, 81–82
  overview, 42–43
  poly(A)-binding protein genes, 82
  primary mesenchyme cell proteins, 59–65
  profilin gene, 84
  proto-oncogenes, 87
  retroposons, 87
  ribosomal protein genes, 82
  snRNA genes, 86
  SpCoel1 gene, 84
  SpEGFI and *SpEGFII* genes, 82
  sperm-associated genes, 83, 85–86
  transcription factor genes, 72–73
  tubulin genes, 73–74
  ubiquitin genes, 78
  vitellogenin genes, 83
  Wnt pathway-related genes, 66–67
S9 gene, gene expression in sea urchin development, 54–55
SHR gene, radial pattern formation in plant ground tissue, 218–219
Smad2 gene, gastrulation regulation in mouse, 137–138, 140
SM30 gene, expression in sea urchin development, 60
SM50 gene, expression in sea urchin development, 56, 60–62
snRNA, gene expression in sea urchin development, 86
Sox genes, gastrulation regulation in mouse, 133, 143
SPATULA gene, gynoecium development role in *Arabidopsis*, 177–180
SpCoel1 gene, expression in sea urchin development, 84
SpEGFI and *SpEGFII* genes, expression in sea urchin development, 82
Sperm-associated genes, expression in sea urchin development, 83, 85–86
Stomates, pattern formation, 18–22
Strongylocentrotus purpuratus, see Sea urchin

T

Transcription factors, gene expression in sea urchin development, 72–73
Transforming growth factor β, gastrulation regulation in mouse, 130–131
Transgenics, gastrulation organizer in mouse embryo, tissue patterning, 140–141
Trichomes
  description, 3
  pattern formation, 22–24
TTG gene, radial pattern formation in plant epidermis, 216–218
Tubulin, gene expression in sea urchin development, 73–74

U

Ubiquitin, gene expression in sea urchin development, 78

V

Vascular tissue, pattern formation in plants
embryonic development
cellular pattern postembryonic perpetuation, 224–225
radial pattern formation, 219–220
gynoecium development genetics in *Arabidopsis,* 164–165
Vitellogenin, gene expression in sea urchin development, 83

W

Wnt pathway genes
gastrulation regulation in mouse, 130, 136, 141
gene expression in sea urchin development, 66–67
wol gene, radial pattern formation in plant vascular tissue, 219–220

Contents of Previous Volumes

Volume 40

1. **Homeobox Genes in Cardiovascular Development**
   Kristin D. Patterson, Ondine Cleaver, Wendy V. Gerber, Matthew W. Grow, Craig S. Newman, and Paul A. Krieg

2. **Social Insect Polymorphism: Hormonal Regulation of Plasticity in Development and Reproduction in the Honeybee**
   Klaus Hartfelder and Wolf Engels

3. **Getting Organized: New Insights into the Organizer of Higher Vertebrates**
   Jodi L. Smith and Gary C. Schoenwolf

4. **Retinoids and Related Signals in Early Development of the Vertebrate Central Nervous System**
   A. J. Durston, J. van der Wees, W. W. M. Pijnappel, and S. F. Godsave

5. **Neural Crest Development: The Interplay between Morphogenesis and Cell Differentiation**
   Carol A. Erickson and Mark V. Reedy

6. **Homeoboxes in Sea Anemones and Other Nonbilaterian Animals: Implications for the Evolution of the Hox Cluster and the Zootype**
   John R. Finnerty

7. **The Conflict Theory of Genomic Imprinting: How Much Can Be Explained?**
   Yoh Iwasa

Volume 41

1 Pattern Formation in Zebrafish—Fruitful Liaisons between Embryology and Genetics
   Lilianna Solnica-Krezel

2 Molecular and Cellular Basis of Pattern Formation during Vertebrate Limb Development
   Jennifer K. Ng, Koji Tamura, Dirk Büscher, and Juan Carlos Izpisúa-Belmonte

3 Wise, Winsome, or Weird? Mechanisms of Sperm Storage in Female Animals
   Deborah M. Neubaum and Mariana Wolfner

4 Developmental Genetics of *Caenorhabditis elegans* Sex Determination
   Patricia E. Kuwabara

5 Petal and Stamen Development
   Vivian F. Irish

6 Gonadotropin-Induced Resumption of Oocyte Meiosis and Meiosis-Activating Sterols
   Claus Yding Andersen, Mogens Baltsen, and Anne Grete Byskov

Volume 42

Cumulative Subject Index, Volumes 20 through 41

Volume 43

1 Epigenetic Modification and Imprinting of the Mammalian Genome during Development
   Keith E. Latham

2 A Comparison of Hair Bundle Mechanoreceptors in Sea Anemones and Vertebrate Systems
   Glen M. Watson and Patricia Mire

3 Development of Neural Crest in *Xenopus*
   Roberto Mayor, Rodrigo Young, and Alexander Vargas

4 Cell Determination and Transdetermination in *Drosophila* Imaginal Discs
  Lisa Maves and Gerold Schubiger

5 Cellular Mechanisms of Wingless/Wnt Signal Transduction
  Herman Dierick and Amy Bejsovec

6 Seeking Muscle Stem Cells
  Jeffrey Boone Miller, Laura Schaefer, and Janice A. Dominov

7 Neural Crest Diversification
  Andrew K. Groves and Marianne Bronner-Fraser

8 Genetic, Molecular, and Morphological Analysis of Compound Leaf Development
  Tom Goliber, Sharon Kessler, Ju-Jiun Chen, Geeta Bharathan, and Neelima Sinha

Volume 44

1 Green Fluorescent Protein (GFP) as a Vital Marker in Mammals
  Masahito Ikawa, Shuichi Yamada, Tomoko Nakanishi, and Masaru Okabe

2 Insights into Development and Genetics from Mouse Chimeras
  John D. West

3 Molecular Regulation of Pronephric Development
  Thomas Carroll, John Wallingford, Dan Seufert, and Peter D. Vize

4 Symmetry Breaking in the Zygotes of the Fucoid Algae: Controversies and Recent Progress
  Kenneth R. Robinson, Michele Wozniak, Rongsun Pu, and Mark Messerli

5 Reevaluating Concepts of Apical Dominance and the Control of Axillary Bud Outgrowth
  Carolyn A. Napoli, Christine Anne Beveridge, and Kimberley Cathryn Snowden

6 Control of Messenger RNA Stability during Development
  Aparecida Maria Fontes, Jun-itsu Ito, and Marcelo Jacobs-Lorena

7 EGF Receptor Signaling in *Drosophila* Oogenesis
  Laura A. Nilson and Trudi Schupbach